Springer Series in Materials Science

Volume 197

For further volumes:
http://www.springer.com/series/856

The Springer Series in Materials Science covers the complete spectrum of materials physics, including fundamental principles, physical properties, materials theory and design. Recognizing the increasing importance of materials science in future device technologies, the book titles in this series reflect the state-of-the-art in understanding and controlling the structure and properties of all important classes of materials.

Bernard Gil

Physics of Wurtzite Nitrides and Oxides

Passport to Devices

Springer

Bernard Gil
Institut de Physique de Montpellier
University of Montpellier 2
Montpellier, Cedex
France

ISSN 0933-033X ISSN 2196-2812 (electronic)
ISBN 978-3-319-06804-6 ISBN 978-3-319-06805-3 (eBook)
DOI 10.1007/978-3-319-06805-3
Springer Cham Heidelberg New York Dordrecht London

Library of Congress Control Number: 2014940318

Printed on acid-free paper

Springer is part of Springer Science+Business Media (www.springer.com)

Preface

This book aims at offering the master students—and probably to other ones—who are studying the physics of wide bandgap semi-conductors, the elements required to rapidly grasp the concept of solid state physics that are needed to start a formation or a research activity at the boarder between physics, chemistry, electrical engineering.

This field has known tremendous developments during the past 20 years, and it will probably continue to be very exciting: so many applications are possible to wide band gap semi-conductors and so few have been satisfied to date.

I have been routinely working on optical properties of wide band gap semi-conductors, of their heterostructures, of their nanostructures since 1991 and I could attend every year a lot of scientific international conferences. Each time was the opportunity for me to meet new faces, sometimes very young ones. Most of these newcomers had received a formation in chemistry, electronics, physics, mathematics, or another one, sometimes including astronomy. They are often launched into the international scientific arena just after having spent a few weeks or months in immersion into a research group, and they have to accommodate a lot of concepts in very different domains. This is not so easy for them. I had a lot of opportunities to chat with many of these newcomers. I got convinced as many of my colleagues are now, too, that there is a need for a general book in which they could find gathered most of the concepts that they need to know. When Claus Ascheron from Springer asked me to write such book, I accepted it without having in mind the challenging proposal he had made to me. You are holding such a book in your hands. The field is wide and necessarily unexhaustively addressed, but I hope this monograph contains a strong enough message for being of value for a lot of my young colleagues.

This book is declined along five chapters: basic symmetry and physical properties linked to it, basics of growth and structural characterization methods, band structure effects and lattice vibrations, optical properties of bulk materials, and finally physics and optical properties of low-dimensional systems. Photonics, quantum optics, plasmonics, and transport properties are not treated; they are from very much specialized areas and are addressed or in the way of being addressed in specialized research books.

This book is the fruit of many collaborations. I would first like to thank the many students that I contributed to train. Many of them now being colleagues in some universities or at the National Centre of Scientific Research, my employing institution. Let me thank Philippe Boring, Pierre Bigenwald, Pierre Lefebvre, Matthieu Moret, Lionel Aigouy, Claude Boemare, Sandra Ruffenach, Andenet Alemu, Magloire Tchounkeu, Francois Demangeot, Cyril Pernot, Abdelhadi Abounadi, Amal Rajira, Tomasz Ochalski, Yves-Mathieu Le-Vaillant, Marian Zamfirescu, Mathieu Gallart, Aurelien Morel, Xue Bing Zhang, Sokratis Kalliakos, Romuald Intartaglia, Benedicte Maleyre, Richard Bardoux, Stephane Faure, Luc Beaur, Daniel Rosales, Julien Selles, Huong Thi Ngo, and Rereao Hahe. I had a chance to work with a lot of handsome colleagues among which are Olivier Briot, Mathieu Leroux, Jean-Yves Duboz, Nicolas Grandjean, Eric Tournie, Benjamin Damilano, Julien Brault, Jean-Michel Chauveau, Christian Morhain, Thierry Bretagnon, Amelie Dussaigne, Jean Massies, Bernard Beaumont, Pierre Gibart, Philippe Vennegues, Pierre Ruterana, Thierry Guillet, Christelle Brimont, Pierre Valvin, Christian L'henoret, Thierry Taliercio, Didier Felbacq, Brahim Guizal, Guillaume Cassabois, Emmanuel Rousseau, Christelle Eve, Jacques Leyzat, Regine Pauzat, Bruno Daudin, …

At the international scale, Izabella Grzegory, Hadis Morkoç, Isamu Akasaki and Shuji Nakamura have been my mentors in the early days of the nitrides. I took advantages of fruitful exchanges with Yahusiko Arakawa, Bo Monemar, Alex Zunger, Hiroshi Amano, Akihiko Yoshikawa, Yasuhi Nanishi, Yoichi Kawakami, Kazamasu Hiramatsu, Katsumi Kishino, Shigefusa Chichibu, Hideto Miyake, Alexey Kavokin, Axel Homann, Martin Stutzmann, Bruno Meyer, Juergen Christen, Jorg Neugebauer, Friedhelm Beschtedt, Alois Krost, Andreas Hangleiter, Martin Feneberg, Ruediger Goldhahn, Charles Foxon, Kevin O'Donnell, Rob Martin, Peter Parbrook, Galia Pozina, Tatiana Shubina, Serguey Ivanov, Valery Yu Davidov, Tanya Paskova, Plamen Paskov, Fernando Ponce, Russell Dupuis, Ted Moutsakas, Steven den Baars, Umesh Misra, James Speck, Zlatko Sitar, Eva Monroy, Xinqiang Wang, Piotr Perlin, Tadeusz Suski, Abderrahmane Kadri, Karima Zitouni, and so many others.

Pierre Bigenwald who carefully browsed the five chapters to expurgate them from typos deserves receiving specific acknowledgments.

Montpellier Bernard Gil

Contents

Chapter 1
Basic Crystallography and Other Properties Linked with Symmetry

This chapter treats of crystallography, basic physical properties linked to the crystal structure. It also gives a few elements of basic group theory.

1.1 The Hexagonal Point Symmetry Deduced from the Shape of Natural Wurtzitic Crystals

Atomic organisation of wurtzitic semi-conductors follows the crystallographic structure named "zincite", that was, to the author's knowledge, first described as red oxide of zinc by American mineralogist Archibald [1] at the dawn on the nineteenth century. Some sulfide minerals also crystallize according to this atomic organisation, among which are greenockite (CdS), first observed in 1840 in Bishopton, Scotland, and named after the landowner Lord Greenock. The term "wurtzite" is correlated with the description of the hexagonal form of zinc sulfide crystals in 1861 and named after French chemist Charles-Adolphe Wurtz. The wurtzite group also includes, rambergite (MnS) discovered in sediments and named at the end of the twentieth century after the Norwegian-Swedish mineralogist, Hans Ramberg. We also wish to indicate that cadmoselite CdSe (artificially synthesized before it was identified as a mineral stone) was discovered in 1957. In the language of semi-conductorists, these materials belong to the group of II–VI compounds: the cation (resp. anion) is a group II (resp. VI) element of Mendeleev's table. Apart from the II–VI family, III-nitrides (III-N's) are also important semi-conductors the cation is a group-III element generally boron, aluminium, gallium or indium. The most energetically favourable atomic organisation for boron nitride is not wurtzite; BN is a layered compound like graphite. The most stable crystalline structure of AlN, GaN and InN semi-conductors is wurtzitic. These materials are not extracted from mines as binary compounds probably due to the high stability of the nitrogen molecule (9.3 eV per bond) or to the chemical reactivity of ammoniac. NH_3 is a very stable molecule: nitrogen single atoms are highly reactive and donot live long without binding to the most abundant

B. Gil, *Physics of Wurtzite Nitrides and Oxides*, Springer Series in Materials Science 197, DOI: 10.1007/978-3-319-06805-3_1,

Fig. 1.1 Photograph of a natural wurtzitic crystal. Note the needle-like shape and the existence of many facets with different orientations illustrating the hexagonal symmetry

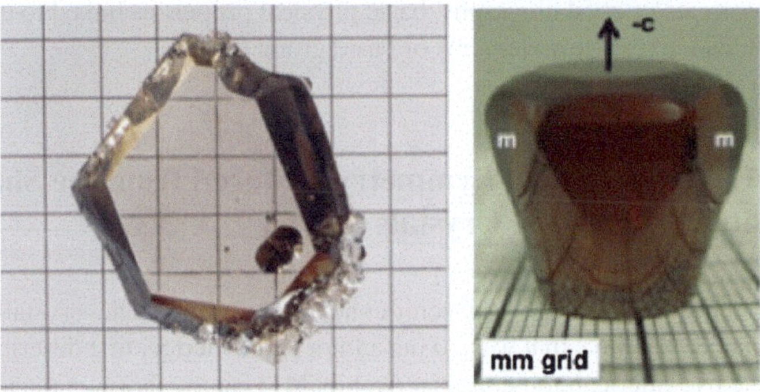

Fig. 1.2 *Left* Photograph of an artificial crystal of bulk gallium nitride (courtesy Dr. I. Grzegory, Polish Academy of Sciences). *Right* Photograph of an aluminum nitride bulk artificial single crystal (courtesy Prof. Zlatko Sitar, North Carolina State University)

elements in nature: H, O, etc. Therefore, the necessary conditions to associate nitrogen and group III element in order to form pure III-Ns are never met in nature, to the best of our knowledge. The existence of AlN was demonstrated in 1862 by Briegler and Geuther [2] and Mallet synthesized it in 1876 [3] with metallic aluminum and sodium carbonate reactant at high temperatures. The growth of GaN was achieved in 1932 by Johnson et al. [4] later achieved, in 1938, the growth of indium nitride was later achieved, in 1938 [5]. The hexagonal symmetry of wurtzite crystals was quantitatively calibrated during the nineteenth century from the orientation of the directions orthogonal to the facets of crystals found in mines. Crystals are most generally found as small needles, exhibiting specific shapes as shown in Fig. 1.1.

Artificially grown bulk semi-conductors like gallium nitride (Fig. 1.2-left) or aluminum nitride (Fig. 1.2-right) are also obtained under faceted crystalline shapes. Although the shape of the crystals presented in Fig. 1.2 (left) and 1.2 (right) are

Fig. 1.3 Photograph of wooden tutorial object designed such as to show a typical repeat of basic building blocks which at the end to form the whole crystal shape

fairly different, they share a common property. The shape of the bunch of directions orthogonal to the crystal facets constitutes some kind of figure of merit.

The figure obtained after the stereographic projection of the ensemble of the directions that are orthogonal to the facets of these crystals is compatible with the symmetry of a regular hexagonal polygon. This figure is a constant quantity typical of the crystalline structure according to the first law of crystallography. It is known as "the law of constant angles between crystal facets", early initiated by Danish Nicolas Steno's seminal work in 1669 and later formulated by French mineralogist Jean-Baptiste Romé de l'Isle in 1772. For wurtzitic semi-conductors, this figure permits to reveal a six-fold symmetry along a given direction, the existence of a three-fold and two-fold symmetries collinear to the six fold one and six symmetry planes parallel to the six-fold symmetry axis. These symmetry elements are generic of the crystalline form named di-hexagonal pyramidal by mineralogists. In terms of modern group theory language, from the shape of these needles, one establishes that the orientation (or point symmetry) of wurtzite crystals is a subgroup (labelled C_6v in Schoenflies notation or P_6mm in the Hermann-Mauguin one) of the full hexagonal symmetry (labelled D_6h in Schoenflies notations or P_6/mmm in the Hermann-Mauguin one). French abbot René-Just Haüy in 1774 carefully observed facets of minerals and proposed the second law of crystallography: crystals consist of three-dimensional stacking at a macroscopic scale of a given microscopic basic building block as tentatively illustrated in Fig. 1.3. This kind of wooden crystal were still used as tutorial examples to teach students in the middle of the twentieth century.

In 1849, French scientist Auguste Bravais postulated the principle of specific three-dimensional translational invariance (existence of translational symmetry operations on the basic building block imagined by Haüy). From this postulate results the notion of three-dimensional periodic crystalline lattice which is the basic principle used to determine the crystallographic structure of materials using radiocrystallography. Before discussing this very recent technique since X-rays were discovered by German physicist Wilhelm Conrad Röntgen in 1895, it is worthwhile allocating some time to simple mathematical analysis of crystal properties in line with their symmetries. We will restrict ourselves to wurtzite and, in particular, catch the opportunity to show how powerful the first laws of crystallography are. From the bunch of angles made by the directions normal to the crystal's facet, we can determine the ratio c/a or a/c of the basic hexagonal building block parameter—along the six-fold symmetry axis—to the one orthogonal to it.

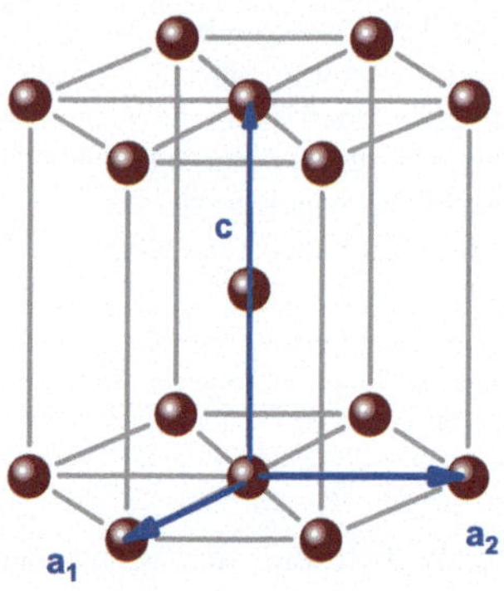

Fig. 1.4 Representation of the crystallographic axes $\vec{a}_1, \vec{a}_2$ and $\vec{c}$ that generate the unit hexagonal cell (*blue lines*). The full hexagonal cell is shown for the completeness

1.2 The Hexagonal Lattice, Its Reticular Planes and Their Description Using Simple Euclidian Geometry

The hexagonal crystallographic system is based on two vectors $\vec{a}_1$ and $\vec{a}_2$ of identical length a at 120° from each other, and a third one, of length c orthogonal to the plane generated by the preceding two, as described by French scientist Auguste Bravais in 1849 and as illustrated in Fig. 1.4.

Lengths a and c being the dimensions of the lattice vectors of the hexagonal cell, we associate them to unit vectors $\vec{i}, \vec{j}$ and $\vec{k}$.

This gives:

$$\vec{i} \cdot \vec{k} = \vec{j} \cdot \vec{k} = 0$$

and

$$\vec{i} \cdot \vec{j} = \cos(120°) = -1/2$$

We now consider the (hkℓ) reticular plane in terms of its so-called Miller indices. This notation was proposed by British mineralogist William Hallowes Miller in 1839.

By definition, the (hkℓ) plane intersects the reticular axes at lengths $\frac{a}{h}$, $\frac{a}{k}$ and $\frac{c}{\ell}$ from the origin as shown in Fig. 1.5.

Letters h, k and ℓ are algebraic integer numbers. There are some specific simple orientations for such planes:

- Miller indices (100) represent a plane parallel to the $(\vec{j}, \vec{k})$ plane,
- Miller indices (010) represent a plan e parallel to the $(\vec{i}, \vec{k})$ plane,
- Miller indices (001) represent a plane parallel to the $(\vec{i}, \vec{j})$ plane.

The notation {hkℓ} is traditionally used to identify all the planes equivalent to (hkℓ) by the symmetry of the lattice. As an example, (100) and (010) planes belong to the {100} family.

Thanks to hexagonal symmetry, and in contrast to the cubic case, $\{100\} \neq \{001\}$.

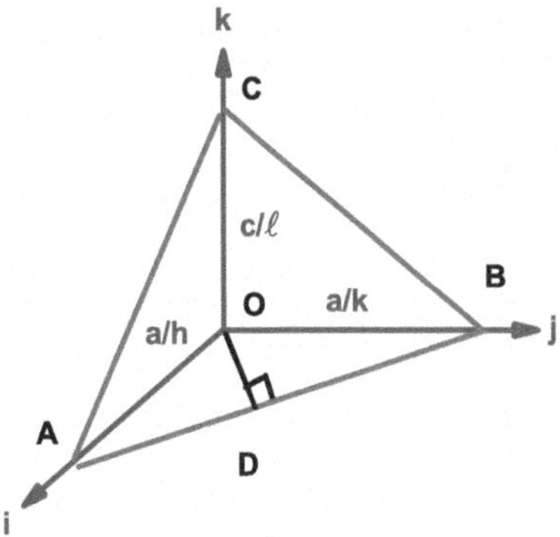

Fig. 1.5 Orientation of the reticular plane (hkℓ) with respect to the crystallographic axes of the hexagonal lattice

Let [hkℓ]—with square brackets—represent a direction in the basis of the direct lattice vectors: [hkℓ] is equivalent to $ha\vec{i}+ka\vec{j}+c\vec{k}$. The notation < hkℓ > is used for all the directions that are equivalent to [hkℓ] by symmetry operations of the crystal.

In the most general case, and for non-cubic crystals, direction [hkℓ] is **NOT** orthogonal to plane (hkℓ).

Given an (hkℓ) plane, we define three unit vectors ($\vec{U}$,$\vec{V}$,$\vec{W}$) orthogonal to each other, two of them ($\vec{U}$ and $\vec{V}$) being vectors of the (hkℓ) plane, whilst the third one ($\vec{W}$) is orthogonal to it.

The $\vec{U}$ direction ($\vec{U}$ is parallel to vector $\vec{AB}$ in Fig. 1.5) is chosen as [-kh0], and thus lies in the ($\vec{i}$,$\vec{j}$) plane. $\vec{U}$ satisfies the sonal equation and corresponds to a crystal direction.

The [-ℓ0h] direction (vector $\vec{AC}$ in Fig. 1.5) connects intersections of the (hkℓ) plane with crystallographic directions $\vec{a}_1$ and $\vec{c}$ at lengths $\frac{a}{h}$ and $\frac{c}{\ell}$ from the origin. It forms with $\vec{U}$ a basis of the (hkℓ) plane. This direction is not orthogonal to $\vec{U}$, which is not the best choice for mathematical calculations. Thus, it is more appropriate to determine, using an orthonormalization procedure a vector $\vec{V}$ lying in the (hkℓ) plane and perpendicular to $\vec{U}$. This vector is parallel to vector $\vec{DO}+\vec{OC}$ with

$$\|\vec{DO}\| = \frac{a\sqrt{3}}{2\sqrt{h^2+hk+k^2}}$$

and $\|\vec{OC}\| = \frac{c}{\ell}$ as indicated in Fig. 1.5.

It also satisfies the zone equation for the (hkℓ) plane, demonstrating that it corresponds to a crystal direction.

To this couple of vectors vectors ($\vec{U}$, $\vec{V}$), we add $\vec{W}$, also obtained after an orthonormalization procedure.

In the hexagonal ($\vec{i}, \vec{j}, \vec{k}$) basis, the coordinates of these unit vectors are:

$$\begin{bmatrix} \vec{U} \\ \vec{V} \\ \vec{W} \end{bmatrix} = \begin{bmatrix} -Ak & Ah & 0 \\ -Aa\ell B_{ac}(2h+k) & -Aa\ell B_{ac}(2k+h) & 2c\frac{B_{ac}}{A} \\ \frac{2}{\sqrt{3}}cB_{ac}(2h+k) & \frac{2}{\sqrt{3}}cB_{ac}(2k+h) & \sqrt{3}\ell a B_{ac} \end{bmatrix} \begin{bmatrix} \vec{i} \\ \vec{j} \\ \vec{k} \end{bmatrix} \quad (1.1)$$

Table 1.1 Values of the ratio c/a for various wurtzitic semi-conductors

Material	BN	GaN	InN	AlN	ZnS	ZnSe	CdS	CdSe	ZnO
c/a	1.64	1.62	1.61	1.6	1.64	1.63	1.62	1.63	1.60

The number of digits is compatible with the determination method

where

$$A = \frac{1}{\sqrt{h^2 + hk + k^2}}$$

and

$$B_{ac} = \frac{1}{\sqrt{3\ell^2 a^2 + 4c^2(h^2 + hk + k^2)}}$$

The angle between the [001] direction and $\vec{W}$ is given as:

$$\Theta = \arccos\left[\frac{1}{\sqrt{1 + \frac{4c^2}{3a^2}\frac{(h^2+hk+k^2)}{\ell^2}}}\right]$$

that depends on the value of the three Miller indices and on the ratio c/a.

At this stage, elementary Euclidian algebraic calculations demonstrate the possibility to determine the c/a ratio in wurtzitic crystals from the orientation of the directions of the crystal facets. The only issue is to have crystals with a large enough number of facets. Thus, we gain access to a well documented set of experimental values of Θ, so that their associated h, k and ℓ numbers can be determined simultaneously with the value of c/a. Table 1.1 summarizes the results of c/a ratio typical of the most common wurtzitic semi-conductors.

1.3 The Four-Index Bravais-Miller Representation of the Orientation of Reticular Planes in Hexagonal Crystals

In the case of the hexagonal (and rhomboidal) lattice systems, one alternative convention is to use a 4-numbers representation (h k i ℓ) for reticular planes, where i $= -$(h + k). In this case, h, k and ℓ are identical to the Miller indices, and i is redundant: four independent vectors cannot generate a three-dimensional space. Figure 1.6 illustrates some typical orientations of (hkℓ) planes that are found in natural crystals.

This four-indices scheme for labelling planes in a hexagonal lattice is very convenient to illustrate the identical nature of reticular planes by permutation of the indices. For example, we better spot the similarity between planes $(11\bar{2}0)$ and $(1\bar{2}10)$ than when they are written as (110) and $(1\bar{2}0)$.

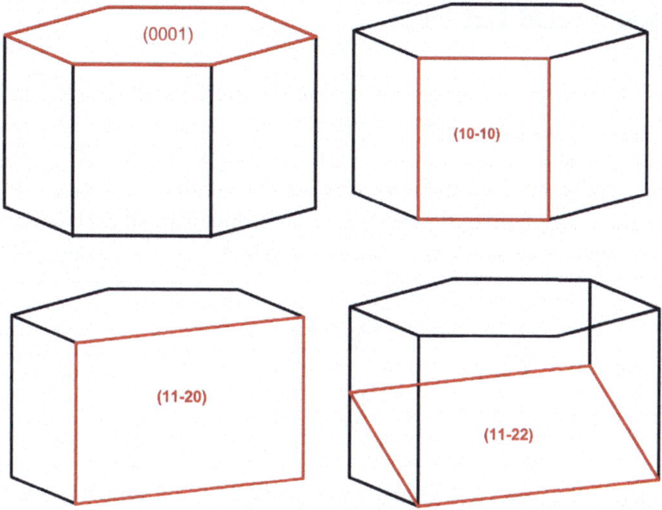

Fig. 1.6 Sketches of typical orientations of reticular planes of interest for wurtzitic semi-conductors. The four-indices Miller-Bravais indexing system is used to represent the orientations

1.4 Representation of Hexagonal Crystal Directions Using Four Indices

In some books of mineralogy, we find another four-indices system for the directions in hexagonal crystals.

Let:

$$\vec{D} = ha\vec{i} + ka\vec{j} + \ell c\vec{k}$$

This direction may be also noted as $[hk.\ell]$ and is fully determined with respect to the hexagonal cell.

The corresponding Weber four indices are defined as: $[hk.\ell] \rightarrow [h'k't\ell']$ and the relationships are:

$$\begin{aligned} h' &= \frac{n(2h-k)}{3} \\ k' &= \frac{n(2k-h)}{3} \\ t &= -(h'+k') \\ \ell' &= n\ell \end{aligned}$$

where n is a factor used to transform the new indices into smaller integers. To avoid confusion, we will not use this notation.

1.5 The Reciprocal Lattice

The reciprocal lattice is not necessary for geometric crystallography. However, it facilitates some calculations, and its utilization is mandatory when studying X-ray diffraction by periodical structures or, out of the context of crystallography, to treat band structure phenomena, or light propagation. The reciprocal lattice is the Fourier transform of the direct lattice. It generates a three-dimensional (3D) space via basis vectors $\vec{a}_1^*$, $\vec{a}_2^*$ and $\vec{c}^*$ that are defined from the basis vectors $\vec{a}_1$, $\vec{a}_2$ and $\vec{c}$ of the direct lattice:

$$\begin{aligned} \vec{a}_i\vec{a}_j^* &= \delta_{ij} \\ \vec{c}.\vec{c}^* &= 1 \\ \vec{a}_i^*.\vec{c}^* &= 0 \end{aligned}$$

One deduces then, that $\vec{a}_1^*$ is orthogonal to both $\vec{a}_2$ and $\vec{c}$, which dictates that:

$$\begin{aligned} \vec{a}_1^* &= \alpha(\vec{a}_2 \wedge \vec{c}) \\ \vec{a}_1 \cdot \vec{a}_1^* &= \alpha\vec{a}_1.(\vec{a}_2 \wedge \vec{c}) = \alpha.V_{ol} = 1 \end{aligned}$$

V_{ol} is the volume $(\vec{a}_1 \wedge \vec{a}_2).\vec{c} = \vec{a}_1.(\vec{a}_2 \wedge \vec{c})$

$$\alpha = \frac{1}{V_{ol}}$$

then

$$\vec{a}_1^* = \frac{\vec{a}_2 \wedge \vec{c}}{V_{ol}}$$

Similarly, one obtains $\vec{a}_2^*$ and $\vec{c}^*$:

$$\vec{a}_2^* = \frac{\vec{c} \wedge \vec{a}_1}{V_{ol}}$$

and

$$\vec{c}^* = \frac{\vec{a}_1 \wedge \vec{a}_2}{V_{ol}}$$

In line with the definitions of the direct and reciprocal lattices, it is possible to generate operations like scalar or vectorial products, using vectors from both spaces:

Let:

$$\vec{R} = r_1\vec{a}_1 + r_2\vec{a}_2 + r_3\vec{c}$$

and

$$\vec{N}^* = n_1\vec{a}_1^* + n_2\vec{a}_2^* + n_3\vec{c}^*$$

Scalar product:

$$\vec{R} \cdot \vec{N}^* = r_1 n_1 + r_2 n_2 + r_3 n_3$$

Let us consider a series of reticular planes $(hk\ell)$ and choose the closest to the origin O.

Let A, B and C be the intersections of this plane with the three axes (see Fig. 1.5).

Vectors $\vec{AB}$ and $\vec{AC}$ both belong to that plane, their directions are respectively [−k h 0] and [$-\ell$ 0 h].

$$\vec{AB} = \vec{AO} + \vec{OB} = -\vec{a}_1\frac{1}{h} + \vec{a}_2\frac{1}{k}$$

$$\vec{AC} = -\vec{a}_1\frac{1}{h} + \vec{c}\frac{1}{\ell}$$

Let $\vec{N}_{hk\ell}^*$ be a vector of the reciprocal lattice:

$$\vec{N}_{hk\ell}^* = h\vec{a}_1^* + k\vec{a}_2^* + \ell\vec{c}^*$$

This vector defines the $[hk\ell]^*$ family in the reciprocal lattice.

The three integer numbers (h, k, et ℓ) being primes between each other, the extremity of $\vec{N}_{hk\ell}^*$ is the first node of the reciprocal lattice from the origin along the $\vec{N}_{hk\ell}^*$ direction.

$$\vec{AB} \cdot \vec{N}_{hk\ell}^* = \left(-\vec{a}_1\frac{1}{h} + \vec{a}_2\frac{1}{k}\right) \cdot (h\vec{a}_1^* + k\vec{a}_2^* + \ell\vec{c}^*) = -\vec{a}_1 \cdot \vec{a}_1^* + \vec{a}_2 \cdot \vec{a}_2^* = 0$$

$$\vec{AC} \cdot \vec{N}_{hk\ell}^* = \left(-\vec{a}_1\frac{1}{h} + \vec{c}\frac{1}{\ell}\right) \cdot (h\vec{a}_1^* + k\vec{a}_2^* + \ell\vec{c}^*) = -\vec{a}_1 \cdot \vec{a}_1^* + \vec{c} \cdot \vec{c}^* = 0$$

The two vectors $\vec{AB}$ and $\vec{AC}$ of the $(hk\ell)$ plane are orthogonal to the vector $\vec{N}_{hk\ell}^*$ of the reciprocal lattice, which gives birth to the very important statement below:

The direction $[hk\ell]^*$ of the reciprocal lattice is orthogonal to the reticular plane $(hk\ell)$ of the direct lattice.

After some simple geometrical calculation, one could also demonstrate that the length of $\vec{N}_{hk\ell}^*$ is connected with the distance $d_{hk\ell}$ between adjacent reticular $(hk\ell)$ planes of the direct lattice by:

$$\vec{N}_{hk\ell}^* \cdot \vec{d}_{hk\ell} = 1$$

$$d_{hk\ell} = \frac{ac\sqrt{3}}{\sqrt{4c^2(h^2 + hk + k^2) + 3a^2\ell^2}}$$

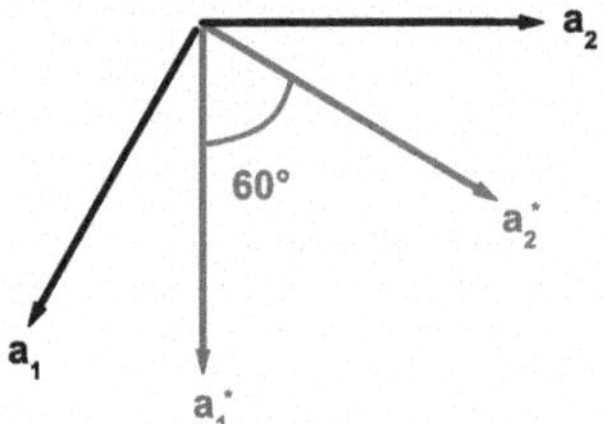

Fig. 1.7 Relative orientations of the vectors of the reciprocal (*starred symbols*) and direct (*unstarred letters*) lattice of the hexagon in the (0001) plane. Vectors c and c* are both orthogonal to that plane

We wish to outline that, in the most general way, direction $[hk\ell]$ *in the basis of the direct lattice vectors IS NOT orthogonal to* $(hk\ell)$ *plane expressed in the basis of the direct lattice vectors, since they do not form an orthogonal basis.*

On Fig. 1.7 are represented the direct and reciprocal lattices basis vectors for the hexagon in the (0001) plane.

Relatively to the crystallographic basis set, the vectors of the reciprocal lattice express as:

$$\begin{bmatrix} \vec{a}_1^* \\ \vec{a}_2^* \\ \vec{c}^* \end{bmatrix} = \begin{bmatrix} \frac{4}{3a} & \frac{2}{3a} & 0 \\ \frac{2}{3a} & \frac{4}{3a} & 0 \\ 0 & 0 & \frac{1}{c} \end{bmatrix} \cdot \begin{bmatrix} \vec{i} \\ \vec{j} \\ \vec{k} \end{bmatrix} \tag{1.2}$$

We note that mathematical calculations can be made using any of these two bases.

$$\vec{N}_{hk\ell}^* = h\vec{a}_1^* + k\vec{a}_2^* + \ell\vec{c}^*$$

may be expressed in terms of direct-lattice basis-vectors $\vec{a}_1$, $\vec{a}_2$ and $\vec{c}$ as:

$$\vec{N}_{hk\ell}^* = h\vec{a}_1^* + k\vec{a}_2^* + \ell\vec{c}^* = \frac{2}{3a^2}(2h+k)\vec{a}_1 + \frac{2}{3a^2}(2k+h)\vec{a}_2 + \frac{\ell}{c^2}\vec{c}$$

that we rewrite versus $(\vec{i}, \vec{j}, \vec{k})$:

$$\vec{N}_{hk\ell}^* = \frac{2}{3a}(2h+k)\vec{i} + \frac{2}{3a}(2k+h)\vec{j} + \frac{\ell}{c}\vec{k}$$

After renormalisation of $\vec{N}_{hk\ell}^*$ with its length, one obtains a unit vector $\frac{\vec{N}_{hk\ell}^*}{\|\vec{N}_{hk\ell}^*\|}$ identical to $\vec{W}$.

Hence, zone indices of the direction perpendicular to plane $(hk\ell)$ are, in suitably normalized triplet form, simply:

$$\left[2h+k, h+2k, \frac{3a^2}{2c^2}\ell\right],$$

that are not, in general, three integer numbers.

1.6 The Orthogonal Basis Set

In order to express tensors and matrices, that represent physical properties of crystals, it is recommended to use a new set of axes: $(\vec{x}, \vec{y}, \vec{z})$ which are always mutually perpendicular. Starting from now, the calculations will be made in the international orthogonal $(\vec{x}, \vec{y}, \vec{z})$ basis. The relationship between the unit vectors of the international basis and those of the hexagonal one are:

$$\begin{bmatrix} \vec{x} \\ \vec{y} \\ \vec{z} \end{bmatrix} = \begin{bmatrix} \frac{2}{\sqrt{3}} & \frac{1}{\sqrt{3}} & 0 \\ 0 & 1 & 0 \\ 0 & 0 & 1 \end{bmatrix} \cdot \begin{bmatrix} \vec{i} \\ \vec{j} \\ \vec{k} \end{bmatrix} \tag{1.3}$$

Vector $\vec{x}$ is orthogonal to $\vec{j}$—and, therefore, $\vec{y}$—and makes a 30° angle with $\vec{i}$.

This basis change leads to the following expressions relating $(\vec{U}, \vec{V}, \vec{W})$ and $(\vec{x}, \vec{y}, \vec{z})$:

$$\begin{bmatrix} \vec{U} \\ \vec{V} \\ \vec{W} \end{bmatrix} = \begin{bmatrix} -\frac{\sqrt{3}}{2}Ak & \frac{1}{2}A(2h+k) & 0 \\ -\frac{\sqrt{3}}{2}Aa\ell B_{ac}(2h+k) & -\frac{3}{2}Aa\ell B_{ac}k & 2c\frac{B_{ac}}{A} \\ AcB_{ac}(2h+k) & \sqrt{3}AcB_{ac}k & \sqrt{3}\ell aB_{ac} \end{bmatrix} \begin{bmatrix} \vec{x} \\ \vec{y} \\ \vec{z} \end{bmatrix} \tag{1.4}$$

The unit vectors of the crystallographic basis and the vectors of the reciprocal lattice express as follow in terms of the unity vectors of the international basis set:

$$\begin{bmatrix} \vec{i} \\ \vec{j} \\ \vec{k} \end{bmatrix} = \begin{bmatrix} \frac{\sqrt{3}}{2} & -\frac{1}{2} & 0 \\ 0 & 1 & 0 \\ 0 & 0 & 1 \end{bmatrix} \cdot \begin{bmatrix} \vec{x} \\ \vec{y} \\ \vec{z} \end{bmatrix} \tag{1.5}$$

and

$$\begin{bmatrix} \vec{a}_1^* \\ \vec{a}_2^* \\ \vec{c}^* \end{bmatrix} = \begin{bmatrix} \frac{2}{a\sqrt{3}} & 0 & 0 \\ \frac{1}{a\sqrt{3}} & \frac{1}{a} & 0 \\ 0 & 0 & \frac{1}{c} \end{bmatrix} \cdot \begin{bmatrix} \vec{x} \\ \vec{y} \\ \vec{z} \end{bmatrix} \tag{1.6}$$

respectively.

1.7 The Determination of the Lattice Parameters by X-ray Diffraction

Soon after the discovery of the X-rays by Wilhelm Conrad Röntgen in 1895, people were hesitating regarding their nature. It was, however, believed that they had very short wavelengths (10^{-9} m). After some measurements, German physicist Arnold

Johannes Wilhelm Sommerfeld had determined the ratio between the wavelengths of visible light and X-rays was close to 10,000. Italian chemist Amedeo Avogadro had claimed in 1811 that two equal volumes of different gases, in identical temperature and pressure conditions contained identical numbers of atoms. One century later, this inspired in 1912 Max von Laüe to suggest from the values of Avogadro's number and molar weights of species, the sizes of atoms and molecules. He then proposed that crystals could be used as diffraction gratings for X-rays. He determined theoretically the way to observe such a diffraction and, in addition, proposed diffraction patterns. The experiment was carried out by German scientists Walter Friedrich and Paul Knipping. After a few initial failures, they met with success on April 23, 1912. X-rays when travelling through the crystal formed the pattern of bright spots that proved Laüe's hypothesis was correct. The explanation of the phenomenon is the following and ruled by two different mechanisms: first, each atom scatters the impinging beam in all directions of the three dimensional space. Then, the scattered propagating waves interfere, thus creating the observed diffraction patterns. British physicist George Paget Thompson (the son of Scottish physicist Joseph John Thompson who discovered the electron in 1897) has produced a theory, treating the scattering of X-rays by an atom with Z electrons like Z scatterings independent of each other. This model, which neglects the influence of the nucleus of this atom, predicts the scattered intensity to be proportional to Z. X-rays poorly detect a light atom, like H or Li, whilst a heavy one like gold for instance is easily spotted, thus explainig its use in contemporary medicine.

1.7.1 Diffraction by a Linear Grating

Regarding the interference pattern, let us consider a reticular plane. In this plane, atoms are by definition regularly arranged according to a two-dimensional lattice. There exists a family of identical planes parallel to the considered one. They are regularly spaced by a quantity we call d here. Each atom behaves like a point source and diffuses X-rays in all dimensions of the 3D space. The scattered beams interfere. The phenomenon is identical to the diffraction of light by a linear grating but it is much more difficult to address because it occurs at three dimensions.

Let us consider a network of parallel slits $\{\ldots, S, S', \ldots\}$ separated by a constant distance d. An incident beam, normal to the plane of the slits—so that, all slits are excited in phase and emit accordingly—is diffracted by the series of slits in a direction making the angle θ with the incident direction.

Figure 1.8 illustrates the geometrical aspects of the diffraction phenomenon in a plane orthogonal to the plane of slits. The plane of the figure then contains both incident and transmitted (or diffracted) beams. The condition for having light in the BB′ plane is that the difference in optical paths between all beams be an integer number times the wavelength, which writes:

$$S'B' - SB = n\lambda$$

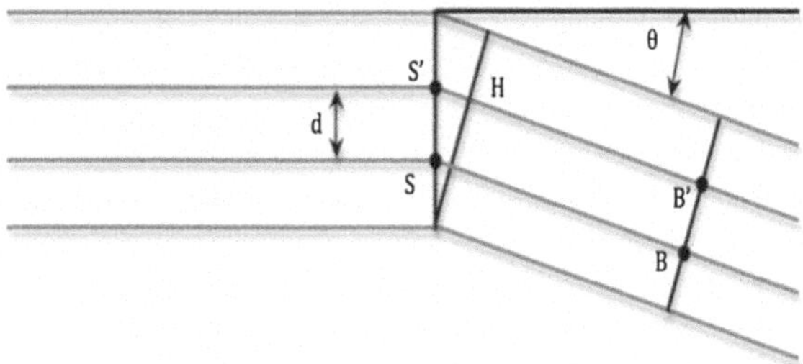

Fig. 1.8 Geometrical aspects of the diffraction phenomenon in a plane orthogonal to the plane of slits

This quantity equals SH may be identified to $d \sin \theta$.
Therefore, we get:

$$d \sin \theta = n\lambda$$

or, alternatively

$$\theta = \arcsin(n\lambda/d)$$

In case the incident beam makes an angle i with the plane's normal, this equation becomes:

$$d(\sin \theta - \sin i) = n\lambda$$

or, alternatively:

$$\theta = \arcsin(n\lambda/d + \sin i)$$

The angle θ varies with the diffraction order n.

1.7.2 *Diffraction by a Linear Lattice, and by a Planar One*

Let us now consider a linear chain of equidistant atoms shined by a monochromatic beam of X-rays. Such a situation is encountered when rotating a crystal around a given crystallographic direction. This configuration is called the rotating-crystal diffraction method. The X-ray detector can be either a cylinder photographic sheet (detector of the old days) or a linear/bi-dimensional array of X-ray detectors (a smarter detection system but more expensive too…). Now, the elementary scattering sources are individual points. In contrast to the linear slits, the scattering is now isotropic around each source. The scattering occurs in the 3D space, and the diffracted beams of different orders n form a series of cones parallel to the linear chain, each one having an apex angle $\phi = \pi/2 - \theta$ forming diffraction surfaces. See Fig. 1.9 (left) and (right).

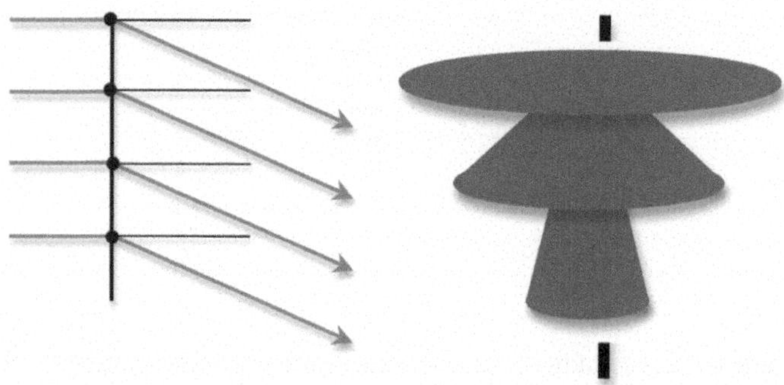

Fig. 1.9 *Left* Geometrical representation of the X-ray diffraction in a plane that contains the linear chain, the incoming beams and the reflected ones. *Right* Geometrical representation of the X-ray diffraction surfaces plane that is parallel to the linear chain. Note the cones have different angles when n changes

Let us now consider a planar two-dimensional periodic arrangement of atoms. One can demonstrate that each plane reflects the incident beam as if it was reflected by a plane mirror whatever the incidence angle, similarly with the laws of light reflection. To avoid confusion, we emphasize here that the origin of this behaviour is correlated to interferences whereas the reflexion of light by a mirror has to be interpreted using Maxwell's equations.

1.7.3 Diffraction by a Three-Dimensional Lattice

Let us consider a family of reticular planes ($hk\ell$) with interplane reticular distance d. Each plane reflects the X-rays, as mentioned above, whatever the incident angle. The beams reflected by all planes interfere, with extinctions at specific angles. A global reflexion occurs for specific angles, what is called selective reflectivity. Let us consider in Fig. 1.10, the optical paths A′B′C′ and ABC of X rays impinging two adjacent planes under an angle θ.

AA′ and CC′ are orthogonal to AB and BC respectively. The incidence and reflexion angles are equal.

The necessary condition to observe a diffracted radiation in the plane orthogonal to BC and B′C′ is:

$$A'B'C' - ABC = n\lambda.$$

Let G (resp. H) be the projection of BC (resp. AB) on A′B′ (resp. B′C′). We remark that:

$$HB = GB' = d\sin\theta$$

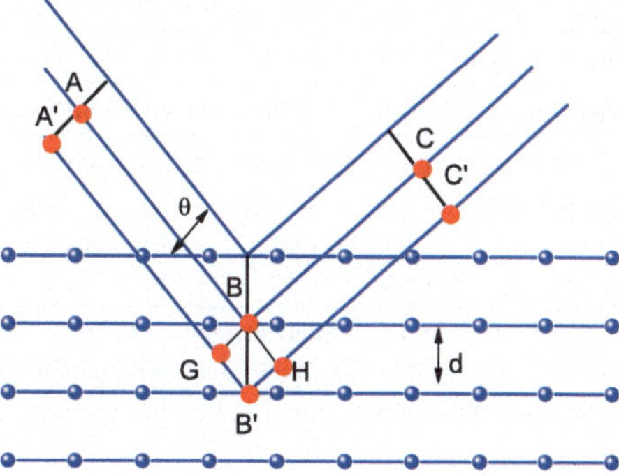

Fig. 1.10 Geometrical sketch showing the orientation of incident and relected beams on a family of reticular planes

Then:

$$\mathrm{A'B'C'} - ABC = 2d \sin \theta$$

Then, the so-called "Bragg relation" is obtained:

$$n\lambda = 2d \sin \theta$$

which is of paramount importance in radio-crystallography. It indicates that, when an X-ray beam encounters a family of reticular planes, it is diffracted under an angle:

$$\theta = \arcsin \frac{n\lambda}{2d}$$

Back to the hexagonal lattice, keeping in mind that the distance $d_{hk\ell}$ between adjacent reticular $(hk\ell)$ planes of the direct lattice is correlated with the norm of the reciprocal lattice vector N^*_{hkl} by:

$$\vec{N}^*_{hk\ell} \cdot \vec{d}_{hk\ell} = 1$$

For any lattice, a simple calculation leads us to express analytically the Bragg relation:

$$\theta_{hk\ell} = \arcsin \left[\frac{n\lambda}{2} \sqrt{\frac{4}{3a^2}(h^2 + hk + k^2) + \frac{\ell^2}{c^2}} \right]$$

X-ray diffraction angles lead to the determination of a and c.

Table 1.2 summarizes the values of c and a measured for wurtzite semi-conductors as well as values of the c/a ratio that should be compared with optical values. We have also included information regarding hexagonal boron nitride and graphite. For these two cases, one notices large values of c with respect to a, due to the loose chemical bound along the six-fold symmetry axis. These compounds are layered compounds. For boron nitride, the wurtzitic phase is metastable.

Table 1.2 Values of the lattice parameters c and a together with the ratio c/a for various wurtzitic semi-conductors

Material	BN	GaN	InN	AlN	ZnS	CdS	ZnO	hex-BN	hex-C
c (10^{-10} m)	4.170	5.185	5.708	4.980	6.260	6.714	5.213	6.661	6.708
a (10^{-10} m)	2.550	3.188	3.539	3.110	3.822	4.136	3.253	2.504	2.461
c/a (X-ray)	1.635	1.627	1.613	1.610	1.638	1.623	1.602	2.66	2.725
c/a (optics)	1.64	1.62	1.61	1.6	1.64	1.62	1.60	–	–

The two right-hand columns correspond to layer compounds, which are found under the crystalline-layered compound hexagonal system; which differ from wurtzite. Values of c/a obtained using both X-rays and optics are given. The number of digits is compatible with the determination method

1.8 The Determination of Space Symmetry by X-ray Analysis

1.8.1 The First Brillouin Zone

German physicist Max Von Laüe has proposed to compute, at a recording distance much larger than the size of the studied crystal, the amplitude of a wave elastically scattered from $\vec{k}$ to $\vec{k}'$:

$$A = \sum_{uvw} exp\left[-i2\pi\vec{\rho}_{uvw}\cdot\Delta\vec{k}\right]$$

The summation is extended over all the sites of the lattice generated by vectors $\vec{a}_1$, $\vec{a}_2$ and $\vec{c}$:

$$\vec{\rho}_{uvw} = u\vec{a}_1 + v\vec{a}_2 + w\vec{c}$$

This summation reaches a maximum when all scattering centres are in phase, i.e. when

$$\vec{\rho}_{uvw}\cdot\Delta\vec{k} = n(\text{with } n \in N)$$

This requires to simultaneously fulfill three equations:

$$\vec{a}_1\cdot\Delta\vec{k} = p$$

$$\vec{a}_2\cdot\Delta\vec{k} = q$$

and

$$\vec{c}\cdot\Delta\vec{k} = r$$

with

$$(p, q, r) \in N^3$$

These three identities are the Laüe equations: they indicate that $\Delta\vec{k}$ is by definition a vector of the reciprocal lattice.

We now write:

$$\Delta\vec{k}^2 = (\vec{k}' - \vec{k})^2 = (\vec{k} - \vec{G})^2 - \vec{k}^2$$

by introducing a vector of the reciprocal lattice $\vec{G}$.

Thanks to energy conservation in case of elastic scattering: $\Delta\vec{k}^2 = 0$ which leads us to rewrite:

$$\Delta\vec{k}^2 = (\vec{k}' - \vec{k})^2 = (\vec{k} - \vec{G})^2 - \vec{k}^2 = \vec{k}^2 - 2\vec{k}\vec{G} + \vec{G}^2 - \vec{k}^2$$

This gives the important relation:

$$2\vec{k} \cdot \vec{G} - \vec{G}^2 = 0$$

We rewrite this equation:

$$2\vec{k} \cdot \vec{G} = \vec{G}^2$$

can be rearranged into:

$$\vec{k} \cdot (\vec{G}/2) = (\vec{G}/2)^2$$

It means that $\vec{k}$ is a half-vector of the reciprocal lattice:

$$\vec{k} = \vec{G}/2$$

The locus of these $\vec{k}$'s is called the Wigner-Seitz cell or the first Brillouin zone of the reciprocal lattice.

1.8.2 The Structure Factor

In the preceding sections, we have considered the interferences made by atoms forming families of reticular planes. From the interferences patterns made by the atomic arrangement and these atoms, a simple geometrical analysis led us to obtain the Bragg relation, correlating a diffraction angle to the wavelength of the X-rays and to the separation between adjacent reticular planes. The pattern obtained by intercepting the diffracted beam with a recording system gives an accurate description of the reciprocal lattice, which basic building block is interpreted in terms of the Wigner-Seitz cell of the reciprocal lattice, often called the first Brillouin zone. Typical diffraction angles being measured for different reticular planes, one can, at the end, reach the value of the lattice parameters. In the case of simple structures, an extended version of the equation is:

$$A = \sum_{uvw} exp\left[-i2\pi\vec{\rho}_{uvw} \cdot \Delta\vec{k}\right]$$

can be alternatively written:

$$A = \int n(\rho) exp\left[-i2\pi\vec{\rho}\cdot\Delta\vec{k}\right] dV$$

where $n(\vec{\rho})$ is the local electronic density.

Let us suppose that each cell contains several atoms (s for example); the nucleus of the jth atom located at position:

$$\vec{\rho_j} = x_j\vec{a}_1 + y_j\vec{a}_2 + z_j\vec{c}$$

relatively to the position of the node:

$$\vec{\rho}_{uvw} = u\vec{a}_1 + v\vec{a}_2 + w\vec{c}$$

$$(x_i, y_i, z_i) \in [0, 1]^3$$

Let the origin be $\vec{\rho_{000}}$ so that we can write the total electronic density in the crystal $n(\vec{\rho})$ as a double summation over all atomic positions in one cell, and through all cells:

$$n(\vec{\rho}) = \sum_{uvw}\sum_{j=1,s} c_j(\vec{\rho} - \vec{\rho_j} - \vec{\rho}_{uvw})$$

where c_j is the electronic density associated with atom j. Then:

$$A_{\Delta\vec{k}} = \sum_{uvw}\sum_{j=1,s}\int c_j(\vec{\rho} - \vec{\rho_j} - \vec{\rho}_{uvw}) exp\left[-i2\pi\vec{\rho}\cdot\Delta\vec{k}\right] dV$$

$$A_{\Delta\vec{k}} = \sum_{uvw}\sum_{j=1,s}\int c_j(\vec{\rho} - \vec{\rho_j} - \vec{\rho}_{uvw}) exp\left[-i2\pi(\vec{\rho} - \vec{\rho_j} - \vec{\rho}_{uvw})\Delta\vec{k}\right]$$
$$exp\left[-i2\pi(\vec{\rho_j} + \vec{\rho}_{uvw})\Delta\vec{k}\right] dV$$

This equation becomes:

$$A_{\Delta\vec{k}} = \sum_{uvw}\sum_{j=1,s} F_j exp\left[-i2\pi(\vec{\rho_j} + \vec{\rho}_{uvw})\cdot\Delta\vec{k}\right] dV$$

where quantities F_j are defined as:

$$F_j = \int c_j(\vec{x}) exp\left[-i2\pi\vec{x}\Delta\vec{k}\right]$$

$$A_{\Delta\vec{k}} = \sum_{uvw} exp\left[-i2\pi\vec{\rho}_{uvw}\cdot\Delta\vec{k}\right]\sum_{j=1,s} F_j exp\left[-i2\pi\vec{\rho}_j\Delta\vec{k}\right]$$

We know, from above, that $\sum_{uvw} exp\left[-i2\pi\vec{\rho}_{uvw}\cdot\Delta\vec{k}\right]$ doesnot vanish—if and only if—$\Delta\vec{k}$ is a vector $\vec{G}$ of the reciprocal lattice.

Then, the quantity:

$$S_{\vec{G}} = \sum_{j=1,s} F_j exp\left[-i2\pi\vec{\rho}_j\vec{G}\right]$$

is called structure factor.

Let

$$\vec{G}_{hk\ell} = h\vec{a}_1^* + k\vec{a}_2^* + \ell\vec{c}^*,$$

then:

$$S_{hk\ell} = \sum_j F_j exp\left[-i2\pi(hx_j + ky_j + \ell z_j)\right]$$

This equation leads to selection rules on (h, k and ℓ), depending on the relative positions of atoms within the cell, when the argument of the complex exponential vanishes. Beyond the scope of point group symmetry, from the relative positions of similar atoms in the cell, are defined complementary symmetry operations within the cell that give access to the space group of the crystal. There are 32 point groups and 230 space groups. Studying them is beyond the scope of this monography.

1.8.3 The Perfect Wurtzite Structure

The wurtzite structure is more complex than this: as demonstrated by W.L. Bragg in 1914, it consists of two interpenetrating Hexagonal Closed Packed sub-lattices, one for each atomic species, offset along the *c* axis by 5/8 of the cell height. The atomic positions in the unit cell are (0, 0, 0) and (2/3, 1/3, 1/2) for the anions, and (0, 0, 3/8) and (2/3, 1/3, 7/8) for the cations and the positions of the atoms in a hexagonal cell are indicated in Fig. 1.11.

It is worthwhile noticing that the structure factor is the sum of contributions from the cations and from the anions:

$$S_{hk\ell} = \left(1 + exp\left[-i\pi\frac{2h + 4k + 3\ell}{3}\right]\right)\left(F_{cation} + F_{anion} exp\left[-i\frac{3\pi\ell}{4}\right]\right)$$

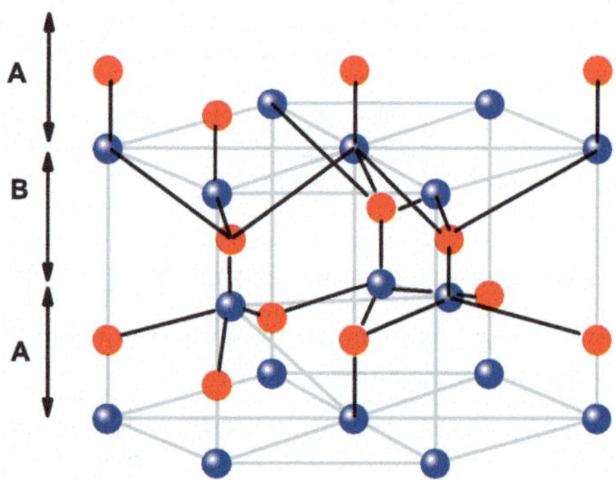

Fig. 1.11 Plot of the positions of the atoms in a hexagonal cell of a wurtzite crystal

We know from the properties of the exponential function that:

$$1 + exp\left[-i\pi\frac{2h+4k+3\ell}{3}\right]$$
$$= 2\cos\pi\frac{2h+4k+3\ell}{6}(\cos\pi\frac{2h+4k+3\ell}{6} - i\sin\pi\frac{2h+4k+3\ell}{6})$$
$$\cos\pi\frac{2h+4k+3\ell}{6} = 0$$

when

$$\pi\frac{2h+4k+3\ell}{6} = p\frac{\pi}{2} \quad \text{with} \quad p \in \mathbf{Z}$$

This equation indicates that extinguished radiations correspond to $(h, k, -(h+k), \ell)$ reticular planes for which $2h + 4k + 3\ell = 3\text{p}$, with $p \in \mathbf{Z}$

This equation is called a selection rule.

On Fig. 1.12 is reported the diffraction pattern recorded by shining a powder of indium nitride with the CuK_α X-ray radiation (λ = 0.15 nm). The powder is constituted with micro-crystals randomly oriented so that artefact selection rules due to specific configuration of the crystals in the context of the geometry of the diffraction experiment can be disregarded.

The intensity of the diffraction pattern corresponding to a given set of Miller indices $\{h, k, \ell\}$ is given by the square of the diffraction amplitude, $\|A_{hk\ell}\|^2$.

Identification of the $(hk\ell)$ sets that lead to a given peak in the context of an overall consistency is a tricky problem. Signatures of specific planes are at last identified, indicating the existence of selection rules. The cell parameters are $a_0 = 3.5390$ Å and $c_0 = 5.7083$ Å. We note that these values are given with a large number of digits, this is not systematic, and have to be correlated with the experiment's temperature (here room temperature), and with doping since the presence of foreign atoms modifies the lattice constants. The corresponding space group indicating the site symmetry for the wurtzite is No. 186 (in International Union of Crystallography classification) or P_{63mc} in Hermann-Mauguin notation or C^3_{6V} in Schoenflies notation.

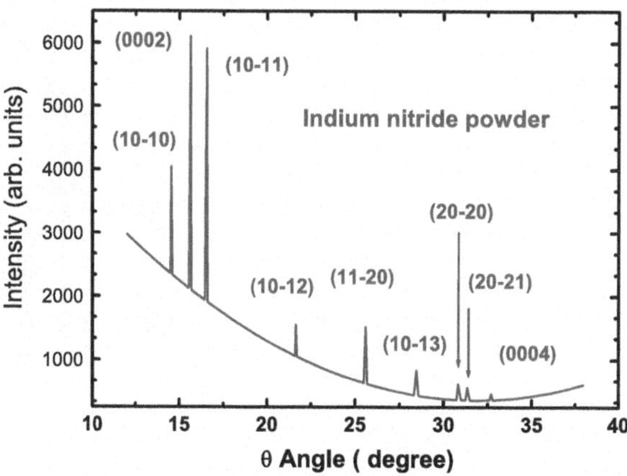

Fig. 1.12 Powder diffraction pattern of indium nitride recorded at room temperature using the CuK_α radiation ($\lambda = 0.15$ nm)

1.8.4 The Internal Displacement Parameter

The internal displacement parameter u is defined as the anion-cation bond length that is also the nearest-neighbour distance in the crystal divided by the c lattice parameter. The interatomic distances are expressed versus this parameter.

In an ideal wurtzite structure represented by four touching hard spheres, the values of the axial ratio and the internal parameter are $c/a = \sqrt{\frac{8}{3}} = 1.633$ and $u = 3/8 = 0.375$, respectively. In the hexagonal basis, the atomic positions in the unit cell are (0, 0, 0) and (2/3, 1/3, 1/2) for the anions, (0, 0, u) and (2/3, 1/3, (u+1/2)) for the cations.

In *Cartesian coordinates*, the crystallographic vectors of wurtzite are:

$$\begin{bmatrix} \vec{a}_1 \\ \vec{a}_2 \\ \vec{c} \end{bmatrix} = a \begin{bmatrix} \frac{\sqrt{3}}{2} & -\frac{1}{2} & 0 \\ 0 & 1 & 0 \\ 0 & 0 & \frac{c}{a} \end{bmatrix} \cdot \begin{bmatrix} \vec{x} \\ \vec{y} \\ \vec{z} \end{bmatrix} \tag{1.7}$$

The atomic positions are (0, 0, 0), and $a(\frac{1}{\sqrt{3}}, 0, \frac{c}{2a})$ for anions.

For cations, positions are $a(\frac{1}{\sqrt{3}}, 0, (\frac{1}{2} - u)\frac{c}{a})$, and $a(0, 0, (1 - u)\frac{c}{a})$.

The nearest-neighbour bond length along the c direction is:

$$b = cu$$

and off-axis nearest-neighbour bond length is, as indicated in Fig. 1.13:

Fig. 1.13 Relative positions of the atoms in the wurtzite structure, directions to nearest neighbours. The c axis is vertical

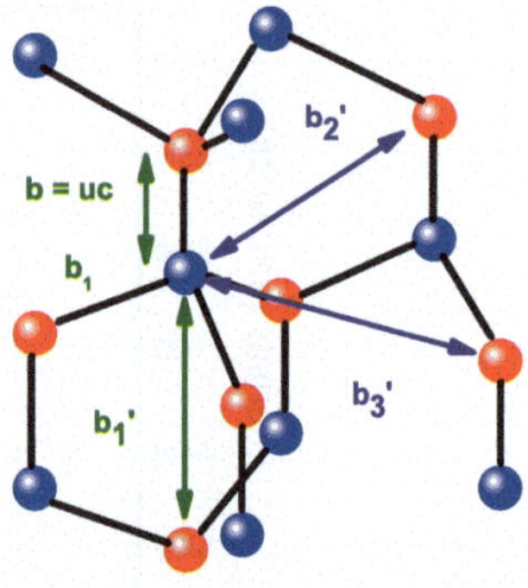

Table 1.3 The evolution of a, c/a and u for wurtzitic III-nitrides

Material	GaN	InN	AlN
a (10^{-10} m)	3.1879	3.5390	3.110
c/a	1.626	1.613	1.610
u	0.377	0.379	0.382

$$b_1 = \sqrt{\frac{a^2}{3} + (1 - u^2)c^2}$$

The four nearest neighbours of each atom form a trigonal pyramid.

We remark that there are three types of second-nearest neighbours designated as:

$$b_1' = (1 - u)c$$

$$b_2' = \sqrt{a^2 + u^2c^2}$$

and

$$b_3' = \sqrt{\frac{a^2}{3} + (\frac{1}{2} - u)^2c^2}$$

In many wurtzitic semi-conductors such as nitrides, experimentally observed $c/$ ratios and values of internal displacement parameter u differ from the ideal values. These discrepancies are due to atoms being real quantum objects, departing more or less from the hard sphere description, either in vacuum or when chemically bonded to other atoms. It should be pointed out that a strong correlation exists between the c/a ratio and the u parameter so that when c/a decreases, u increases in a manner to keep the four tetrahedral distances nearly constant through a distortion of tetrahedral angles. Table 1.3 illustrates the evolution of c/a and u for wurtzitic III-nitrides.

Fig. 1.14 Relative ordering of the atomic planes in the wurtzite structure along the c direction

Table 1.4 The evolution of spontaneous polarization P_{sp}, ratio c/a and internal displacement parameter u for wurtzitic semi-conductors

Material	GaN	InN	AlN	ZnO	BeO
P_{sp} (Cm^{-2})	−0.0339	−0.0413	−0.0898	−0.057	−0.045
c/a	1.626	1.613	1.610	1.60	1.622
u	0.377	0.379	0.382	0.382	0.378

1.9 The Spontaneous Polarization Along the c Axis

In Fig. 1.14, we present a sketch of the atomic planes along the six-fold symmetry axis, which consists of successive stacking of planes that only contain cations or anions, the so-called ABAB stacking.

The important point to outline from the figure is the non-overlap of the center of gravity of reticular planes of cations (positive charges) and anions (negative charges) that produces an electronic dipole along the c direction of the crystal. There is a spontaneous polarization internal to the wurtzitic crystal. This quantity is not easy to measure in bulk crystals, due to the polar nature of the surface. This surface easily traps impurities so that the total polarization drop through the crystal vanishes. Impurities, topological defects may create electric fields and compensate this quantity too. This quantity has been computed by several group of theorists, using very sophisticated approaches in the context of the quantum mechanics description of the chemical bonds. These so-called ab-nitio calculations have furnished some numerical values for these quantities. It can be seen in Table 1.4 that the spontaneous polarization P_{sp} in III-Ns (resp. II-Os) has a negative value. The conventional [0001] direction taken as positive goes from the group III- (respectively group II-) element atom to the group V- (respectively group VI-) element and opposes to P_{sp}. In nitrides for instance, the negative value indicates that the electric dipole is oriented from the N atom towards the group III-element (Al, Ga, or In) one.

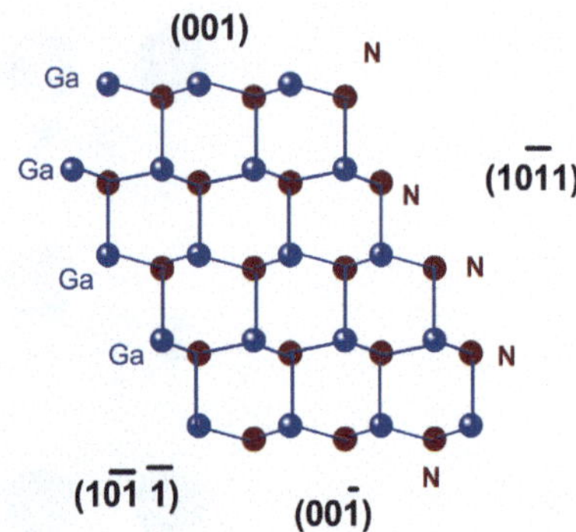

Fig. 1.15 relative orientation of the atomic stackings and non equivalence of the (0001) and (000$\bar{1}$) surface. By convention, the (0001) plane only contains metallic atoms

It is worthwhile noticing, from Fig. 1.15, that when cutting the [0001] bonds above a group III- (respectively group II-) element, we define the (0001) surface only composed with metallic atoms. If cutting the [0001] bonds below the reticular plane of group V- (respectively group VI-) element, the $(0, 0, 0, \bar{1})$ surface contains now non metallic atoms. The chemical composition of the $(0, 0, 0, 1)$ surface differs from that of the $(0, 0, 0, \bar{1})$ surface. These two surfaces are not equivalent. One can anticipate that they will behave very differently, when studying, for instance, their responses to chemical aggressions. Let's define the metallic surface as the (0001). One can easily demonstrate that the $(3, 0, \bar{3}, \bar{1})$ surface only contains metallic atoms whereas $(3, 0, \bar{3}, 1)$ does not contain any. In contrast, some surfaces such as $(1, 1, \bar{2}, 2)$ contain both kinds of atoms.

There are no real correlations between the values of P_{sp} and the c/a ratio, the internal displacement parameter. This indicates that P_{sp} is a real effect, deriving from the quantum nature of the chemical bonds, that depends on the specific stacking of the different atoms. The important point to outline here is that, thanks to the quantum origin of the value spontaneous polarization, its value in an alloy like $In_{1-x}Ga_xN$ is not obtained by averaging the spontaneous polarization in both binaries, since local modifications of the crystal composition due to chemical disorder in the cation lattice may lead to significant departure of the real value from the predicted one.

In the case of nitride reticular planes having arbitrary $(hk\ell)$ orientations, according to the relation that connects the $(\vec{U}, \vec{V}, \vec{W})$ vectors and the international basis set, the spontaneous polarization has components in the $\vec{V}$ and $\vec{W}$ directions only. These components are obtained by simple matrix algebra as follows:

$$\begin{bmatrix} (P_{sp})_U \\ (P_{sp})_V \\ (P_{sp})_W \end{bmatrix} = \begin{bmatrix} -\frac{\sqrt{3}}{2}Ak & \frac{1}{2}A(2h+k) & 0 \\ -\frac{\sqrt{3}}{2}Aa\ell B_{ac}(2h+k) & -\frac{3}{2}Aa\ell B_{ac}k & 2c\frac{B_{ac}}{A} \\ AcB_{ac}(2h+k) & \sqrt{3}AcB_{ac}k & \sqrt{3}\ell aB_{ac} \end{bmatrix} \begin{bmatrix} 0 \\ 0 \\ P_{sp} \end{bmatrix} \quad (1.8)$$

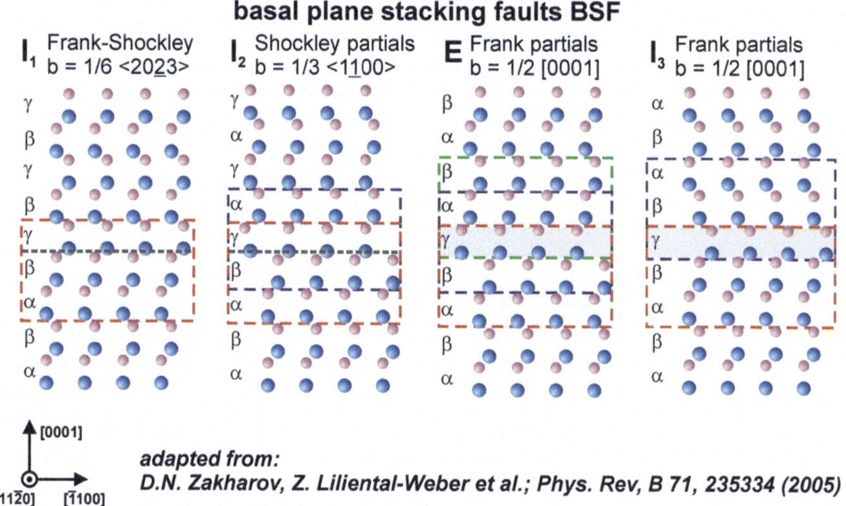

Fig. 1.16 Basal stacking faults along the [0001] direction. Courtesy Gordon Schmidt, Frank Bertram and Juergen Christen, University Otto von Gericke of Magdeburg

1.10 Defects in the Lattice

There are sometimes some departures from the perfect crystal arrangements. Such departures constitutes defects which may have deleterious influence on optical and transport properties. Among such defects are basal stacking faults. In Fig. 1.16 are represented basal stacking faults which are some of the possible lattice defects. Studying defects is out of the scope of this book.

1.11 Piezoelectric Effects in Wurtzitic Semi-conductors

As discussed in the preceding section, the non-overlap of positive and negative atomic charges produces a polarization field along the six-fold symmetry axis of wurtzitic semi-conductors. This effect is sometimes called the pyroelectric effect because its magnitude may change when temperature of the crystal changes, leading to change of the c/a ratio for instance. Among the thirty-two possible point groups required to describe the orientational symmetries for bulk crystals, only ten are pyro-electric crystals, i.e. they exhibit a spontaneous polarisation field of given orientation with respect to the crystallographic axes. These point groups are in general moderate symmetry sub-groups of the seven groups that describe the full symmetry of the seven crystallographic systems. Among the twenty-one non centro-symmetric groups, ten do not exhibit inversion symmetry with respect to a center point of the cell. In addition, they also possess a complementary property that does not exist in the eleven

Table 1.5 Piezoelectric and pyroelectric properties of the 32 point groups

32 symmetry classes	
12 non piezo-electric classes	20 piezo-electric classes
O and the 11 centro-symmetric classes:	10 non pyroelectric or non polar classes:
$O, C_i, C_{2h}, D_{2h}, D_{4h}, C_{3i}, D_{3d}, C_{6h}, D_{6h}, T_h, O_h$	$D_2, D_{2d}, D_3, D_4, S_4, D_6, C_{3h}, D_{3h}, T, T_d$
	10 pyroelectric or polar classes:
	$C_1, C_2, C_3, C_4, C_6, C_s, C_{2v}, C_{3v}, C_{4v}, C_{6v}$

centro-symmetric groups and group O—this one, due to redunding symmetry effects: upon application of a strain field, a complementary polarization may be induced in the crystal. This strain-induced polarization is called piezo-electric polarization and a crystal behaving accordingly is named piezo-electric. Symmetry considerations and utilization of group representations have enabled researchers to determine whether a crystal is piezo-electric, pyro-electric, both or none. The results are compiled in Table 1.5. Some cubic crystals like the zinc-blende semi-conductors (T_d symmetry) are piezo-electric crystals like rhomboedral quartz (D_3 symmetry). Quartz is not pyro-electric whilst the most symmetrical wurtzite (C_{6v} symmetry) exhibits spontaneous and may be piezo-electrically polarized. The impossibility to tune the spontaneous polarization in wurtzite crystals—we remind that it is oriented along the six-fold symmetry axis—is a rigid situation, compensated by a flexible situation regarding the piezo electric one. The orientation, magnitude and sign of the latter can be tuned almost at will (Table 1.6).

Given a strain field described by a two-dimensional strain tensor which components are ϵ_{xx}, ϵ_{yy}, ϵ_{zz}, ϵ_{xz}, ϵ_{xz} and ϵ_{xy} the components of the piezo electric polarization are given by:

$$\begin{bmatrix} P_x \\ P_y \\ P_z \end{bmatrix} = \begin{bmatrix} 0 & 0 & 0 & 0 & 2e_{15} & 0 \\ 0 & 0 & 0 & 2e_{15} & 0 & 0 \\ e_{13} & e_{13} & e_{33} & 0 & 0 & 0 \end{bmatrix} \begin{bmatrix} \epsilon_{xx} \\ \epsilon_{yy} \\ \epsilon_{zz} \\ \epsilon_{yz} \\ \epsilon_{xz} \\ \epsilon_{xy} \end{bmatrix} \tag{1.9}$$

where quantities e_{ij} are the components of the piezo-electric tensor.

In the case of nitride reticular planes having arbitrary (h, k, ℓ) orientations, according to the relation that connects the $(\vec{U}, \vec{V}, \vec{W})$ vectors and the international basis set, the piezo-electric polarization has components in the $\vec{U}$, $\vec{V}$ and $\vec{W}$ directions.

These components are obtained by simple matrix algebra as follows:

$$\begin{bmatrix} (P_{pz})_U \\ (P_{pz})_V \\ (P_{pz})_W \end{bmatrix} = \begin{bmatrix} -\frac{\sqrt{3}}{2}Ak & \frac{1}{2}A(2h+k) & 0 \\ -\frac{\sqrt{3}}{2}A a \ell B_{ac}(2h+k) & -\frac{3}{2}A a \ell B_{ac}k & 2c\frac{B_{ac}}{A} \\ AcB_{ac}(2h+k) & \sqrt{3}AcB_{ac}k & \sqrt{3}\ell a B_{ac} \end{bmatrix} \begin{bmatrix} P_x \\ P_y \\ P_z \end{bmatrix} \tag{1.10}$$

Table 1.6 The components of the piezo electric tensor for wurtzitic semi-conductors

Material	GaN	InN	AlN
e_{13} (Cm^{-2})	−0.338	−0.413	−0.533
e_{33} (Cm^{-2})	−0.372	−0.454	−0.623
e_{15} (Cm^{-2})	−0.167	−0.112	−0.351

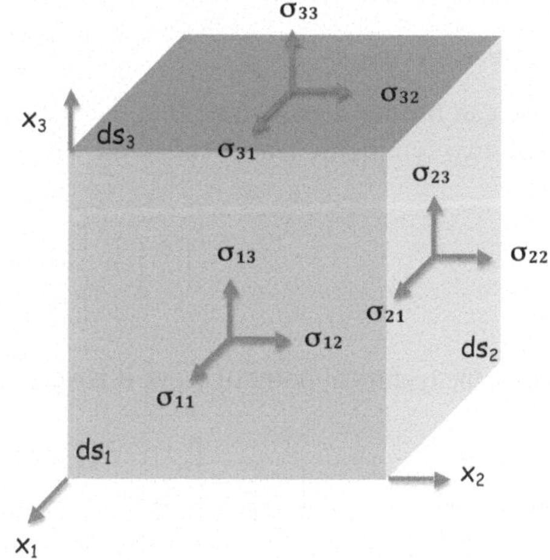

Fig. 1.17 relative orientation of the components of the stress tensor

1.12 Stresses and Strains

1.12.1 The Stress Tensor

Let us consider an elastic solid at the equilibrium with the directions orthogonal to its facets parallel to the $\vec{x}$, $\vec{y}$ and $\vec{z}$ directions that we propose to represent as $\vec{x_1}$, $\vec{x_2}$ and $\vec{x_3}$ respectively. It is well known from continuous media mechanics that side i of facets located at the origin O, which elementary surface is $ds_i = dx_j dx_k$ experiences an elastic force:

$$d\vec{F}_i = \vec{\sigma}_i ds_i = \vec{\sigma}_i dx_j dx_k$$

Here, the (i, j, k) triplet designs a circular permutation through triplet (1, 2, 3). The components of such a force along the ℓ direction is:

$$dF_{i\ell} = \sigma_{i\ell} dx_j dx_k$$

as shown on Fig. 1.17.

We obtain 3 equations $d\vec{F}_i = \vec{\sigma}_i ds_i = \vec{\sigma}_i dx_j dx_k$ and 9 equations $dF_{i\ell} = \sigma_{i\ell} dx_j dx_k$. The scalar quantities $\sigma_{i\ell}$ are the components of the stress at the origin O.

$$|\sigma_{i\ell}| \simeq |\sigma_{i\ell}^0| + \frac{|\partial\sigma_{i\ell}^0|}{\partial x_i} dx_i$$

In the context of homogeneous stress:

$$|\sigma_{i\ell}| \simeq |\sigma_{i\ell}^0| + \frac{|\partial\sigma_{i\ell}^0|}{\partial x_i} dx_i = |\sigma_{i\ell}^0|$$

One can further demonstrate that the scalar quantities $\sigma_{i\ell}$ are the components of a rank-two symmetric tensor called the stress tensor that we write as follows:

$$\left[\vec{\vec{\sigma}}\right] = \begin{bmatrix} \sigma_{11} & \sigma_{12} & \sigma_{13} \\ \sigma_{12} & \sigma_{22} & \sigma_{23} \\ \sigma_{13} & \sigma_{23} & \sigma_{33} \end{bmatrix} \tag{1.11}$$

From the technical point of view, it is very convenient to write the stress tensor:

$$\left[\vec{\vec{\sigma}}\right] = \begin{bmatrix} \sigma_{11} & \sigma_{12} & \sigma_{13} \\ \sigma_{12} & \sigma_{22} & \sigma_{23} \\ \sigma_{13} & \sigma_{23} & \sigma_{33} \end{bmatrix} = \begin{bmatrix} \sigma_1 & \sigma_6 & \sigma_5 \\ \sigma_6 & \sigma_2 & \sigma_4 \\ \sigma_5 & \sigma_4 & \sigma_3 \end{bmatrix} \tag{1.12}$$

where couples of indices 11, 22, 33, 23, 13, and 12 are one by one represented by one index running from 1 to 6.

Alternatively, the stress tensor may be written under the useful form below:

$$\left[\vec{\vec{\sigma}}\right] = \begin{bmatrix} \sigma_1 & \sigma_6 & \sigma_5 \\ \sigma_6 & \sigma_2 & \sigma_4 \\ \sigma_5 & \sigma_4 & \sigma_3 \end{bmatrix} = \begin{bmatrix} \sigma_1 \\ \sigma_2 \\ \sigma_3 \\ \sigma_4 \\ \sigma_5 \\ \sigma_6 \end{bmatrix}. \tag{1.13}$$

1.12.2 The Strain Tensor

Let us consider two close points A and B of a solid at the equilibrium in the absence of any mechanical perturbation. We write their coordinates relatively to the origin O, in terms of the orthogonal vectors $\vec{x}$, $\vec{y}$ and $\vec{z}$ as follows:

$$\left[\overrightarrow{OA}\right] = \begin{bmatrix} x_1 \\ x_2 \\ x_3 \end{bmatrix} \text{ and } \left[\overrightarrow{OB}\right] = \begin{bmatrix} x_1 + dx_1 \\ x_2 + dx_2 \\ x_3 + dx_3 \end{bmatrix} \tag{1.14}$$

That gives:

$$\left[\overrightarrow{AB}\right] = \left[\vec{ds}\right] = \begin{bmatrix} dx_1 \\ dx_2 \\ dx_3 \end{bmatrix} \tag{1.15}$$

Under the influence of the stress, A shifts to A' and B shifts to B' as follows

$$\left[\overrightarrow{AA'}\right] = \left[\vec{u}\right] = \begin{bmatrix} u_1 \\ u_2 \\ u_3 \end{bmatrix} \text{ and } \left[\overrightarrow{BB'}\right] = \left[\vec{u} + d\vec{u}\right] = \begin{bmatrix} u_1 + du_1 \\ u_2 + du_2 \\ u_3 + du_3 \end{bmatrix} \tag{1.16}$$

The coordinates of A' and B' are:

$$\left[\overrightarrow{OA'}\right] = \begin{bmatrix} x_1 + u_1 \\ x_2 + u_2 \\ x_3 + u_3 \end{bmatrix} \tag{1.17}$$

and

$$\left[\overrightarrow{OB'}\right] = \begin{bmatrix} u_1 + du_1 + x_1 + dx_1 \\ u_2 + du_2 + x_2 + dx_2 \\ u_3 + du_3 + x_3 + dx_3 \end{bmatrix} \tag{1.18}$$

Then $\overrightarrow{ds}$ is transformed into $\overrightarrow{ds}'$

$$\left[\overrightarrow{A'B'}\right] = \left[\vec{ds}'\right] = \begin{bmatrix} du_1 + dx_1 \\ du_2 + dx_2 \\ du_3 + dx_3 \end{bmatrix} \tag{1.19}$$

We now write for every component i:

$$du_i = \frac{\partial u_i}{\partial x_1} dx_1 + \frac{\partial u_i}{\partial x_2} dx_2 + \frac{\partial u_i}{\partial x_3} dx_3$$

or, by introducing a (3×3) matrix $[\beta]$ which components are $\beta_{ij} = \frac{\partial u_i}{\partial x_j}$:

$$\left[\vec{du}\right] = [\beta] . \left[\vec{dx}\right] \tag{1.20}$$

Here, $[\beta]$ is the matrix representation of the strain tensor $\vec{\vec{\beta}}$.

$[\beta]$ is the sum of an antisymmetric matrix $[\varpi]$ plus a symmetric one $[\epsilon]$.

$$\begin{bmatrix} \beta_{11} & \beta_{12} & \beta_{13} \\ \beta_{21} & \beta_{22} & \beta_{23} \\ \beta_{31} & \beta_{32} & \beta_{13} \end{bmatrix} = \begin{bmatrix} 0 & \varpi_{12} & \varpi_{13} \\ -\varpi_{12} & 0 & \varpi_{23} \\ -\varpi_{13} & -\varpi_{23} & 0 \end{bmatrix} + \begin{bmatrix} \epsilon_{11} & \epsilon_{12} & \epsilon_{13} \\ \epsilon_{21} & \epsilon_{22} & \epsilon_{23} \\ \epsilon_{31} & \epsilon_{32} & \epsilon_{13} \end{bmatrix} \tag{1.21}$$

Fig. 1.18 Sketch of a stretching under the effect of the diagonal components of the strain tensor

Fig. 1.19 Sketch of a shear (*right*) acting on an unstrained sample (*left*) under the effect of the off-diagonal components of the strain tensor

where

$$\varpi_{ij} = \frac{1}{2}\left(\frac{\partial u_i}{\partial x_j} - \frac{\partial u_j}{\partial x_i}\right)$$

and

$$\epsilon_{ij} = \frac{1}{2}\left(\frac{\partial u_i}{\partial x_j} + \frac{\partial u_j}{\partial x_i}\right)$$

One can demonstrate that the antisymmetric tensor has the symmetry of a pseudo-vector which can be represented by a three-component axial vector of components ϖ_{23}, ϖ_{31}, and ϖ_{12}. This contribution shall be disregarded here.

The symmetric tensor represents the variation of the lengths between the different points of the solid.

The diagonal elements ϵ_{ii} are the stretching components of the strain field.

On Fig. 1.18 is illustrated a typical stretching of a barrel: the length is increased along one direction, reduced along the remaining two. On Fig. 1.19 is illustrated a typical shear of a barrel:

The off-diagonal elements ϵ_{ij} with $i \neq j$ are the shear components of the strain field. The symmetric tensor represents the variation of the lengths between the different points of the solid.

One may have to represent the symmetrical strain tensor as a six-component vector:

$$\begin{bmatrix} \epsilon_1 \\ \epsilon_2 \\ \epsilon_3 \\ \epsilon_4 \\ \epsilon_5 \\ \epsilon_6 \end{bmatrix} = \begin{bmatrix} \epsilon_{11} \\ \epsilon_{22} \\ \epsilon_{33} \\ 2\epsilon_{23} \\ 2\epsilon_{31} \\ 2\epsilon_{12} \end{bmatrix}. \tag{1.22}$$

1.12.3 The Stiffness and Compliance Tensors

Hooke's law is a well known proportionality relation that correlates the deformation experienced by a solid with the stress it is submitted to. In our case, a tensorial relation connects the rank-two strain tensor with the rank-two stress one. It writes:

$$\left[\overset{\rightrightarrows}{\epsilon}\right] = \left[\overset{\rightrightarrows}{S}\right] \cdot \left[\overset{\rightrightarrows}{\sigma}\right] \tag{1.23}$$

where $\overset{\rightrightarrows}{S}$ *is the compliance tensor*, a rank-four tensor.

Reversely, one can write:

$$\left[\overset{\rightrightarrows}{\sigma}\right] = \left[\overset{\rightrightarrows}{C}\right] = \left[\overset{\rightrightarrows}{\epsilon}\right] \tag{1.24}$$

where $\overset{\rightrightarrows}{C}$ *is the stiffness tensor*, a rank-four tensor.

Both $\overset{\rightrightarrows}{S}$ and $\overset{\rightrightarrows}{C}$ have $3^4 = 81$ independent components—all are scalar numbers—that are represented by S_{ijkl} or C_{ijkl} with each of the four indices running from 1 to 3.

Fortunately, the number of these components can be reduced to 36, thanks to symmetry considerations.

Components of both $\overset{\rightrightarrows}{S}$ and $\overset{\rightrightarrows}{C}$ are governed by the same symmetry rules.

Let us here take $\overset{\rightrightarrows}{C}$ for the sake of the illustration.

One can demonstrate that $C_{ijkl} = C_{jikl} = C_{ijlk} = C_{klij}$.

Then, it is possible to represent any of these rank-four tensors using a (6×6) matrix.

We remind the reader that both the stress and strain tensors can be represented by 6-component vectors.

The relationships between the indices of the six-component vectors and those of the (3×3) symmetrical tensors are the following: the indices of the six-component vectors run from 1 to 6 when where couples of indices are 11, 22, 33, 23, 13, and 12 respectively.

The components of the stress tensor are: $\sigma_m = \sigma_{ij}$.

The components of the strain tensor and of its representative vector are, in addition, connected by a complementary relation:

$$\epsilon_m = f(m)\epsilon_{ij}$$

with:

$$f(m) = 1 \quad \text{when m} \in \{1, 2, 3\}$$
$$f(m) = 2 \quad \text{when m} \in \{4, 5, 6\}$$

Here, we have to write the following relationships:

$$C_{mn} = g(m, n)C_{ijkl}$$

with:

$$g(m, n) = 1 \quad \text{when } (m, n) \in (\{1, 2, 3\}, \{1, 2, 3\})$$
$$g(m, n) = 2 \quad \text{when } (m, n) \in \{(1, 2, 3\}, \{4, 5, 6\}) \quad \text{or when } (m, n) \in (\{4, 5, 6\}, \{1, 2, 3\})$$
$$g(m, n) = 4 \quad \text{when } (m, n) \in (\{4, 5, 6\}, \{4, 5, 6\})$$

Then, the compliance and stiffness rank-four tensors can be represented using (6 × 6) matrices of the kind below:

$$\left[\vec{\vec{S}}\right] = \begin{bmatrix} S_{11} & S_{12} & S_{13} & S_{14} & S_{15} & S_{16} \\ S_{12} & S_{22} & S_{23} & S_{24} & S_{25} & S_{26} \\ S_{13} & S_{23} & S_{33} & S_{34} & S_{35} & S_{36} \\ S_{14} & S_{24} & S_{34} & S_{44} & S_{45} & S_{46} \\ S_{15} & S_{25} & S_{35} & S_{45} & S_{55} & S_{56} \\ S_{16} & S_{26} & S_{36} & S_{46} & S_{56} & S_{66} \end{bmatrix}; \quad \left[\vec{\vec{C}}\right] = \begin{bmatrix} C_{11} & SC_{12} & C_{13} & C_{14} & C_{15} & C_{16} \\ C_{12} & C_{22} & C_{23} & C_{24} & C_{25} & C_{26} \\ C_{13} & C_{23} & C_{33} & C_{34} & C_{35} & C_{36} \\ C_{14} & C_{24} & C_{34} & C_{44} & C_{45} & C_{46} \\ C_{15} & C_{25} & C_{35} & C_{45} & C_{55} & C_{56} \\ C_{16} & C_{26} & C_{36} & C_{46} & C_{56} & C_{66} \end{bmatrix} \tag{1.25}$$

The relations between the strain and the stress now write:

$$[\vec{\epsilon}] = \left[\vec{\vec{S}}\right] \cdot [\vec{\sigma}] \quad \text{and} \quad [\vec{\sigma}] = \left[\vec{\vec{C}}\right] \cdot [\vec{\epsilon}] \tag{1.26}$$

We wish here to insist on the fact that these objects are representations in the Voigt's notation of tensors of the three-dimensional space and that we can not applicate to them classical algebraic calculations, typical of Euclidian spaces like matrix equations for rotation of axes. In such cases, one has to use relations for tensors.

1.12.4 The Stiffness and Compliance Tensors in Wurtzitic Semi-conductors

It may be demonstrated that, in line with the six-fold symmetry along the $\vec{z}$ ($\vec{x_3}$ axis), directions $\vec{x_1}$ and $\vec{x_3}$ are equivalent. Then, the compliance and stiffness tensors take the simplified forms given below:

$$\left[\vec{\vec{S}}\right]=\begin{bmatrix} S_{11} & S_{12} & S_{13} & 0 & 0 & 0 \\ S_{12} & S_{11} & S_{13} & 0 & 0 & 0 \\ S_{13} & S_{13} & S_{33} & 0 & 0 & 0 \\ 0 & 0 & 0 & S_{44} & 0 & 0 \\ 0 & 0 & 0 & 0 & S_{44} & 0 \\ 0 & 0 & 0 & 0 & 0 & S_{66} \end{bmatrix}; \quad \left[\vec{\vec{C}}\right]=\begin{bmatrix} C_{11} & C_{12} & C_{13} & 0 & 0 & 0 \\ C_{12} & C_{11} & C_{13} & 0 & 0 & 0 \\ C_{13} & C_{13} & C_{33} & 0 & 0 & 0 \\ 0 & 0 & 0 & C_{44} & 0 & 0 \\ 0 & 0 & 0 & 0 & C_{44} & 0 \\ 0 & 0 & 0 & 0 & 0 & C_{66} \end{bmatrix} \tag{1.27}$$

Adding the complementary equation:

$$S_{66}=2(S_{11}-S_{12})$$

$$C_{66}=\frac{1}{2}(C_{11}-C_{12})$$

Then, the wurtzite symmetry prescripts five components to be different and independent for stiffness and compliance tensors.

We would like to indicate that:

$$[\vec{\epsilon}]=\left[\vec{\vec{S}}\right]\cdot\left[\vec{\vec{C}}\right]\cdot[\vec{\epsilon}] \tag{1.28}$$

The matrix product of the compliance and stiffness tensors gives the identity (6 × 6) matrix:

$$\left[\vec{\vec{S}}\right]\cdot\left[\vec{\vec{C}}\right]=[1] \tag{1.29}$$

The C_{ij}s can be expressed as functions of the S_{ij}s and vice versa:

$$S_{11}=\frac{C_{33}C_{11}-C_{13}^2}{(C_{11}-C_{12})\left[C_{33}(C_{11}+C_{12})-2C_{13}^2\right]}$$

$$S_{12}=\frac{C_{13}^2}{(C_{11}-C_{12})\left[C_{33}(C_{11}+C_{12})-2C_{13}^2\right]}$$

$$S_{13}=-\frac{C_{13}}{C_{33}(C_{11}+C_{12})-2C_{13}^2}$$

$$S_{33}=\frac{C_{11}+C_{12}}{C_{33}(C_{11}+C_{12})-2C_{13}^2}$$

$$S_{44}=\frac{1}{C_{44}}$$

The recommended values of the stiffness coefficients of nitride semi-conductors are given in Table 1.7.

Table 1.7 Values of the stiffness coefficients of nitride semi-conductors (at room temperature)

Material	GaN	InN	AlN
C_{11} (GPa)	390 ± 15	190	410 ± 10
C_{12} (GPa)	145 ± 20	104	149 ± 10
C_{13} (GPa)	106 ± 20	121	99 ± 4
C_{33} (GPa)	398 ± 20	182	389 ± 10
C_{44} (GPa)	105 ± 10	100	125 ± 5

1.12.5 The Energy of a Strained Crystal

Let us consider an unstrained crystal, initially shaped as a unit cube, experiencing a small homogeneous strain field of components represented by ϵ_i in the context of matrix notation.

When the strain changes from ϵ_i to $\epsilon_i + d\epsilon_i$, one can demonstrate that the work due to the stress components σ_i acting on the cube faces is:

$$dW = \sum_i \sigma_i d\epsilon_i$$

that we re-write, using Einstein's notation to get rid of the summation symbol on index repeated twice:

$$dW = \sigma_i d\epsilon_i$$

This can be rearranged into:

$$dW = C_{ij}\epsilon_j d\epsilon_i$$

still using Einstein's notation.

After integration, taking into account $C_{ji} = C_{ij}$, one arrives at:

$$W = \frac{1}{2} C_{ij}\epsilon_j\epsilon_i$$

W has the dimension of a pressure (GPa) and represents the strain energy per unit volume of the crystal.

1.13 Basic Elements of Group Theory

1.13.1 The Concept of Algebraic Groups

- A group G is a collection of elements $\{A, B, C, \ldots\}$ which are inter-related according to certain rules.

Table 1.8 Multiplication table of group G

G	E	A	B	$\dots$	U	V
E	E	A	B	$\dots$	U	V
A	A	AA	AB	$\dots$	AU	AV
B	B	BA	BB	$\dots$	BU	BV
$\vdots$	$\vdots$	$\vdots$	$\vdots$	$\ddots$	$\vdots$	$\vdots$
U	U	UA	UB	$\dots$	UU	UV
V	V	VA	VB	$\dots$	VU	VV

- The product of two elements of the group must be an element of the group: $\forall(A, B) \in G \times G$; $AB = C$, with $C \in G$. When $AB = BA$ the group is called Abelian.

There exists one identity element E:
$E \in G$ and $\forall(A) \in G$; AE = EA = A. E—comes from German word Einheit—is called identity element.

- The associative law of multiplication must hold:

$$\forall(A, B, C) \in G \times G \times G;\ (\mathrm{AB})\mathrm{C} = \mathrm{A}(\mathrm{BC})$$

- Any element A of a group must have a reciprocal element. This reciprocal element may be noted A^{-1}.

$$\forall(A) \in G, \exists B \in G \text{ so that } \mathrm{AB} = \mathrm{BA} = \mathrm{E}.$$

- The reciprocal of a product of two or more elements is equal to the product of the reciprocals in reverse order.

$$(ABC..XY)^{-1} = Y^{-1}X^{-1}B^{-1}A^{-1}$$

Groups may have a finite number of elements (they are called finite groups) or unlimited numbers of elements (infinite groups). The number of elements of a group, generally represented G is called the order of the group.

- The multiplication table of the symmetry elements (Table 1.8) is a table of h rows and h columns. At the intersection of column X and row Y is the product XY. Each row and each column in the group multiplication table lists each of the group elements once and only once: this is the rearrangement theorem. Two rows may not be identical, nor may be two columns. Thus, each row and each column is a rearranged list of the group elements.
  When groups are Abelians, the multiplication table is symmetric: AB = BA.
- A group has subgroups, labelled H, which are collections of elements of G that are themselves forming a group. E belongs to all subgroups of G. E alone is a

subgroup of G. Let h be the order of H. Then, as prescripted by the theorem of Lagrange, g/h is an integer number.

- A group may be separated into various smaller sets of elements called classes. Let A and X be two elements of a group G and B a third element of G, so that $B = X^{-1}AX$. B, the similarity transform of A by X, is the conjugate of A. Three important properties may be derived from this definition: each element is itself-conjugate; and if A is conjugate of B, then B is conjugate of A; and if A is conjugated of B and C, then B and C are conjugate with each other. *A complete set of elements conjugate one to another is called a class of the group.*
  The orders of classes must be integral factors of the order of the group. The identity constitutes a class. In an Abelian group, each element constitutes its own class. Classes are symmetry operations of the same kind.
- Two groups G and G' are isomorphic if there exists a function that sets up a one-to-one correspondence between the elements of the groups in a way that respects the given group operations. Such a function is called an isomorphism. From the standpoint of group theory, isomorphic groups have the same properties (same order, same multiplication table,...) and need not be distinguished.
  An isomorphism from a group (G, $\cdot$) to itself is called an automorphism of this group. Thus, it is a bijection $f : G \to G$ such that $\mathrm{f(A)} \cdot \mathrm{f(B)} = \mathrm{f(A \cdot B)}$, $\forall (A, B) \in G \times G$.
  An automorphism always maps the identity to itself. The image under an automorphism of a conjugacy class is always a conjugacy class (the same or another). The image of an element has the same order as that element.
- Besides the one-to-one correspondence between two groups, many-to-one correspondence may exist. The groups are said to be homomorphic. The isomorphism preserves the structure of the original group, but a homomorphism causes some of the structure of the original group to be lost. Both properties are reflected in the behaviour of multiplication tables. The orders of the two homomorphic groups may be different.
- The **direct product** of groups G and H which orders are g and h respectively is a group of order $g \cdot h$, noted $G \times H$. Its elements are, by construction, the products of all the elements of G by all the elements of H. If G has p classes and H has p' classes, then $G \times H$ has $p.p'$ classes. A condition for a group to be a direct product of two groups is that neither its order nor its number of classes are prime numbers.

1.13.2 Representations of Finite Groups by Matrices

Let us consider a n-dimensional vectorial space E_n with its eigenvectors $\{\vec{e_n}\}$.

Let $X = \sum_j X_j\vec{e_j} \equiv X_j\vec{e_j}$ (with Einstein's notation for sommation) a vector of E_n.

Let the operator $\mathbf{A}$ so that $Y = \mathbf{A}X$.

This may be written $Y_k = A_{kj}X_j$.

The $(n \times n)$ matrix with elements A_{kj} is a representation of operator $\mathbf{A}$ in the basis $\{\vec{e_n}\}$. k (resp. j) is the column (resp. row) index.

The trace of the matrix representative of operator $\mathbf{A} - \sum_k A_{kk}$—is an invariant under any basis transformation. *It is called the character of matrix A.*

The full set of the matrices representating in the basis $\{\vec{e_n}\}$ all the operators **A** of a n-dimensional group G forms a group and forms a n-dimensional representation of the operators **A** of the n-dimensional group G. Any representation of a finite group G is equivalent to a unitary representation of this group (all matrices are unitary matrices).

Note: When two operators **A** and **B** are represented by matrices M_A and M_B respectively: $\mathbf{A} \neq \mathbf{B} \nrightarrow M_A \neq M_B$. Isomorphisms and homomorphisms may exist between G and the group of matrices M representing the operators of G. Homomorphisms often occurs: several operators of G are represented by the same matrix.

The abstract **basis of irreducible representations** is of paramount importance for representing symmetries. Let $\Gamma_\nu(A)$ and $\Gamma_\mu(A)$ be two matrix representations of a group G. These representations are of dimensions ν and μ respectively. If for each element of the group we define a matrix of dimension $\nu + \mu$ as:

$$\Gamma(A) = \begin{array}{|c|c|} \hline \Gamma_\nu(A) & 0 \\ \hline 0 & \Gamma_\mu(A) \\ \hline \end{array} = \Gamma_\nu(A) \oplus \Gamma_\mu(A)$$

The matrices Γ form a representation of G. This representation leads to block-diagonal matrices. It is said to be reducible; it is the direct sum of representations Γ_ν and Γ_μ. The summation differs from a classical summation; it is a summation over two different vectorial spaces ϵ_ν and ϵ_μ.

$$\Gamma = \Gamma_\nu \bigoplus \Gamma_\nu \quad \text{and} \quad \epsilon = \epsilon_\nu \bigcup \epsilon_\mu.$$

A representation with a block-diagonal shape is a **reducible representation**. Once the ultimate sizes of the blocks are reached, i.e. when no similarity operation is susceptible to reduce the sizes of such blocks, the representation is named **irreducible**.

There is an infinite number of possible representations for a group, just like an idea may be formulated using as many languages as one can invent. The number of irreducible representations of a group is limited: it is equal to the number of classes of the group. The number of irreducible representations of an Abelian group is equal to the number of its elements.

Let us compare the full set of irreducible representations to the finite number of Chinese ideograms. Any language spoken by a person can be written with these Chinese ideograms. An assembly of people speaking different tongues can hardly communicate. But writing their sayings using the Chinese ideograms allows anyone knowing Chinese ideograms to understand every word. Since it is possible to associate many sounds to a Chinese ideogram, each of the irreducible representations has an infinite number of possible basis vectors. Any representation can be expanded along the set of irreducible representations by manipulating with an ad-hoc algebra the representation characters to reduce with the characters of the irreducible rep-

Table 1.9 Table of characters of group G

G	C_1	C_2	C_3	C_4
Γ_1	χ_{11}	χ_{12}	χ_{13}	χ_{14}
Γ_2	χ_{21}	χ_{22}	χ_{23}	χ_{24}
Γ_3	χ_{31}	χ_{32}	χ_{33}	χ_{34}
Γ_4	χ_{14}	χ_{24}	χ_{34}	χ_{34}

resentations. We will develop this quantitatively. To achieve that, it is necessary to construct the table of character of the group.

1.13.3 Character Tables and Irreducible Representations

The **table of characters** is a table with as many rows as irreducible representations (Γ_is), and as many columns as classes (C_js). Let n be this number, it is a $n \times n$ table sketched in Table 1.9 when $n = 4$.

Quantity χ_{ij} is the character of matrices representing elements of class C_j in irreducible representation Γ_i. these quantities may be complex numbers.

Trick: a quick method to calculate the character of a matrix representing an operator **A** in a given representation is to consider the number of basis vectors unchanged under **A** (character 1) diminished from the number of basis vectors transformed into their opposite under **A** (character -1). The vectors transformed differently under **A** have a vanishing contribution to the characters of the representation.

There are a few orthogonality relations that help to construct the table of irreducible representations:

$$\sum_{\mathbf{A}} \chi_j(\mathbf{A})\chi_i^*(\mathbf{A}) = g\delta_{ij}$$

Let class $C_{\mathbf{A}}$ of **A** have $N_{\mathbf{A}}$ elements. Then, the equation above becomes:

$$\sum_{C_A} N_A \chi_j(C_A)\chi_i^*(C_A) = g\delta_{ij}$$

$$\sum_{C_A} N_A \chi_j(C_A)\chi_i^*(C_B) = g\delta_{AB}$$

- Given an irreducible representation, the sum of the squared modulii of characters through all operators **A** is the order of the group:

$$\sum_{\mathbf{A}} |\chi_j(\mathbf{A})|^2 = g$$

– The sum of the squared modulii of characters of the identity through all irreducible representations (Γ's) is the order of the group:

$$\sum_{\Gamma} |\chi_{j1}|^2 = g$$

Since χ_{j1} is the dimension n_j of irreducible representation Γ_j, the sum of the squared dimensions of the irreducible representations is the order of the group.

$$\sum_{\Gamma} |n_j|^2 = g$$

This permits us to decompose a reducible representation Γ into irreducible representations. Starting from:

$$\Gamma = \sum_{i} a_i \Gamma_i$$

and

$$\chi(\Gamma) = \sum_{i} a_i \chi(\Gamma_i)$$

The number of times a_i the representation Γ_i appears in the decomposition of Γ is:

$$a_i = 1/g \sum_{\mathbf{A}} \chi_{\Gamma(\mathbf{A})} \chi^*_{i(\mathbf{A})}$$

$$a_i = 1/g \sum_{C_A} NA \chi_{\Gamma}(C_A) \chi^*_i(C_A)$$

The direct product of two representations is a representation which has for each class and each irreducible representation a character with value the product of the characters for this class and this irrediucible representation. It is sometimes an irreducible representation but not always. The notation is: $\Gamma = \Gamma_1 \bigotimes \Gamma_2$. The multiplication table is a $(n \times n)$ table with n rows and n columns. At the intersection of ith row and jth column is the product $\Gamma_i \bigotimes \Gamma_j$ expressed as an expansion of irreducible representations. An example is sketched below (Table 1.10) in case of four irreducible representations.

It is important to note there are relationships between wave functions of the Schrödinger equation and irreducible representations.

One can demonstrate that the non-degenerate solutions of the Schrödinger equation are basis functions of unidimensional irreducible representations of the group that describe the system under examination.

A k-degenerate wave function generates a representation of dimension k which can be irreducible or reducible.

Table 1.10 Multiplication table of the irreducible representations of group G

Multiplication table	Γ_1	Γ_2	Γ_3	Γ_4
Γ_1	Γ_1	Γ_2	Γ_3	Γ_4
Γ_2	Γ_2	$\Gamma_2 \otimes \Gamma_2$	$\Gamma_2 \otimes \Gamma_3$	$\Gamma_2 \otimes \Gamma_4$
Γ_3	Γ_3	$\Gamma_3 \otimes \Gamma_2$	$\Gamma_3 \otimes \Gamma_3$	$\Gamma_3 \otimes \Gamma_4$
Γ_4	Γ_4	$\Gamma_4 \otimes \Gamma_2$	$\Gamma_4 \otimes \Gamma_3$	$\Gamma_4 \otimes \Gamma_4$

Any wave function can be decomposed as a linear combination of irreducible representations of the group of the Schrödinger equation:

$$\Psi(\vec{r}) = \sum_{P} \sum_{n=1,n_p} \Psi_n^P(\vec{r})$$

where $\Psi_n^P(\vec{r})$ is the nth basis function of state P of irreducible representation Γ_P, of degeneracy n_p. To go further, if the calculation of symetrized wave functions is necessary, one has to determine the matrix that represents the action of each symmetry operator in the basis set chosen to treat the problem one wants to solve. Then, one applicates **projection operators** to the basis set in order to obtain a new basis which basis functions are those of the irreducible representations.

A projector is an operator which applicated to an arbitrary function f transforms it specifically. To each irreducible representation Γ_μ of dimension n_μ can associated a projector P^μ which is built as an appropriate linear combination of the symmetry operations **A**. Group theory permits to demonstrate that operator P^μ is defined as a summation through all symmetry operations weighted by the character of the class this operation has in irreducible representation Γ_μ:

$$P^\mu = \sum_{\mathbf{A}} \chi^\mu(\mathbf{A})\mathbf{A}$$

Then:

$$P^\mu f_i^\nu = 0 \quad \text{when } \mu \neq \nu, \forall i, i = 1, \ldots, n_\nu$$

and

$$P^\mu f_i^\mu = g/n_\mu f^\mu \quad \text{when } \mu = \nu.$$

1.13.4 The Point Group C_{6v}

Each of the following are symmetry elements of group C_{6v} and constitute a class:

- The identity.
- Two six-fold symmetry axis parallel to the [001] direction: one corresponds to a rotation of $2\pi/6$ and the second to a rotation of $-2\pi/6$ (or $10\pi/6$).

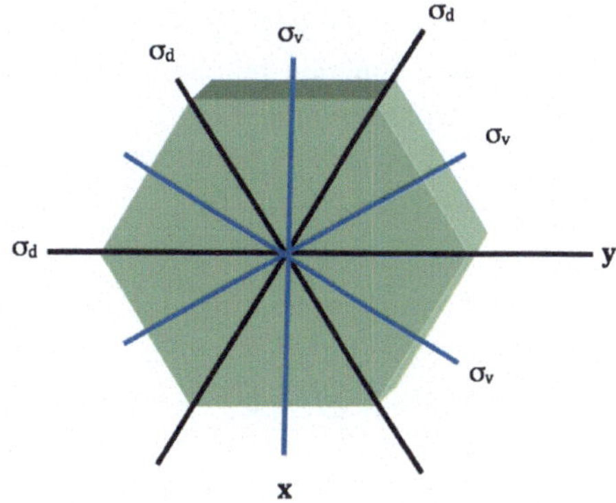

Fig. 1.20 Geometrical representation of some of the reflection planes of C_{6v} relatively to the hexagon

- A rotation of $2\pi/3$ around the [001] direction rotation of $-2\pi/3$ (or $4\pi/3$) around the [001] direction.
- A rotation of π around the [001] direction rotation.
- Three reflection planes making $\pi/3$ with each other. These planes are generated by $(\vec{y}, \vec{z})$ and are noted σ_d.
- Three reflection planes making $\pi/3$ with each other, orthogonal to the precedent three. These planes are generated by $(\vec{x}, \vec{z})$ and are noted σ_v. The geometrical representation of these reflection planes relatively to the hexagon is shown in Fig. 1.20.

1.13.5 Application of Group Theory to the Calculation of Integrals

Let integral:

$$\int \Psi_\alpha^* F_\beta \Psi_\gamma d\tau$$

where Ψ_α and Ψ_γ are wave vectors corresponding to eigenvalues E_α and E_γ of a quantum problem. The quantity F_β is an operator, the elementary volume of the integration space is $d\tau$.

Let us suppose that the relationship between the quantifies (Ψ_α, Ψ_γ, F_β) and irreducible representations (Γ_α^*, Γ_γ, Γ_β) has been established.

The product $\Psi_\alpha^* F_\beta \Psi_\gamma$ transforms like $\Gamma_\alpha^* \bigotimes \Gamma_\gamma \bigotimes \Gamma_\beta$

If, using an appropriate basis transformation, we are able to reduce the product representation into:

$$\Gamma_\alpha^* \bigotimes \Gamma_\gamma \bigotimes \Gamma_\beta = \Gamma_\delta \bigoplus \Gamma_\mu \bigoplus \Gamma_\nu \bigoplus \dots$$

Then, the product $\Psi_\alpha^* F_\beta \Psi_\gamma = \varphi_\delta + \varphi_\mu + \varphi_\nu + \cdots$
where φ_δ is a function belonging to the functional space which basis generates irreducible representation Γ_δ.

The use of projectors permits us to establish that, if Γ_δ does not appear in the decomposition of $\Gamma_\alpha^* \bigotimes \Gamma_\gamma \bigotimes \Gamma_\beta$, there is no term φ_δ in the above expansion. Then:

$$P^\delta \Gamma_\alpha^* \bigotimes \Gamma_\gamma \bigotimes \Gamma_\beta = 0$$

In particular, if $\Gamma_\alpha^* \bigotimes \Gamma_\gamma \bigotimes \Gamma_\beta \notin \Gamma_1$ then $\int \Psi_\alpha^* F_\beta \Psi_\gamma d\tau$ $P^1 \Gamma_\alpha^* \bigotimes \Gamma_\gamma \bigotimes \Gamma_\beta = 0$ or

$$P^1 \Psi_\alpha^* \bigotimes \Psi_\gamma \bigotimes \Psi_\beta = 0$$

that may also be written after introducing the expression of P^1:

$$\sum_{\mathbf{A}} \Psi_\alpha^* \bigotimes \Psi_\gamma \bigotimes F_\beta = 0$$

Conclusion:

$$\mathit{if} \quad \Gamma_\alpha^* \bigotimes \Gamma_\gamma \bigotimes \Gamma_\beta \notin \Gamma_1 \quad \mathit{then} \quad \int \Psi_\alpha^* F_\beta \Psi_\gamma d\tau = 0$$

This identity constitutes the conditions for matrix elements to vanish.

1.13.5.1 Selection Rules for Optical Transitions

Time-dependent perturbation theory and Fermi Golden rule permit to calculate the radiative recombination rate ω for a transition between an initial level $|i\rangle$ and a final level $|f\rangle$ as proportional to the square of the matrix element between levels $|i\rangle$ and $|f\rangle$ under the action of operator H. This writes:

$$\omega \sim |\langle i|H|f\rangle|^2$$

In the frame-work of the dipolar interaction, H transforms like a vector of the three-dimensional space. It will transform like Γ_1 or Γ_5 depending on the photon polarization field, as indicated by the tables of C_{6v}. Then, straightforward application of multiplication table permits to calculate the selection rules between any initial level $|i\rangle$ and any final level $|f\rangle$ for both polarizations of the electric field of the photon.

1.13.6 Group Theory and Perturbations

We consider a crystal with initial Hamiltonian H_0 under an external perturbation V.

$$H = H_0 + V.$$

The Hamiltonian H_0 (resp. external perturbation V) has symmetry G_0 (resp. G_1). G_1 symmetry is in general lower than G_0 one.

We will consider two possible situations here.

- *Case a*: $G_0 = G_1$, the perturbation does not change the symmetry of the crystal.

The eigen-values of the Schrödinger equations are already classified, according to the irreducible representations of G_0. The classification and degeneracies of energy levels in H are identical to those of H_0, the eigenvalues are shifted by $V^{\mu} = \langle f^{\mu}|Vf^{\mu}\rangle$. The perturbation does not lead to a dissociation of the degenerated levels except in case of accidental degeneracy when the wave functions of the un-perturbed problem transform like a reducible representation of G_0.

We now consider two wave functions $|i\rangle$ and $|j\rangle$, both transforming like an irreducible representation Γ_{μ} of G_0. In addition, let: $E_i = \langle f_i|H_0|f_i\rangle$ and $V_{ij} = \langle f_i|V|f_j\rangle$.

H_0 and V transform like Γ_1 of G_0. Then, matrix element $\langle f_i|\Gamma_1|f_j\rangle$ transforms like $\Gamma_{\mu} \otimes \Gamma_{\mu} = \Gamma_1 \oplus \ldots$ Then, both $E_i \neq 0$ and $V_{ij} \neq 0$.

1.13.6.1 Levels of Identical Symmetry are Coupled by a Non-symmetry-Breaking Perturbation

Eigenstates are obtained as the solution of the (2×2) Hamiltonian:

$$\begin{vmatrix} E_1 + V_{11} - E & V_{12} \\ V_{12}^* & E_2 + V_{22} - E \end{vmatrix}$$

when $E_1 = E_2$, the degeneracy is lifted except if, simultaneously, $E_1 + E_2 + V_{11} + V_{22} = 0$ and $V_{12} = 0$, a possible but very restrictive situation.

If $E_1 = E_2$ and wave functions ($|i\rangle$, $|j\rangle$) belong to different irreducible representations of G_0, $V_{12} = 0$; then, the degeneracy of the two levels is lifted via V_{11} and V_{22}.

- *Case b*: $G_0 > G_1$, the perturbation reduces the symmetry of the crystal.

In that case, G_1 is a sub-group of G_0; the eigen values of H have to be classified, according to the irreducible representations of G_1. Each irreducible representation Γ of G_0 is a representation γ of G_1. That representation γ may be reducible or irreducible in G_1.

$$\Gamma = \sum_i a_i \gamma_i$$

Irreducible representation Γ of G_0 must be equivalent to γ of G_1 for all symmetry operations g belonging to both G_0 and G_1. The characters of representations are in the following relationship:

$$\chi[\Gamma(g)] = \sum_i a_i \chi[\gamma_i(g)]$$

The relationships between the Γs of G_0 and the γs of all possible G_1s are given in the compatibility table.

1.13.7 *Angular Momenta and Group Theory: Simple and Double Groups*

When an atom is located in the free space, the group of symetrie of the Schrödinger equation belongs to the group of rotations, and its wave functions are the spherical harmonics $Y^{\ell}_{m_\ell}$. The ensemble of the rotations in the three-dimensional space formq an infinite group, with an infinite number of classes (all the rotations of any angle around an axis of arbitrary orientation). A rotation of the function of an angle α around z—equivalent to rotating the axes of $-\alpha$ on the $Y^{\ell}_{m_\ell}$—writes:

$$\begin{pmatrix} e^{-i\ell\alpha} & 0 & & & & \\ 0 & e^{-i(\ell-1)\alpha} & 0 & & & \\ & 0 & \ddots & & & \\ & & & \ddots & 0 & \\ & & & 0 & e^{i(\ell-1)\alpha} & 0 \\ & & & & 0 & e^{i\ell\alpha} \end{pmatrix}$$

The character is:

$$\chi^{\ell}(\alpha) = \frac{\sin\left[\left(\ell + \frac{1}{2}\right)\alpha\right]}{\sin[\alpha/2]}$$

Angular momentum algebra indicates this equation holds for any value of the total angular momentum $\vec{J} = \vec{L} + \vec{S}$, where $\vec{S}$ is the spin operator.

We remark that:

$$\chi^{J}(\alpha + 2\pi) = \chi^{J}(\alpha)(-1)^{2J}$$
$$\chi^{J}(\alpha + 4\pi) = \chi^{J}(\alpha)$$

We remark that, if J is half-integer, we will have to consider for identity a rotation of 4π. In this specific case, a supplementary operation occurs, as well as supplementary irreducible representations. These 4π and 2π symmetries double the number of elements and add complementary irreducible representations. **The double group represents the symmetries of Fermions** (J is half-integer) while **the simple group represents the symmetries of Bosons** (J is integer).

For the simple group, we had the relation: $\sum_{\Gamma} |n_j|^2 = g$.

For the double group, we now have: $\sum_{\Gamma} |n_j|^2 = 2g$.

In the specific case of C_{6v}, we have a simple group twelve-fold with six classes and six irreducible representations: $12 = 1 + 1 + 1 + 1 + 2^2 + 2^2$.

The double group C_{6v} is twenty-four-fold with 9 classes and 9 irreducible representations: $24 = 1 + 1 + 1 + 1 + 2^2 + 2^2 + 2^2 + 2^2 + 2^2$.

Note 1: The symmetry operations are not always rotations. When the point group of the crystal contains a symmetry plane, or other symmetry operations like reverse rotations, we use the isomorphism between this point group and the holoaxial group of the same singony.

Note 2: Identical values of total angular momenta can be obtained in very different conditions. For example, $J = 1$ is obtained either from a spherical harmonics $Y^{\ell}_{m_\ell}$ with $\ell = 1$, odd function in real space or by coupling two J= 1/2 spins giving an even function in real space. It is, then, mandatory to take the partial parity of the wave function into account. The symmetry of an angular momentum with even spatial parity is noted D_J^+ whilst the symmetry of an angular momentum with even spatial parity is noted D_J^- in the compatibility tables with the full rotation group.

1.13.8 Character Tables, Compatibility Table and Multiplication Tables

In this section, we give the tables required to handle group theory when dealing with C_{6v} and its subgroups. The notations are the following ones:

- E represents the identity,
- C_n represent rotations of $2\pi/n$ around the six-fold axis, σ_v and σ_d,
- Operations of the double group are over-lined.—Some wave functions of angular moment $J = +3/2$ and $+1/2$ are represented as $|J, m_J\rangle$ in the columns of the wave functions.

Character table and basis functions for C_{6v}

C_{6v}	E	$\bar{E}$	$(C_2, \bar{C}_2)$	$2C_3$	$2\bar{C}_3$	$2C_6$	$2\bar{C}_6$	$(3\sigma_d, 3\bar{\sigma}_d)$	$(3\sigma_v, 3\bar{\sigma}_v)$	basis functions
Γ_1	1	1	1	1	1	1	1	1	1	z
Γ_2	1	1	1	1	1	1	1	-1	-1	
Γ_3	1	1	-1	1	1	-1	-1	1	-1	$x^3 - 3xy^2$
Γ_4	1	1	-1	1	1	-1	-1	-1	1	$y^3 - 3yx^2$
Γ_5	2	2	-2	-1	-1	1	1	0	0	(x, y)
Γ_6	2	2	2	-1	-1	-1	-1	0	0	$\Gamma_3 \otimes \Gamma_5$
Γ_7	2	-2	0	1	-1	$\sqrt{3}$	$-\sqrt{3}$	0	0	$\lvert 1/2, \pm 1/2\rangle$
Γ_8	2	-2	0	1	-1	$-\sqrt{3}$	$\sqrt{3}$	0	0	$\Gamma_3 \otimes \Gamma_7$
Γ_9	2	-2	0	-2	2	0	0	0	0	$\lvert 3/2, \pm 3/2\rangle$

Multiplication table for C_{6v}

	Γ_1	Γ_2	Γ_3	Γ_4	Γ_5	Γ_6	Γ_7	Γ_8	Γ_9
Γ_1	Γ_1	Γ_2	Γ_3	Γ_4	Γ_5	Γ_6	Γ_7	Γ_8	Γ_9
Γ_2		Γ_1	Γ_4	Γ_3	Γ_5	Γ_6	Γ_7	Γ_8	Γ_9
Γ_3			Γ_1	Γ_2	Γ_6	Γ_5	Γ_8	Γ_7	Γ_9
Γ_4				Γ_1	Γ_6	Γ_5	Γ_8	Γ_7	Γ_9
Γ_5					$\Gamma_1+\Gamma_2+\Gamma_6$	$\Gamma_3+\Gamma_4+\Gamma_5$	$\Gamma_7+\Gamma_9$	$\Gamma_8+\Gamma_9$	$\Gamma_7+\Gamma_8$
Γ_6						$\Gamma_1+\Gamma_2+\Gamma_6$	$\Gamma_8+\Gamma_9$	$\Gamma_7+\Gamma_8$	$\Gamma_7+\Gamma_8$
Γ_7							$\Gamma_1+\Gamma_2+\Gamma_5$	$\Gamma_3+\Gamma_4+\Gamma_6$	$\Gamma_5+\Gamma_6$
Γ_8								$\Gamma_1+\Gamma_2+\Gamma_5$	$\Gamma_5+\Gamma_6$
Γ_9									$\Gamma_1+\Gamma_2+\Gamma_3+\Gamma_4$

Full rotation compatibility table for C_{6v}

D_0^+	Γ_1	D_0^-	Γ_2
D_1^+	$\Gamma_2+\Gamma_5$	D_1^-	$\Gamma_1+\Gamma_5$
D_2^+	$\Gamma_1+\Gamma_5+\Gamma_6$	D_2^-	$\Gamma_2+\Gamma_5+\Gamma_6$
D_3^+	$\Gamma_2+\Gamma_3+\Gamma_4+\Gamma_5+\Gamma_6$	D_3^-	$\Gamma_1+\Gamma_3+\Gamma_4+\Gamma_5+\Gamma_6$
$D_{1/2}^+$	Γ_7	$D_{1/2}^-$	Γ_7
$D_{3/2}^+$	$\Gamma_7+\Gamma_9$	$D_{3/2}^-$	$\Gamma_7+\Gamma_9$
$D_{5/2}^+$	$\Gamma_7+\Gamma_8+\Gamma_9$	$D_{5/2}^-$	$\Gamma_7+\Gamma_8+\Gamma_9$

Compatibility table for C_{6v} and some of its subgroups

C_{6v}	Γ_1	Γ_2	Γ_3	Γ_4	Γ_5	Γ_6	Γ_7	Γ_8	Γ_9
C_{2v}	Γ_1	Γ_3	Γ_2	Γ_4	$\Gamma_2+\Gamma_4$	$\Gamma_1+\Gamma_3$	Γ_5	Γ_5	Γ_5
$C_s(E=x)$	Γ_1	Γ_2	Γ_1	Γ_2	$\Gamma_1+\Gamma_2$	$\Gamma_1+\Gamma_2$	$\Gamma_3+\Gamma_4$	$\Gamma_3+\Gamma_4$	$\Gamma_3+\Gamma_4$
$C_s(E=y)$	Γ_1	Γ_2	Γ_2	Γ_1	$\Gamma_1+\Gamma_2$	$\Gamma_1+\Gamma_2$	$\Gamma_3+\Gamma_4$	$\Gamma_3+\Gamma_4$	$\Gamma_3+\Gamma_4$

1.13.9 The Translation Group

The properties of crystals are unchanged under translations $\vec{\tau}_n = n_1\vec{a}_1 + n_2\vec{a}_2 + n_3\vec{a}_3$ where the n_i's are integers and the $\vec{a}_i$'s are the primitive translation vectors. This may be written as $\{E|\vec{\tau}_n\}$ using the symbol of Seitz.

Obviously, from the definition above, $\{E|\vec{\tau}_n + \vec{\tau}_{n'}\} = \{E|\vec{\tau}_{n+n'}\}$ and as $(\mathbb{Z}, +)$ form an Abelian group, the ensemble of translations $\{E|\vec{\tau}_n\}$ constitutes an Abelian finite translation group with $N_1N_2N_3$ elements.

The cyclic boundary conditions for crystals of finite dimensions $N_1\vec{a}_1$, $N_2\vec{a}_2$ and $N_3\vec{a}_3$ write

$$\{E|\vec{\tau}_{N_1,0,0}\} = \{E|\vec{\tau}_{0,N_2,0}\} = \{E|\vec{\tau}_{0,0,N_3}\} = \{E|\vec{0}\}$$

Further defining a vector of the reciprocal lattice $\vec{N}^*_{hkl} = h\vec{a}^*_1 + k\vec{a}^*_2 + l\vec{a}^*_3$, and using cyclic boundary conditions leads to:

$$\vec{k} = (h/N_1)\vec{a}^*_1 + (k/N_2)\vec{a}^*_2 + (l/N_3)\vec{a}^*_3$$

$\{E|\vec{\tau}_n\}$ may be represented by $e^{-i\vec{k}.\vec{\tau}_n}$.

1.13.10 The Space Group

Let $\vec{t}_1, \vec{t}_2, \ldots \vec{t}_j$ be the positions of identical atoms in the unit cell.

For the wurtzite C_{6v} structure, $\vec{t}_1 = \vec{0}$, and $\vec{t}_2 = 2/3\vec{a}_1 + 1/3\vec{a}_2 + \vec{c}/2$ for the anions; and $\vec{t}_3 = 3/8\vec{c}$ and $\vec{t}_4 = 2/3\vec{a}_1 + 1/3\vec{a}_2 + 7/8\vec{c}$ for the cations.

The elements of the space group write $\{\mathbf{A}|\vec{\alpha}\}$ where $\mathbf{A}$ is a symmetry operator of the point group and $\vec{\alpha} = \vec{\tau}_n + \vec{t}$. In this notation, $\vec{t}$ is a fractional translation operator. Let

$$\{\mathbf{A}|\vec{\alpha}\} \cdot t_j = \vec{\tau}_{n'} + \vec{t}_{j'}$$

where $\vec{t}_{j'}$ represents the position of an atom in the unit cell—$\vec{t}_{j'}$ may coincide with $\vec{t}_j$—and $\vec{\tau}_{n'}$ is a suitable translation vector. Since $\{\mathbf{A}|\vec{\alpha}\}$ is a symmetry operation of the crystal, $\vec{t}_{j'}$ and $\vec{t}_j$ correspond to similar atoms.

In summary, operator $\{\mathbf{A}|\vec{\alpha}\}$ contains the operators of the point group A, and accounts for the translational symmetry at the scale of the crystal via $\vec{\alpha}$. The translational symmetry operator $\vec{\alpha}$ has two contributions: an intra-elementary cell contribution—$\vec{t}$—and an inter-elementary cell—$\vec{\tau}$.

It can be easily verified that

$$\{\mathbf{A}|\vec{\alpha}\} \cdot \{\mathbf{B}|\vec{\beta}\} = \{\mathbf{AB}|S\vec{\beta} + \vec{\alpha}\}$$

and that:

$$\{\mathbf{A}|\vec{\alpha}\}^{-1} = \{\mathbf{A}^{-1}| - \mathbf{A}^{-1}\vec{\alpha}\}.$$

1.13.10.1 The Ensemble of Operators $\{A|\vec{\alpha}\}$ Constitutes the Space Group of the Crystal

The wurtzite space group is identified as $P6_3mc$ or C_{6v}^4 in the international tables.

References

1. B. Archibald, Am. Mineral. J. **1**, 96 (1810)
2. F. Briegler, A. Geuther, Ann. Chem. **123**, 228 (1862)
3. J.W. Mallet, J. Chem. Soc. **30**, 349 (1876)
4. W.C. Johnson, J.B. Parson, J.B., M.C Crew. J. Phys. Chem. **36**, 2651 (1932)
5. R. Juza, H. Hahn, Zeitschrift fur anorganische und allgemeine Chemie **239**, 282 (1938)

Further Reading

6. *Crystals and Crystal Structures* by R.J.D Tilley, Wiley, New York (2006), ISBN: 0470018208
7. *Physical Properties of Crystals: their Representation* by Tensors and Matrices by J.F. Nye, Oxford University Press (re-edited in 2004), ISBN: 0198511655
8. *Nitride semiconductors and Devices* by H. Morkoç, Springer-Verlag, Berlin, Heidelberg, New York, (1999), ISBN: 354064038x
9. *Zinc Oxide, Fundamentals Materials and Device Technology* by H. Morkoç and Umit Ozgür, Wiley-VCH, Berlin, (2009), ISBN: 9783527408139
10. *Wide Band Gap Semiconductors: Fundamental Properties and Modern Photonic and Electronic Devices*, K. Takahashi, A. Yoshikawa, and A. Sandhu ed., Springer Verlag, Berlin Heidelberg New York, (2007), ISBN: 103540472347
11. *Symmetry in the Solid State* by R. S. Knox and A. Gold, W.A. Benjamin, New York, (1964)
12. *Chemical application of group theory* by F. Albert Cotton, Wiley Inter Science, New York, (1971)
13. *Group theory and quantum Mechanics* by M. Tinkham, Mc Graw-Hill Book Company, New York, (1964)
14. *Group theory* by E. P. Wigner, Academic Press, New York, (1959)
15. *Group theory and its application to physical problems* by M. Hamermesh, Addison-Wesley Publishing Company, (1964)
16. *Group theory in quantum mechanics* by V. Heine, Dover Publications Inc, (1993), ISBN: 9780486675855
17. *The application of group theory in physics* by G.Ya Lyubarskii, Pergamon Press, London (1960)
18. *Properties of the thirty-two points groups* by G. F. Koster, J. O. Dimmock, R. G. Wheeler and H. Statz, The MIT Press, Cambridge, Massachussetts, (1963)

Chapter 2
Basics of Growth and Structural Characterization

In this chapter we review some growth methods like crystallization or epitaxy and give some elements relative to structural characterization methods.

2.1 Growth of Bulk Crystals

There exists a large variety of growth methods: growth from a solution, growth from a vapor phase, epitaxial growth. Most of these growth mesthods require very sophisticated facilities and have been detailed in the specialized books including those edited by Hadis Morkoc, or by Dirk Ehrentraut, Elke Meissner and Michal Bockowski for nitrides, or the one edited by Klingshirn, Meyer, Waag Hoffmann and Geurts concerning zinc oxide. There are two main routes. The first one concerns the growth of bulk materials that can be further utilized as substrates after being cut, polished up to a surface with optimized performances (flatness, decontamination of foreign species). The second one concerns the epitaxial growth on a substrate prepared so that it surface is made epi-ready. Both methods have advantages and drawbacks.

In the case of nitrides it is extremely complicated to get free nitrogen atoms so that $Ga + N \rightarrow GaN$. The reason is the very high stability of the nitrogen or ammonia molecules which have to be dissociated to furnish the nitrogen atom susceptible to combine with gallium to form GaN. Thus tricky technological approaches have been proposed so that the crystallization $Ga + 1/2N_2 \rightarrow GaN$ can be triggered and realized. The melting temperature of gallium nitride is extremely high (2,493 °C). The nitrogen pressure needed for congruent melting of GaN is 6 GPa. Since it was impossible to crystallize GaN from the melt, other crystal growth methods were developed. One of them was crystallization from the high temperature liquid solution of gallium under high nitrogen pressure as sketched in Fig. 2.1. The constituents of the material to be crystallized are dissolved in a suitable solvent and crystallization occurs as the solution becomes critically supersaturated. The supersaturation can

B. Gil, *Physics of Wurtzite Nitrides and Oxides*, Springer Series in Materials Science 197, DOI: 10.1007/978-3-319-06805-3_2,

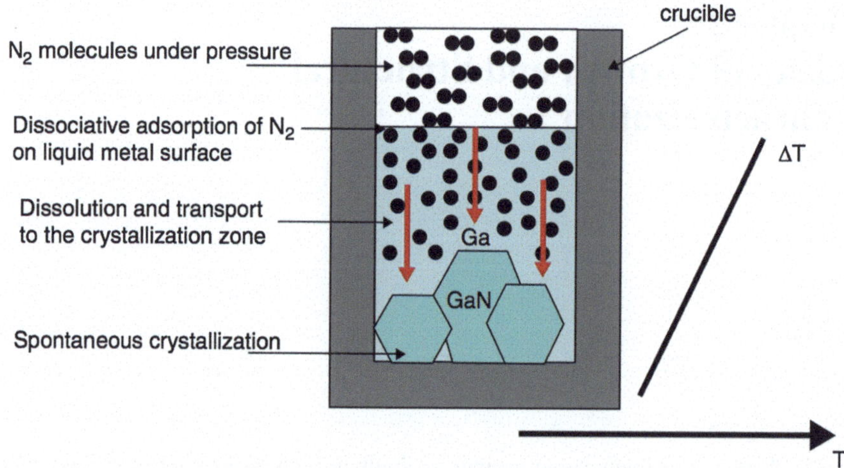

Fig. 2.1 Schematic illustration of the crystallization process in a temperature gradient by the high pressure solution growth method; three basic stages of the method are marked. After [12, 18]

mainly be promoted by cooling the solution or by a transport process in which the solute flows from a hotter to a cooler region. The main advantage of the solution growth is that the growth temperature is lower than that required for growth from the melt. This advantage often results in a better crystal quality with respect to point defects, dislocation density, and low angle grain boundaries, compared to crystals grown directly from their melts. The molecular nitrogen being under pressure is dissociated on the hot liquid gallium surface and the atomic nitrogen is dissolved in the liquid metal. Then, the convective flow of gallium, due to applied temperature gradients, is exploited in order to provide atomic nitrogen to the supersaturated zone of the solution, where the GaN crystallization takes place. The growth of very high quality bulk GaN then slowly occurs for pressures of 2 GPa and at "gentle" temperatures of about 1,800 °C, which is quite far from the melting temperature of GaN. A typical GaN crystal obtained by this method is presented in Fig. 1.2a of the preceding chapter.

The Ammono-thermal growth, is an analogue of hydrothermal growth widely used for the quartz mass production; it is a solvothermal method using NH_3 as a fluid. A closed system, so-called autoclave, is employed rugged enough to bear high pressure under elevated temperature and extremely aggressive environments. We disregard discussing the complex chemistry of these growth processes which is reviewed in [3, 5]; both chapters of the research textbook [12]. This is a very complex growth technology that permits to obtain large size high quality crystals under high growth rate conditions. Mineralizers (ionic additives) are essential to amplify the solubility of the solute in the solvent, and they determine the chemistry of crystal growth. Since NH_3 dissociates into NH_4^+ and NH_2^-, mineralizers that generate NH_4^+ (like NH_4Cl) are called acidic mineralizers whereas those which generate NH_2^-

(like KNH_2) are called basic mineralizers. Mineralizers which do not dominantly generate either NH_4^+ or NH_2^- are called neutral mineralizers. An advantage of the acidic mineralizer NH_4Cl over basic mineralizers such as KNH_2 or $NaNH_2$ is the high solubility in NH_3 around room temperature. Thus, a larger amount of GaN would be dissolvable in ammonia containing an acidic mineralizer in comparison to a basic mineralizer.

There are other interesting approaches for growth of bulk nitrides including the sodium flux method. Using such an approach, the growth of GaN crystals occurs from the Na flux using for example sodium azide Na_3N as sodium and atomic nitrogen source. A high Na content is required to form GaN and the formation of GaN requires less amount of Na when the temperatures rises. The GaN size also increases with decreasing Na content in the Ga-Na solution. For a detailed description of this growth method we refer the readers to: *A brief review on the Na-flux Method Toward Growth of Large size* GaN *Crystal* by Ehrentraut et al. [12].

Other growth methods are reviewed for AlN in "*The growth of bulk aluminum nitride*" by Ronny Kirste and Zlatko Sitar in Gil [15].

The growth of bulk zinc oxide to be further used as substrate can be quite easily achieved using solvothermal method at temperatures of about (800 °C) as reviewed in "*Growth*", by Klingshirn [13].

All these growth methods for bulk crystals require a seed to optimize the growth. These stories always begin from the growth of very tiny bulk single crystals which are carefully examined so that the best of them are then selected and used as a seeds for a second growth step which is supposed to give larger mono crystals, and the process repeats for a third growth protocol, and a fourth one, and so on if needed. When large enough boules of materials are obtained, they can be sawed to get crystal surfaces with specific orientations. Then, they are mechanically polished and chemically etched to be used as substrates for epitaxial growth. In general one uses them for homo-epitaxial growth when the chemical composition of the substrate material is very closed to the composition of the multilayer stacking deposited to form a more or less sophisticated artificial heterostructure with advanced physics and device properties. Imaging techniques like Atomic Force Microscopy can be used, as illustrated in Fig. 2.2, to measure the morphology of a bulk ZnO (10–12)-oriented surface processed after epitaxial growth of a quantum well. The Fig. 1a shows the ZnO crystal structure cut along the (10–12) semipolar plane. The rectangular in-plane unit cell is indicated. The in-plane lattice parameters are 3.25 and 7.67 Å along [1–210] and [−1011], respectively. The angle between the (10–12) and the (0001) planes is 42.77 °. The surface is atomically flat and atomic steps of 1.9 Å are clearly observed, which corresponds to the one monolayer height along the growth plane shown in Fig. 2.2.

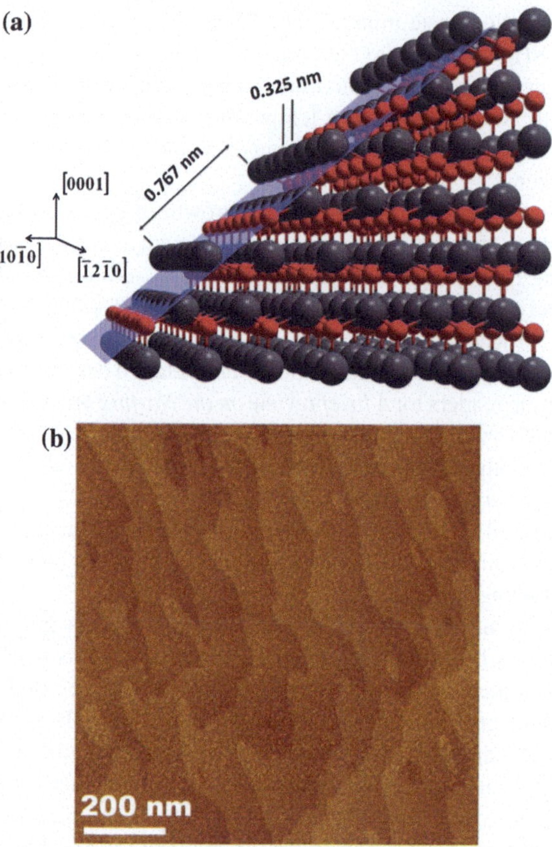

Fig. 2.2 **a** ZnO crystal structure cut parallel to the (10–12) plane. The zinc atoms and the oxygen atoms are represented by large *grey spheres* and small *red spheres*, respectively. The in-plane unit cell parameters are indicated. **b** $1 \times 1\,\mu^2$ atomic force microscopy image taken from a semipolar multiple quantum well surface. The height scale in the image is 2 nm. After Chauveau et al. [2]

2.2 Principle of Epitaxial Growth Methods

The epitaxial growth methods with growth rate of qualitatively speaking about $1\,\mu$/h which are very useful for realizing thin heterostructures with dimensions controlled at the monolayer scale. They take advantage of utilization of small quantities of very high purity precursors and permit to dope specifically some given layers of complex heterostructures to form light emitting diodes, lasers, field effect transistors,.... The growth conditions are not ruled by thermodynamics, but by the diffusion of chemical species at the growth surface. They are out of equilibrium growth mechanisms. The growths are achieved in temperature conditions that are depending on the material considered. They may drastically differ from a material to another one: Indium nitride will be decently grown for temperatures ranging between 550 and 650 °C for avoiding its thermal degradation while about 650 °C are appropriate for growing high quality GaN, and 1,150 °C is the growth temperature range adapted for the high quality epitaxy of AlN. The long standing lack of high electronic and structural quality, large size bulk substrates, their still high costs have required that people to deposit the nitrides and oxides on a large variety of foreign, lattice-mismatched substrates like Saphir, 6HSiC, and even more exotic materials as reviewed in Morkoc [11]. From the lattice mismatch, from the differences in thermal expansion coefficients of the epilayer and of the substrate, materials have long been obtained with huge densities of dislocations. Surprisingly the optical properties were astonishingly very good.

2.2.1 Hydride Vapor Phase Epitaxy

We begin by this epitaxial method for the following reason that it leads to very high growth rate compared with the other epitaxial method and permits to growth materials that will be used as growth substrates. Maruska and Tiejten grew the first single-crystal epitaxial GaN thin films by vapor transport. In their method, HCl vapor flowing over a Ga melt caused the formation of GaCl which was transported downstream. At the substrate, the GaCl mixed with NH_3 leading to the chemical reaction:

$$GaCl + NH_3 \rightarrow GaN + HCl + H_2$$

The growth rate was quite high; about 0.5 μ/min. Tanya Paskova and Bo Monemar report in Gil [16], growth rate of 80 μ/h in "HVPE GaN quasi substrates for nitride device structures". This growth approach is very convenient for preparing nitride substrates that can be further used (and are) as substrates for epitaxial techniques very suited for growing quantum wells, laser diodes, and transistors like Molecular Beam Epitaxy (MBE) or Metal-Organic Vapor Phase Epitaxy (MOVPE). Indeed one can growth very thick layers of material which can be removed from the substrates by an epitaxial lift-off technique before to be process like a piece of bulk material to render the surface epi-ready (appropriate to epitaxial growth).

2.2.2 Metal-Organic Vapor Phase Epitaxy

The principle of the method is to pyrolise in a the flow of a carrier gas, different precursors supplying the constituents of the semiconductors which impeg the substrate surface where they may combine with each other in order to contribute to the growth of the material. In general molecules formed by trimethyl or triethyl radicals and the metal elements are used as metal element precursors. The most word widely used nitrogen precursor is ammonia NH_3. The real situation is more complex than what states the equation below:

$$Ga + NH_3 \rightarrow GaN + 3/2H_2$$

There are a lot of physical parameters to control in order to optimize the growth conditions: among which are for instance the interaction of the residuals of the decomposition of the metal organic species with the surface, the interaction of the un-decomposed fraction of ammonia with the growth surface, the impact of the nature of the carrier gas (H_2 or N_2), the partial pressure of the different actors that contribute to the growth scenario, the total pressure in the growth chamber, the growth temperature,.... All of them which were quantitatively reviewed with great details by Olivier Briot in "MOVPE growth of nitrides" [17].

In the case of zinc oxide, most oxygen precursors are quite reactive, leading to efficient prereactions in the gas phase. This is particularly true when pure oxygen is used. In this case, the reactor pressure has to be reduced drastically, making the growth of ZnO difficult. This is the reason why groups have also focused on alternative precursors for oxygen, such as butanole iso-propanole, or N_2O (nitrous oxide or laughing gas). Dimethyl zinc, di-ethyl-zinc as well as zinc acetyl acetonate are used as zinc precursors. Bis-cyclo-pentadienyl-magnesium (Cp_2Mg) is an often used Mg precursor for the growth of ZnMgO. Since Cp_2Mg is very reactive, reacting readily with oxygen in the gas phase, growth of ZnMgO growth particularly tricky. (The wurtzite phase is the stable crystalline phase up to about 25 % Mg.) The two oxides CdO and ZnO are quite dissimilar materials in terms of their temperature stability, similar to Indium in InGaN, Cd in ZnCdO can only be incorporated at relatively low growth temperatures. A very detailed review can be found in "*Growth*", by Andreas Waag, in Klingshirn et al. [13].

2.2.3 Molecular Beam Epitaxy

An MBE system is a refined form of ultrahigh vacuum evaporation. Elements are heated in furnaces or Knudsen cells and directed beam of atoms or molecules are condensed onto a heated single crystal where they react. Because it is a UHV-based technique it has the advantage of being compatible with a wide range of surface analysis techniques. Figure 2.3 schematically illustrates the cross sectional diagram of a MBE system typical one those used in Hadis Morkoc's laboratory at the Virginia Commonwealth University. This system designed for the growth of nitrides uses ammonia as nitrogen precursor but nitrogen plasma is often used as nitrogen source. The real time control of growth is made possible by in-situ Reflection High Energy Electron Diffraction (RHEED). An electron beam issued from the RHEED gun in Fig. 2.3 glancing the surface of the crystal is diffracted by the atoms this surface. Thanks to the wave like property of electrons, the diffracted electrons interfere according to the crystal structure and spacing of the atoms at the sample surface and the wavelength of the incident electrons, creating specific diffraction patterns according to the surface features of the sample on the RHEED screen. The electron diffraction pattern recorded permits to characterize the crystallography of the sample surface through analysis of the diffraction patterns.

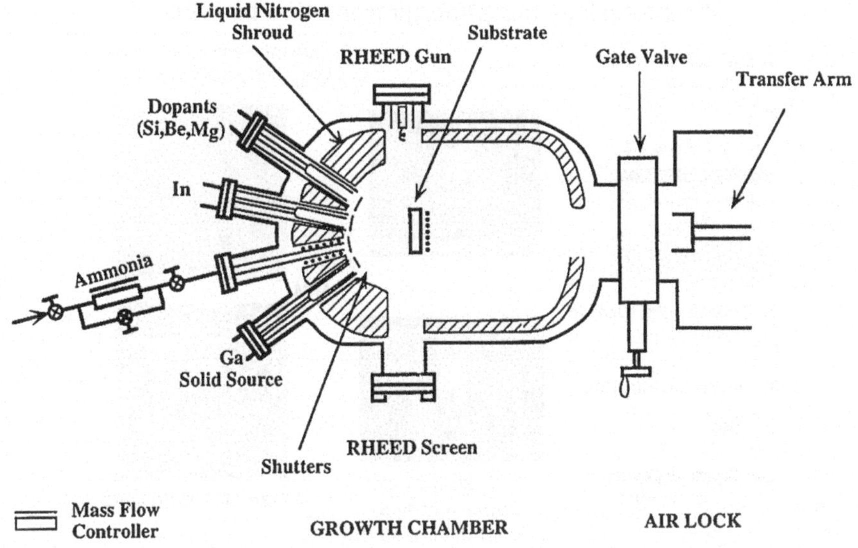

Fig. 2.3 Schematic cross sectional diagram of a MBE system typical one those used in Hadis Morkoc's Laboratory at the Virginia Commonwealth University. Courtesy Hadis Morkoc

In Fig. 2.4 are reported some typical RHEED patterns, AFM patterns recorded in case of the growth of GaN on an $Al_{0.5}Ga_{0.5}N$ template. The growth conditions are qualitatively sketched in the left-hand part of the figure, showing the correlation between growth conditions and modification of the RHEED diffraction pattern. Right hand side are given the corresponding AFM image recorded after the growth out of the growth chamber. There is an obvious and clear correlation between the surface morphology and the diffraction pattern of the RHEED electrons. A transition between a two-dimensional growth regime and three dimensional one (clearly corrugated surface) observed beyond the growth of the critical thickness for coherent growth of GaN on $Al_{0.5}Ga_{0.5}N$ is measured in situ during the growth. An alternative method to calibrate the growth is to follow the RHEED intensity oscillations when time passes in order to probe the growth rate at the monolayer scale. Provided that the growth rate and the temperature correspond to a 2D layer by layer nucleation growth mode, well resolved RHEED oscillations can be observed. In Fig. 2.5 are plotted such oscillations recorded in the case of GaN, $Al_{0.33}Ga_{0.67}N$ and AlN growth on sapphire substrate. The alloy composition can be easily determined by comparing the oscillation frequencies associated with the binary and the ternary alloy.

An MBE system for the growth of ZnO-based materials usually uses metallic sources for Zn, Mg, Cd, etc., being evaporated from effusion cells usually equipped with Pyrolitic boron nitride (PBN) ceramic crucibles stable up to evaporation temperatures of 1,000 °C. There exists a variety of possibilities concerning the oxygen source for MBE. Water is not necessarily a contaminant in the case of ZnO MBE. Pure molecular oxygen does not work well, because the reaction rate with Zn is too small. Another possibility is the use of oxygen containing molecules such as

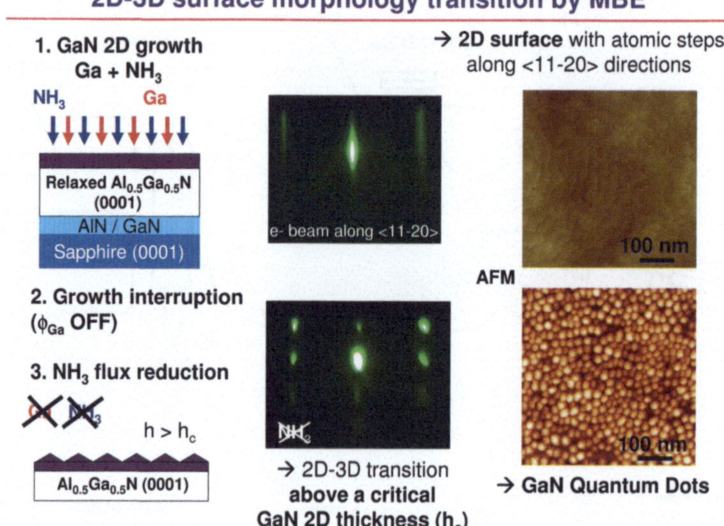

Fig. 2.4 Typical RHEED patterns (*middle*), AFM patterns (*right*) recorded in case of the growth of GaN on an $Al_{0.5}Ga_{0.5}N$ template. The growth conditions are qualitatively sketched in the *left-hand* part of the figure, showing the correlation between growth conditions and modification of the RHEED and AFM patterns. A transition between a two-dimensional growth regime and three dimensional one (clearly corrugated surface) observed beyond the growth of the critical thickness for coherent growth of GaN on $Al_{0.5}Ga_{0.5}N$ is measured in situ during the growth. Courtesy Julien Brault, Benjamin Damilano and Jean Massies

water (H_2O) or hydrogen peroxide (H_2O_2). In the case of H_2O_2, one has to take into account the instability of this molecule. Alternative precursors also include ozone (O_3). When cooled down, ozone can be stored in the liquid state and used as a very efficient precursor. However, safety measures are to be taken into account, since this precursor then is explosive. A very detailed review can be found in "*Growth*", by Andreas Waag, in Klingshirn et al. [13].

2.2.4 The Growth of (001)-Oriented GaN on Sapphire Using a Low-Temperature-Grown Thin Buffer Layer

The crystalline quality of the nitride has long been very poor. In 1983, Yoshida et al. [10] deposited an AlN buffer layer on the sapphire substrate prior to the GaN growth. This technique appeared to be very efficient for improving the overall material quality. It was later adapted to the MOVPE growth by Amano et al. [1] and Nakamura et al. [9] demonstrated that a GaN low temperature buffer layer could be employed with similar success. The growth protocol is illustrated in Fig. 2.6.

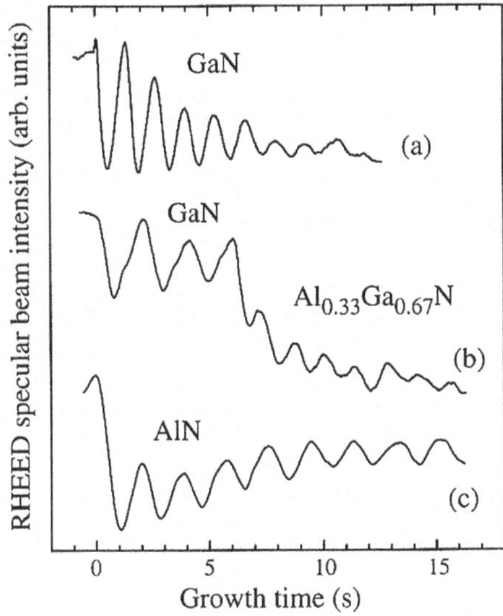

Fig. 2.5 Typical RHEED intensity oscillations for GaN, $Al_{0.33}Ga_{0.67}N$ and AlN on saphirre. Courtesy Nicolas Grandjean

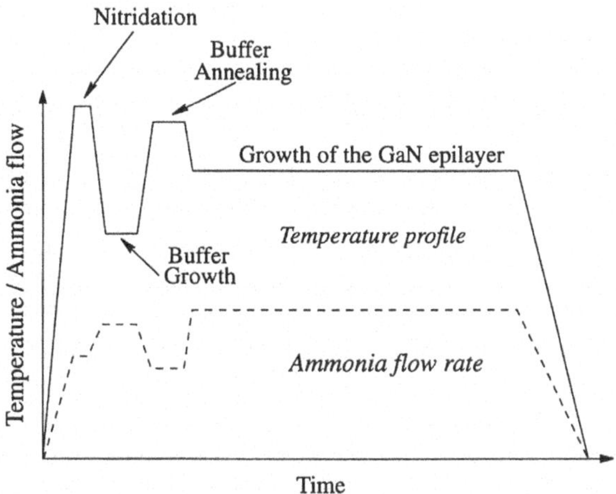

Fig. 2.6 Sketch of the double step growth process: temperature and precursors flow versus time. Courtesy Olivier Briot

The growth protocol has to begin by a pre-treatment of the substrate at high temperature (above 1,000 °C) to reconstruct the substrate's surface and to improve it. Different groups use slightly different protocols under either an ammonia or an hydrogen ambient, different temperatures and treatment times.

Then the substrate temperature is cooled down to a temperature ranging between 500 and 800 °C at which the low temperature buffer layer is deposited. The buffer

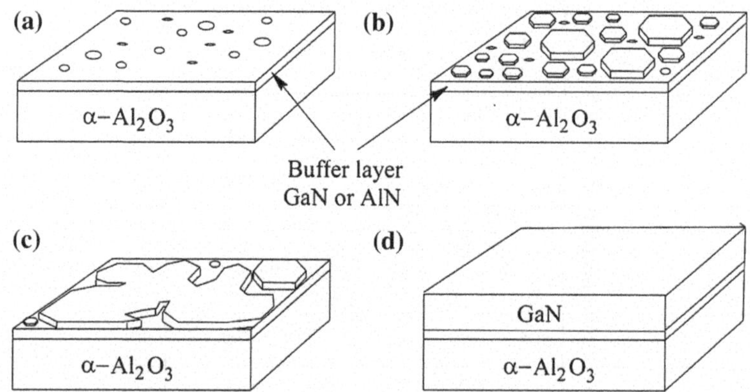

Fig. 2.7 Sketch of the double step growth process: temperature and precursors flow versus time. Courtesy Olivier Briot

thickness is of paramount importance as well as its growth conditions. All the GaN properties are optimized for a GaN buffer thickness of 200–250 Å while thicknesses ranging from 300 to 1,000 Å have been proposed for AlN buffer layers. This is very empirical. We want to remark that various stackings of low temperature grown nitride buffer layers were also proposed.

The buffer layer is then submitted to a heat treatment to improve its crystalline quality that is very poor. Its temperature is ramping elevated under an ammonia flow to stabilize its surface.

Finally the growth of the high temperature and high quality GaN film is realized. The role of the buffer layer is interpreted as follows: When the GaN layer is grown on the buffer a high density of nucleation is obtained (Fig. 2.7a) and small hexagonal islands are formed. Then at an early stage of the island growth, the lateral growth mode is enhanced leading to a progressive coalescence of islands (Fig. 2.7b, c) up to a full coverage of the buffer layer by the high temperature GaN film.

In Fig. 2.8 is reported the interpretation of HIramatsu et al. [7] which proposed after Transmission Electron Microscopy experiment that the AlN buffer layer is crystallized by solid phase epitaxy during the heat treatment into a highly oriented columnar structure. GaN nuclei grow on top of such columns and the high density of GaN nucleation centers is directly linked to the high density of AlN columns.

The columns have disordered orientations: they are tilted with respect to the growth direction and twisted around it (see Fig. 2.9).

The tilt can be calibrated by a rocking curve diffraction experiment around the (0002) X-ray diffraction peak; the twist can be calibrated via a rocking curve experiment around the (10–10) diffraction peak. Measuring twist ((which is in general worse than tilt) is difficult and requires to work in grazing incidence configuration as sketched top in Fig. 2.10. Tilt is often evaluated from other asymmetrical diffraction peaks, using semi-empirical models as illustrated bottom in Fig. 2.10 for an indium nitride crystal. An X-Ray diffraction pattern contains a lot of information

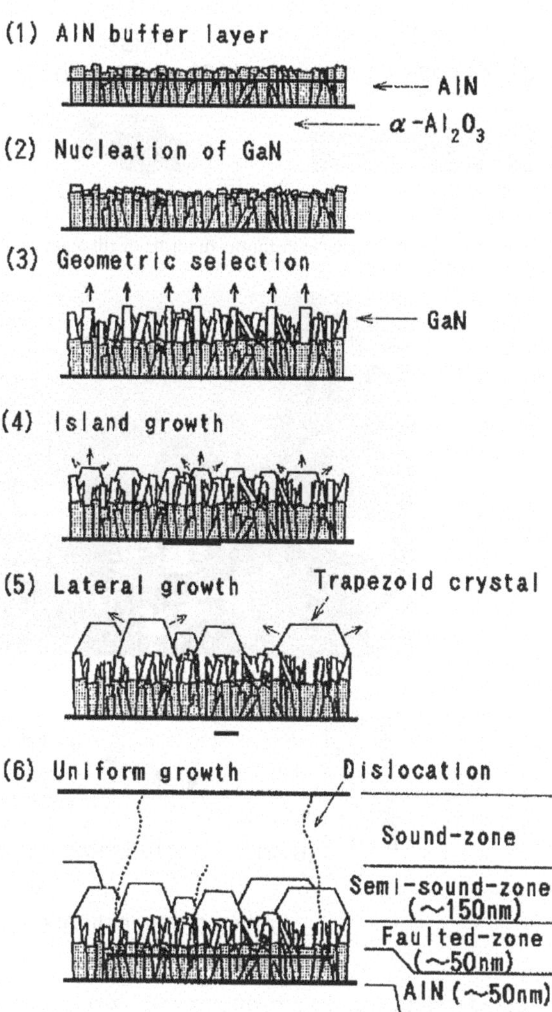

Fig. 2.8 Schematic diagrams showing the growth process of GaN on the AlN buffer layer as the cross sectional views. After [7]

which processing is a real issue. As it is a representation in the reciprocal space of an objet in the real one, the Fourier theorem teaches us that the (0002) X-ray diffraction contains an information regarding the average vertical coherence length (~the layer thickness) whilst (10–10) diffraction peak contains information regarding the average grain size. To get such information requires a line shape fitting which is far out of the scope of this book. As the base of the columns gets larger, thanks to efficient lateral growth, the number of columns emanating at the growth surface gradually decreases as the film thickness increases. The relative tilt and twist between the columns decreases as the film increases expressing that the film continues to evolve

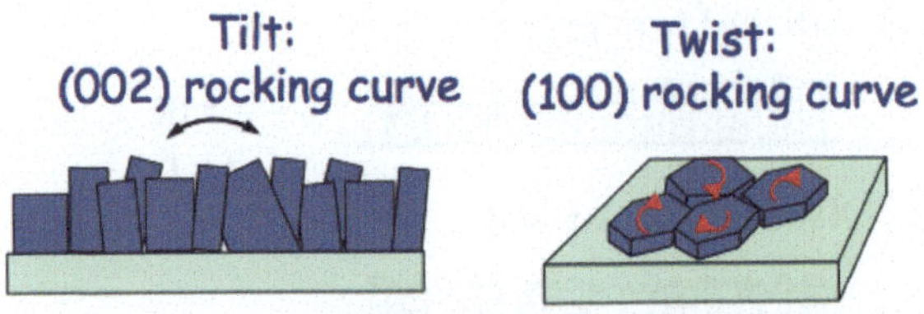

Fig. 2.9 Relative orientations of the GaN nano columns in terms of tilt and twist

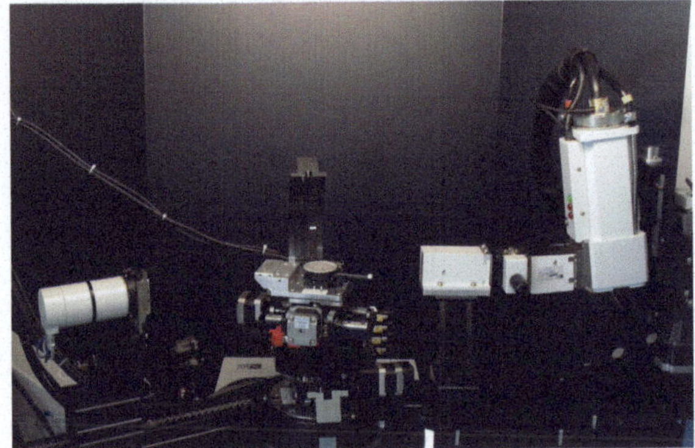

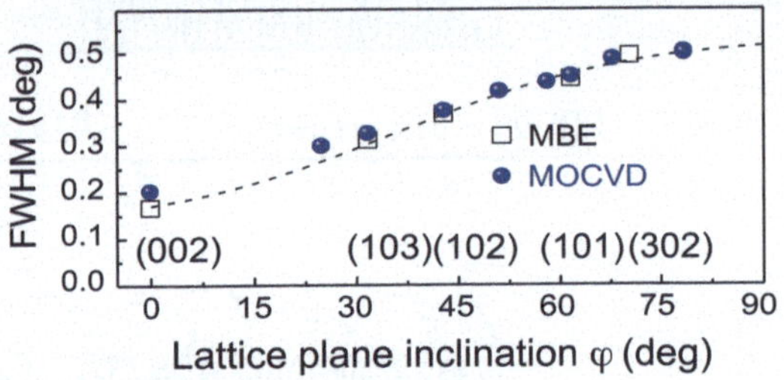

Fig. 2.10 *Top* experimental configuration of an X-ray diffractometer for measuring twist-related information. Courtesy Matthieu Moret. *Bottom* evolution of the full width at half maximum of the X-ray diffraction pattern as a function of the inclination angle for thin InN films deposited on a GaN template by MBE and MOVPE. Courtesy Matthieu Moret

towards a better ordered structure as the growth proceeds. To quantitatively Fig. 2.11 is reported the evolution of the density of dislocation versus growth thickness as reported in by Morkoc [8].

In Fig. 2.11 is reported the evolution of the density of dislocation versus growth thickness as reported in Morkoc [8].

High crystalline quality layers superstore fringes around the main diffraction peak. Such short period fringes are called Pendellosung fringes; they are observed in case of high optical coherence at lateral dimensions comparable to the diameter of the X-ray beam; their periodicity is the thickness of the film. These fringes permits to measure for example the thickness of a quantum well layer as we be shown later. Figure 2.12 illustrates the observation of Pendellosung fringes in a InN films of improved thickness coherence when chemically processed.

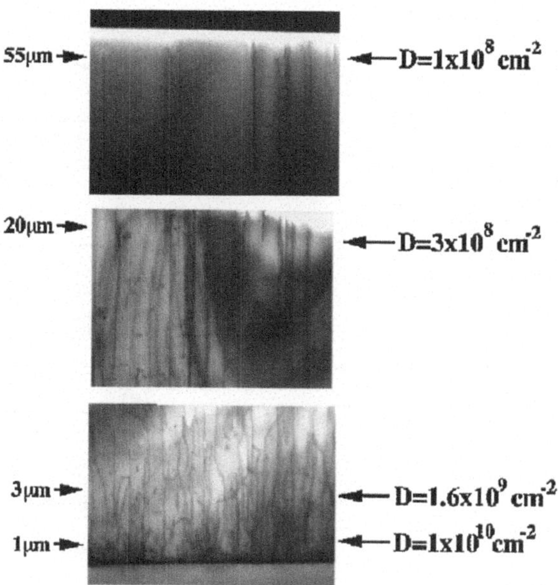

Fig. 2.11 Determination of density of threading dislocations in HVPE GaN layer. After Morkoc [8] and courtesy of J. Jasinski and Z. Liliental-Weber

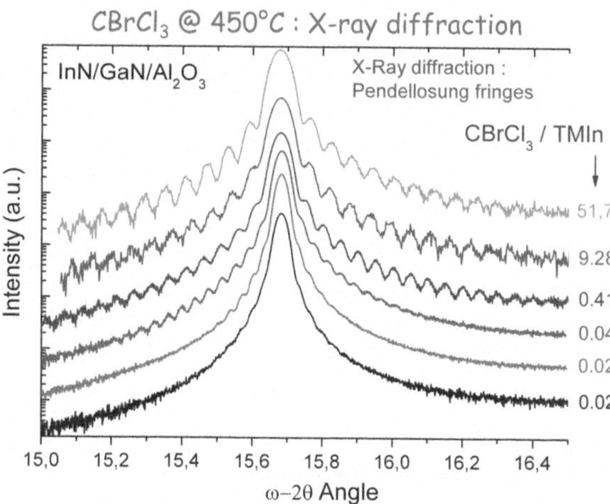

Fig. 2.12 Progressive observation of Pendellosung Fringes for indium films of thickness coherence controlled by halide mediated growth courtesy of Olivier Briot

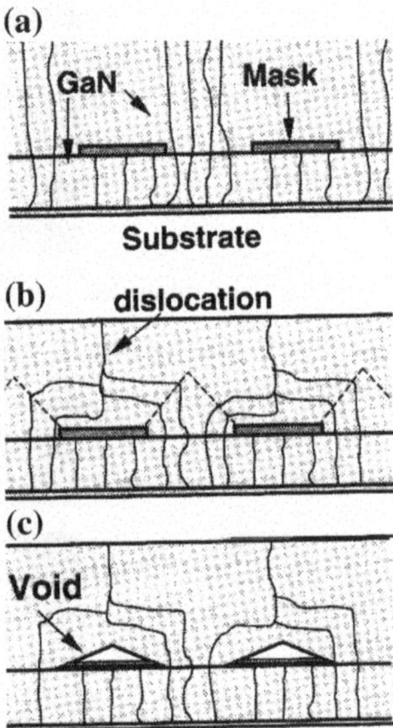

Fig. 2.13 Reduction of the dislocation density using different ELO techniques Hiramatsu in [6]

2.3 Epitaxial Lateral Overgrowth Techniques

Epitaxial lateral overgrowth (ELO) is a technique to grow low-defect-density thin films on lattice-mismatched substrates. The principle of the method consist in first texturing the substrate surface. Then the growth of the semiconductor occurs both growing vertically and laterally at the same time. There are many different possibilities for designing the substrate that are reviewed in [6]. The principle is to reduce the density of dislocations at the active part of the substrate. The basic ELO idea is to block the dislocations under a series of SiO_2 stripes, and to obtain a substrate surface which high dislocation density regions and low dislocation density regions (Fig. 2.13a). Another possibility is to obtain Facet-initiated ELO (FIELO) (Fig. 2.13b) or PENDEO Epitaxy (Fig. 2.13c).

2.4 Epitaxial Growth of Heterostructures

The semiconductors are in general lattice-mismatched and when all mono-layers of the heterostructures are assembled coherently, each of them experiences a strain field in its growth plane. The optimized growth conditions are sometimes different for the different compounds and this may complicate the growth. The deformations can be evaluated by assuming lattice matching to a growth substrate, or by global minimization of the total elastic energy stored in the whole heterostructure. X-ray diffraction can be used to determine the strains state of the different layers of an heterostructure. In Fig. 2.14a we show the $2\theta/\omega$ X-ray diagram taken from a 2.4 nm

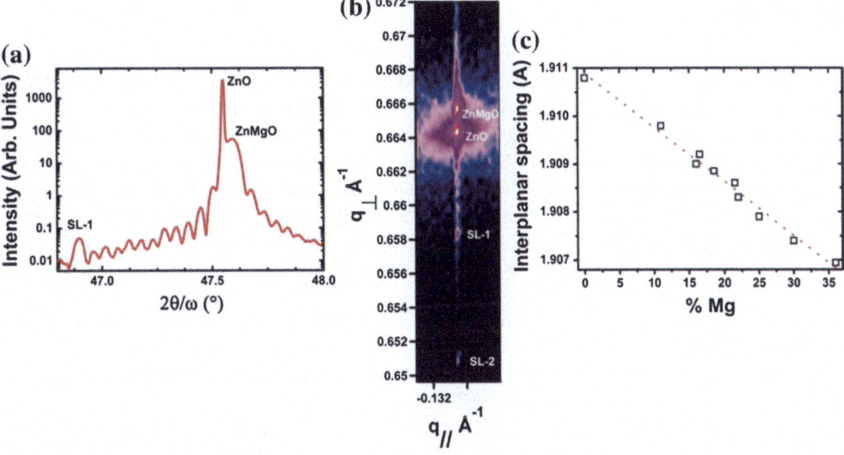

Fig. 2.14 **a** The $2\theta/\omega$ X-ray diagram taken from a 2.4 nm wide ZnO-ZnMgO quantum well grown on a (10–12) ZnO substrate. The ZnO, (Zn, Mg)O peaks as well as a superlattice order one (SL-1) are indicated. **b** Reciprocal space map taken from a MQW sample around taken around the (10–13) reciprocal node. **c** Out of plane lattice parameter of the (Zn, Mg)O barriers deduced as a function of the Mg content. After Built-in electric field in ZnO based semipolar quantum wells grown on (10–12) ZnO substrates by Chauveau et al. [2]

wide ZnO-ZnMgO quantum well grown on a (10–12) ZnO substrate. The position of the diffraction peak related to the (10–12) ZnO sub-strate is measured at 47.548°, corresponding to a reticular distance of 1.9108 Å. The superlattice orders are clearly seen (only one order labeled SL-1 is shown for clarity give a precise value of the thicknesses).

Pendellosung fringes are also observed, confirming the flatness of the QW interfaces and of the interface between the (Zn, Mg)O layer and the ZnO substrate. A reciprocal space mapping (RSM) in asymmetrical geometry is shown in Fig. 2.14b taken around the (10–13) reciprocal node. The ZnO and (Zn, Mg)O diffraction peaks are indicated. Note that the Pendellosung fringes and the superlattice satellites (marked as SL-1 and SL-2) are still clearly observed, even in this asymmetrical configuration. Moreover, the (Zn, Mg)O and the ZnO layers exhibit the same in-plane lattice parameter, i.e., the same $q_{//}$. This proves that no plastic relaxation occurs in a 180 nm thick $Zn_{0.83}$-$Mg_{0.17}O$ layer. The (Zn, Mg)O is fully strained while the ZnO QW is fully unstrained. The out of plane lattice parameter of the barriers $d_{(10-12)}$ was measured precisely for the whole series as a function of the Mg content. The results are shown in Fig. 2.13c. The lattice parameter slowly decreases with increasing Mg content because the minimum elastic energy stored in a (Zn, Mg)O/ZnO heterostructure is expected to be reached along this orientation.

References

1. H. Amano, N. Sawaki, I. Akasaki, Y. Toyoda, Metalorganic vapor-phase epitaxial-growth of a high-quality GaN film using an Al buffer layer. Appl. Phys. Lett. **48**, 383 (1986)
2. J.-M. Chauveau, Y. Xia, I. Ben Taazaet-Belgacem, M. Teisseire, B. Roland, M. Nemoz, J. Brault, B. Damilano, M. Leroux, B. Vinter, Built-in electric field in ZnO based semipolar quantum wells grown on (101–2) ZnO substrates. Appl. Phys. Lett. **103**, 262104 (2013)
3. R. Doradzinski, R. Dwilinski, J. Garczynski, L.P. Sierzputowski, Y. Kanbara, *Ammonothermal Growth of GaN Under Ammono-Basic Conditions* (Springer, Berlin, 2010)
4. B. Gil, *III-Nitride Semiconductors and their Modern Devices*. (Oxford University Press, Oxford, 2013), pp. 121–146
5. T. Hashimoto, S. Nakamura, *A Pathway Toward Bulk Growth of GaN by the Ammonothermal Method* (Springer, Berlin, 2010)
6. K. Hiramatsu, Epitaxial lateral overgrowth techniques used in group III nitride epitaxy Hiramatsu. J. Phys. Condens. Matter **13**, 6962 (2001)
7. K. Hiramatsu, S. Itoh, H. Amano, I. Akasaki, N. Kuwano, T. Shiraishi, K. Oki, Growth mechanism of GaN grown on sapphire with A1N buffer layer by MOVPE. J. Cryst. Growth **115**, 628 (1991)
8. H. Morkoc, Comprehensive characterization of hydride VPE grown GaN layers and templates. Mater. Sci. Eng. R **33**, 135 (2001)
9. S. Nakamura, GaN growth using GaN buffer layer. J. Appl. Phys. Part 2–Lett. **10A**, L1705 (1991)
10. S. Yoshida, S. Misawa, S. Gonda, Improvements on the electrical and luminescent properties of reactive molecular-beam epitaxially grown GaN films by using AlN-coated sapphire substrates. Appl. Phys. Lett. **42**, 427 (1983)
11. H. Morkoc, *Nitride Semiconductors and Devices* (Springer, Berlin, 1999)
12. D. Ehrentraut, E. Meissner, M. Bockowski, *Technology of Gallium Nitride Crystal Growth* (Springer, Berlin, 2010)
13. C.F. Klingshirn, B.K. Meyer, A. Waag, A. Hoffmann, J.M. Geurts, *Zinc Oxide* (Springer, Berlin, 2010)
14. I. Grzegory, High pressure growth of bulk GaN from solutions in gallium. J. Phys. Condens. Matter. **13**, 6875 (2001)
15. B. Gil (ed.), *III-Nitride Semiconductors and their Modern Devices* (Oxford University Press, Oxford, 2013)
16. B. Gil (ed.), *Low Dimensional Nitride Semiconductors* (Oxford University Press, Oxford, 2002)
17. B. Gil (ed.), *Group III-Nitride Semiconductor Compounds* (Oxford University Press, Oxford, 1998)
18. H. Morkoç, *Handbook of Nitride Semiconductors and Devices, GaN-based Optical and Electronic Devices* (Wiley, New York, 2009)

Chapter 3
Electrons and Phonons in Wurtzitic Semi-conductors

We review the basic theory of electronic and lattice dynamics in crystals and applicate it to wurtzite semiconductors. The semiclassical theory of the dielectric constant, of paramount importance to treat optical properties is introduced as well as the **k.p** method for computing electronic states in the neighborhood of a band structure extremum.

3.1 Electrons in a Periodic Potential

3.1.1 The Born-Oppenheimer Adiabatic Approximation

The description of the energetic states of a solid starts with writing the Hamiltonian **H** of this solid. Let the n_e electrons (resp. $\overrightarrow{N_k}$ nuclei) of mass m_0 (resp. M_k) have coordinates $\overrightarrow{r_e}$ (resp. $\overrightarrow{R_k}$). The nuclei charge is Z_k. The Hamiltonian **H** is the sum of a kinetic energy operator **T** and an interaction operator **V**, accounting for the Coulomb interaction between particles:

$$\mathbf{H} = \mathbf{T} + \mathbf{V}$$

with:

$$\mathbf{T} = \mathbf{T}_e + \mathbf{T}_k$$
$$\mathbf{V} = \mathbf{V}_{ee} + \mathbf{V}_{ek} + \mathbf{V}_{kk}$$

The kinetic energy operators are expressed as follows:

$$\mathbf{T}_e = -\sum_{j=1}^{n_e} \frac{\hbar^2}{2m_0} \frac{\partial^2}{\partial r_j^2}$$

B. Gil, *Physics of Wurtzite Nitrides and Oxides*, Springer Series in Materials Science 197, DOI: 10.1007/978-3-319-06805-3_3,

and

$$\mathbf{T}_k = -\sum_{J=1}^{N_k} \frac{\hbar^2}{2M_j} \frac{\partial^2}{\partial R_j^2}$$

The mutual repulsive Coulomb interaction between electrons writes:

$$\mathbf{V}_{ee} = \frac{1}{2} \frac{e^2}{4\pi\epsilon} \sum_{j,i \neq j} \frac{1}{|\overrightarrow{r_i} - \overrightarrow{r_j}|}$$

where factor $\frac{1}{2}$ appears to avoid double-counting interactions.

The repulsive nucleus–nucleus Coulomb interaction operator writes:

$$\mathbf{V}_{kk} = \frac{1}{2} \frac{e^2}{4\pi\epsilon} \sum_{J,I \neq J} \frac{Z_I Z_J}{|\overrightarrow{R_I} - \overrightarrow{R_J}|}$$

while the attractive electron–nucleus Coulomb interaction operator is:

$$\mathbf{V}_{ek} = -\frac{e^2}{4\pi\epsilon} \sum_{i,J} \frac{Z_J}{|\overrightarrow{r_i} - \overrightarrow{R_J}|}$$

In the absence of external potential, the crystal states are obtained as solutions of the Schrödinger equation:

$$\mathbf{H}\Psi(\{\overrightarrow{r_i}\}, \{\overrightarrow{R_J}\}) = E\Psi(\{\overrightarrow{r_i}\}, \{\overrightarrow{R_J}\})$$

The wave-function $\Psi(\{\overrightarrow{r_i}\}, \{\overrightarrow{R_J}\})$ of the interacting electrons and nuclei is represented versus the whole electrons $\{\overrightarrow{r_i}\}$ and nuclei $\{\overrightarrow{R_J}\}$ positions. It has $3(n_e + N_K)$ degrees of freedom and, as soon as the crystal gets a reasonable size, the system cannot be solved, even using the most powerful computers.

Physical approximations are required to keep the problem tractable. The first one is the Born-Oppenheimer approximation. It allows us to separate the electron problem from the nucleus motion. It is motivated by the consideration that, protons or neutrons constituting the nucleus, are $\approx$1,840 times heavier than electrons. When the atomic number of the element Z_k increases, neutrons are more numerous than protons: they compensate, via short range strong-interaction processes, the long range proton–proton Coulomb repulsion. This way, the nucleus is more stable; we remind the reader that, for high values Z_k, the chemical elements are all unstable with time, often leading to radioactive fission. Thus, the ratio $\frac{M_k}{m_e}$ increases rapidly up to 10^4–10^5 and a description of the crystal states with electrons rapidly moving into a quasi-static lattice was proposed. The wave-function writes:

$$\Psi(\{\overrightarrow{r_i}\}, \{\overrightarrow{R_J}\}) = \chi(\overrightarrow{R_J})\Phi(\{\overrightarrow{r_i}\}, \{\overrightarrow{R_J}\})$$

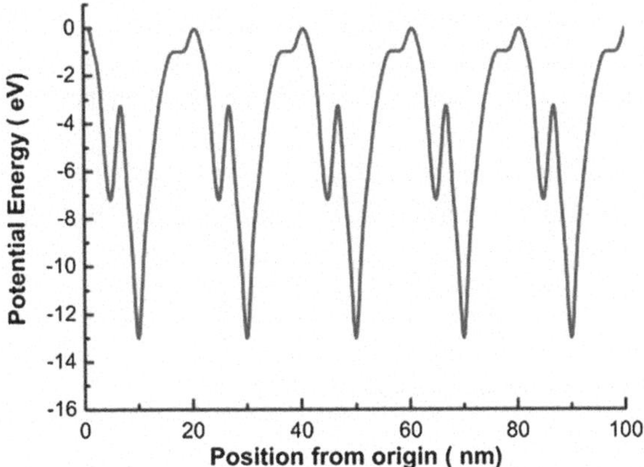

Fig. 3.1 Sketch of a periodic potential (five repeats) for a one-dimension crystal

Given a momentary full set of nuclei configuration $\{\overrightarrow{R_J}\}$, neglecting the $\{\overrightarrow{R_J}\}$-dependence of the electronic wave-function $\Phi(\{\overrightarrow{r_i}\}, \{\overrightarrow{R_J}\})$ with respect to the kinetic operator $\mathbf{T}_k$, one can express the electron eigenstates as:

$$\left[\mathbf{T}_e + \mathbf{V}_{ee} + \mathbf{V}_{ek}(\{\overrightarrow{R_J}\})\right] \Phi_n(\{\overrightarrow{r_i}\}, \{\overrightarrow{R_J}\}) = E_n \Phi_n(\{\overrightarrow{r_i}\}, \{\overrightarrow{R_J}\})$$

Similarly, the nucleus problem writes:

$$\left[\mathbf{T}_k + \mathbf{V}_{kk} + \mathbf{E}_n(\{\overrightarrow{R_J}\})\right] \chi_{n\nu}(\{\overrightarrow{R_J}\}) = E_{nv}(\{\vec{R_J}\}) \cdot \chi_{n\nu}(\{\overrightarrow{R_J}\})$$

In the above equations, n and ν are quantum numbers for electrons and nuclei: they will describe respectively electronic and phonon states.

$\mathbf{V}_{ek}(\{\overrightarrow{R_J}\})$ is a potential possessing the crystal periodicity, such as the one illustrated in Fig. 3.1 for a one-dimension case. One can, for instance, imagine that the potential minima are located at the positions of the nuclei.

We note that the quantity $\mathbf{E}_{ek}(\{\overrightarrow{R_J}\})$ acts as an effective potential of the adiabatic electron-nuclei interaction, that depends on the electronic subsystem occupied. Together with the repulsive Coulomb interaction between nuclei, it provides an effective potential the electrons undergo:

$$V_n(\{\overrightarrow{R_J}\}) = V_{kk}(\{\overrightarrow{R_J}\}) + E_n(\{\overrightarrow{R_J}\})$$

That leads us to express the Schrödinger equation for the nuclei motion in the crystal under the form:

$$\left[\mathbf{T}_k + \mathbf{V}_n(\{\overrightarrow{R_J}\})\right] \chi_{n\nu}(\{\overrightarrow{R_J}\}) = E_{n\nu}(\{\overrightarrow{R_J}\}) \chi_{n\nu}(\{\overrightarrow{R_J}\})$$

The electron energy spectrum describes the electron motion in an effective potential $\mathbf{V}_0(\{\overrightarrow{R_J}\})$ created by the nuclei, having the spatial crystal periodicity. The effective potential associated with the electronic ground state is called the Born-Oppenheimer surface.

The expression of the eigen-values $E_{n\upsilon}(\{\overrightarrow{R_J}\})$ indicates that any modification of the positions $\overrightarrow{R_J}$ produced, for instance, by submitting the crystal to a deformation will alter them.

3.1.2 The One-Electron Approximation

In this section, we re-handle the equation:

$$\left[\mathbf{T}_e + \mathbf{V}_{ee} + \mathbf{V}_{ek}(\{\overrightarrow{R_J}\})\right]\Phi_n(\{\overrightarrow{r_i}\},\{\overrightarrow{R_J}\}) = E_n(\{\overrightarrow{R_J}\})\Phi_n(\{\overrightarrow{r_i}\},\{\overrightarrow{R_J}\})$$

that is transformed as:

$$\left[-\sum_{j=1}^{n_e}\frac{\hbar^2}{2m_0}\frac{\partial^2}{\partial r_j^2} + \frac{1}{2}\frac{e^2}{4\pi\epsilon}\sum_{j,i\neq j}\frac{1}{|\overrightarrow{r_i}-\overrightarrow{r_j}|} - \frac{e^2}{4\pi\epsilon}\sum_{j,J}\frac{Z_J}{|\overrightarrow{r_j}-\overrightarrow{R_J}|}\right]$$
$$\Phi_n(\{\overrightarrow{r_i}\},\{\overrightarrow{R_J}\}) = E_n(\{\overrightarrow{R_J}\})\cdot\Phi_n(\{\overrightarrow{r_i}\},\{\overrightarrow{R_J}\})$$

or, by extracting the j-summation:

$$\sum_{j=1}^{n_e}\left[-\frac{\hbar^2}{2m_0}\frac{\partial^2}{\partial r_j^2} + \frac{1}{2}\frac{e^2}{4\pi\epsilon}\sum_{i\neq j}\frac{1}{|\overrightarrow{r_i}-\overrightarrow{r_j}|} - \frac{e^2}{4\pi\epsilon}\sum_{J}\frac{Z_J}{|\overrightarrow{r_j}-\overrightarrow{R_J}|}\right]$$
$$\Phi_n(\{\overrightarrow{r_i}\},\{\overrightarrow{R_J}\}) = E_n(\{\overrightarrow{R_J}\})\cdot\Phi_n(\{\overrightarrow{r_i}\},\{\overrightarrow{R_J}\})$$

This Hamiltonian is the sum of n_e one-electron Hamiltonians:

$$\left[-\frac{\hbar^2}{2m_0}\frac{\partial^2}{\partial r^2} + \frac{1}{2}\frac{e^2}{4\pi\epsilon}\sum_{\overrightarrow{r_j}\neq\vec{r}}\frac{1}{|\vec{r}-\overrightarrow{r_j}|} - \frac{e^2}{4\pi\epsilon}\sum_{J}\frac{Z_J}{|\vec{r}-\overrightarrow{R_J}|}\right]$$
$$\Phi_n(\{\vec{r}\},\{\overrightarrow{R_J}\}) = E_n(\{\overrightarrow{R_J}\})\cdot\Phi_n(\{\vec{r}\},\{\overrightarrow{R_J}\})$$

and describes the energy spectrum of the electron moving in an effective potential $\mathbf{V}_{eff}$ due to $n_e - 1$ other electrons and N_k ions.

$$V_{eff}(\vec{r}, \{\overrightarrow{R_J}\}) = \frac{1}{2}\frac{e^2}{4\pi\epsilon}\sum_{\overrightarrow{r_j}\neq\vec{r}}\frac{1}{|\vec{r}-\overrightarrow{r_j}|} - \frac{e^2}{4\pi\epsilon}\sum_J \frac{Z_J}{|\vec{r}-\overrightarrow{R_J}|}$$

The main difficulty comes from the first term of this effective potential, the Coulomb interaction:

$$\frac{1}{2}\frac{e^2}{4\pi\epsilon}\sum_{\overrightarrow{r_j}\neq\vec{r}}\frac{1}{|\vec{r}-\overrightarrow{r_j}|}$$

Starting from Lord Sommerfeld free-electron model, several approximations have been proposed.

3.1.3 The Free Electron Model

This model ignores all interactions and leads to a simple series of un-coupled Schrödinger equations:

$$-\sum_{j=1}^{n_e}\left[\frac{\hbar^2}{2m_0}\frac{\partial^2}{\partial r_j^2}\right]\Phi(\{\overrightarrow{r_j}\}) = E\cdot\Phi(\{\overrightarrow{r_j}\})$$

The eigen-values are obtained as

$$E = \sum_{j=1}^{n_e}\frac{\hbar^2}{2m_0}k_j^2$$

They are continuous in **k** space and form a parabolic band of energy versus the real quantum vector **k**. The wave-functions are obtained as products of the wave-functions of individual electrons:

$$\Phi(\{\overrightarrow{r_i}\}) = \prod_{j=1}^{j=n_e}\phi_j(\overrightarrow{r_j})$$

with:

$$\phi_\ell(\overrightarrow{r_\ell}) = \frac{1}{\sqrt{V}}exp\left[i\overrightarrow{k_\ell}\cdot\overrightarrow{r_\ell}\right]$$

V being the volume of the crystal.

At this stage, we have a clear description of the energy spectrum and wave-functions for a gas of free spin-less non-interacting electrons. A representation of the eigenstates of such a gas versus the continuous ensemble of quantum numbers k is given in Fig. 3.2.

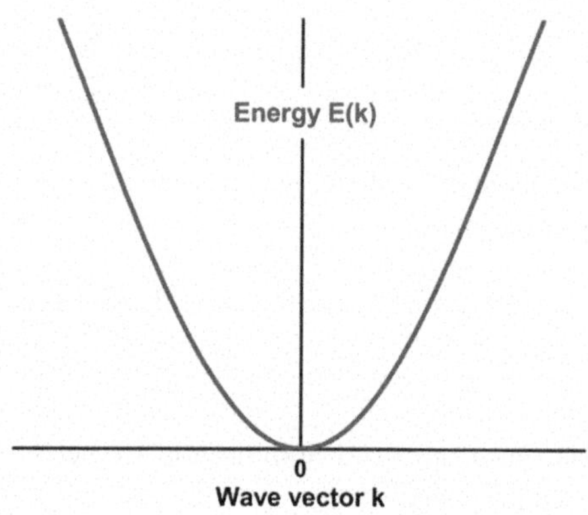

Fig. 3.2 Plot of the eigenvalues $E(\vec{k})$ for a gas of non-interacting electrons in reciprocal space. The wave-vector $\vec{k}$ is real and acts as a quantum number

Equation

$$E_{\vec{k}} = \sum_{j=1}^{n_e} \frac{\hbar^2}{2m_0} k_j^2$$

indicates that the linearly-independent solutions are obtained only if the k_j's take positive values.

Thus, the number of electrons states—neglecting spin—with kinetic energy smaller than $E_{\vec{k}}$ is the number of points in the positive octant of the sphere of radius $\tilde{k}$ where $E_{\vec{k}} = \frac{\hbar^2}{2m_0}\tilde{k}^2$.

For large $\tilde{k}$, the number of points is simply the volume: $\frac{1}{8}\frac{4}{3}\pi\tilde{k}^3$

So, the total number of electrons having an energy smaller than $E_{\vec{k}}$ is:

$$N(E_{\vec{k}}) = V\frac{\pi}{6}\left[\frac{8m_0E_{\vec{k}}}{(2\pi\hbar)^2}\right]^{3/2} = \frac{V}{6\pi^2}\left(\frac{2m_0E_{\vec{k}}}{\hbar^2}\right)^{3/2}$$

The number of states per unit of energy is called the density of states $n(E)$ and is related to $N(E)$ through:

$$\int_0^{E_{\vec{k}}} n(E)dE = N(E_{\vec{k}})$$

$$n(E_{\vec{k}}) = (\frac{dN}{dE})_{E_{\vec{k}}}$$

$$N(E_{\vec{k}}) = V\frac{1}{4\pi^2}\left(\frac{2m_0}{\hbar^2}\right)^{3/2} E_{\vec{k}}^{1/2}$$

Including the spin requires to double the number of wave-functions: each wave-function $\phi_\ell(\vec{r_\ell})$ has to be multiplied by a spinor function which does not operate in real space, but in the spin space. There are two eigen-values for the spin state $\frac{\hbar}{2}$ and $-\frac{\hbar}{2}$ corresponding to spin functions $\begin{bmatrix} 1 \\ 0 \end{bmatrix}$ and $\begin{bmatrix} 0 \\ 1 \end{bmatrix}$ respectively. Two spin-dependant global electron wave-functions are built from each $\phi_\ell(\overrightarrow{r_\ell})$:

$$\psi_\ell(\overrightarrow{r_\ell}) = \phi_\ell(\overrightarrow{r_\ell}) \cdot \begin{bmatrix} 1 \\ 0 \end{bmatrix}$$

and

$$\psi'_\ell(\overrightarrow{r_\ell}) = \phi_\ell(\overrightarrow{r_\ell}) \cdot \begin{bmatrix} 0 \\ 1 \end{bmatrix}$$

Last, the electron obeys Fermi-Dirac statistics that applies to half-integer angular momentum particles: two electrons DONOT dwell in the same state, and the global wave-function HAS TO be anti-symmetric when permuting the two particles. One could demonstrate that the symmetry-compatible wave-function can be written as a determinant:

$$\Phi_D(\{\overrightarrow{r_i}\}) = \frac{1}{\sqrt{2n_e!}} \begin{bmatrix} \psi_1(\overrightarrow{r_1}) & \psi_1(\overrightarrow{r_2}) & \dots & \psi_1(\overrightarrow{r_{2n_e}}) \\ \psi_2(\overrightarrow{r_1}) & \psi_2(\overrightarrow{r_2}) & \dots & \psi_2(\overrightarrow{r_{2n_e}}) \\ \vdots & \vdots & \ddots & \vdots \\ \psi_{2n_e}(\overrightarrow{r_1}) & \psi_{2n_e}(\overrightarrow{r_2}) & \dots & \psi_{2n_e}(\overrightarrow{r_{2n_e}}) \end{bmatrix}$$

Exchanging two columns or rows changes the sign of the whole Φ_D. In addition, when two wave-functions are identical, the global vanishes, satisfying the Fermi exclusion principle. Two electrons may, of course, have the same single electron wave-function $\phi_\ell(\overrightarrow{r_\ell})$, but they **MUST** have different spin states.

The density of states including spin is then:

$$N(E_{\vec{k}}) = \frac{V}{2\pi^2} \left(\frac{2m_0}{\hbar^2} \right)^{3/2} E_{\vec{k}}^{1/2}$$

Let us now consider the quantum states of an electron moving in a spatially-periodic potential.

3.1.4 The Effect of a Periodic Lattice: The Bloch Theorem

Still neglecting electron–electron interactions, the one-electron Hamiltonian is now written as:

$$\left[-\frac{\hbar^2}{2m_0} \frac{\partial^2}{\partial r^2} - \frac{e^2}{4\pi\epsilon} \sum_J \frac{Z_J}{|\vec{r} - \overrightarrow{R_J}|} \right] \Phi_n(\vec{r}) = E_n \cdot \Phi_n(\vec{r})$$

or, alternatively:

$$\left[-\frac{\hbar^2}{2m_0}\frac{\partial^2}{\partial r^2}+U(\vec{r})\right]\Phi_n(\vec{r})=E_n\cdot\Phi_n(\vec{r})$$

where the effective potential $U(\vec{r})$, due to the crystal lattice, is invariant under translation by a lattice vector $\overrightarrow{L}$.

$$U(\vec{r})=U(\vec{r}+\overrightarrow{L})$$

Being solid-state physicists, our convention concerning the reciprocal lattice differs from the crystallographists'. In this chapter, and by contrast with the preceding one, the reciprocal lattice vectors $\overrightarrow{a_j^*}$ are generated by vectors $\overrightarrow{a_i}$:

$$\overrightarrow{a_j^*}.\overrightarrow{a_i}=2\pi\delta_{ij}$$

where δ_{ij} is the Kronecker symbol.

For every vector $\vec{g}$ in the reciprocal lattice: $e^{i\vec{g}.(\vec{L}+\vec{r})}=e^{i\vec{g}.\vec{r}}$. We are allowed to write:

$$U(\vec{r})=\sum_{\vec{g}}U_{\vec{g}}.e^{i\vec{g}.\vec{r}}$$

where the $U_{\vec{g}}$'s are the Fourier transforms of $U(\vec{r})$.

$$U_{\vec{g}}=\frac{1}{V}\int U(\vec{r})e^{-i\vec{g}.\vec{r}}d^3r$$

Let us evaluate the impact of this periodic potential, when treated as a perturbation to the free electron state.

We start from the functions basis $\phi_\ell(\overrightarrow{r_\ell})=\frac{1}{\sqrt{V}}\exp\left[i\overrightarrow{k_\ell}\cdot\overrightarrow{r_\ell}\right]$.

We look for matrix elements of the effective potential $U(r)$ linking different un-perturbated states:

$$U_{\overrightarrow{kk'}}=\frac{1}{V}\int U(\vec{r})e^{-i(\vec{k}-\overrightarrow{k'}).\vec{r}}d^3r$$

We rewrite it as:

$$U_{\overrightarrow{kk'}}=\frac{1}{V}\int\sum_{\vec{g}}U_{\vec{g}}\cdot e^{i\vec{g}\cdot\vec{r}}e^{-i(\vec{k}-\overrightarrow{k'})\cdot\vec{r}}d^3r$$

and:

$$U_{\overrightarrow{kk'}} = \frac{1}{V} \sum_{\vec{g}} \int U_{\vec{g}} \cdot e^{-i(\vec{k}-\overrightarrow{k'}-\vec{g}).\vec{r}} d^3 r$$

This quantity vanishes except if $\exists \vec{g}$ so that $\vec{k} - \vec{k}' = \vec{g}$. Then,

$$U_{\overrightarrow{kk'}} = U_{\vec{g}} = U_{\vec{k}-\overrightarrow{k'}}$$

The non-vanishing matrix elements correspond to states which wave-vectors differ exactly by one vector $\vec{g}$ of the reciprocal lattice.

Using second-order perturbation theory, the wave-function becomes:

$$\Psi_{\vec{k}} = \phi_{\vec{k}} + \sum_{\vec{g}}{}' \frac{U_{\vec{g}}}{\hbar^2 \left[\frac{\vec{k}^2-(\vec{k}+\vec{g})^2}{2m_0}\right]} \phi_{(\vec{k}+\vec{g})}$$

firstly rearranged into:

$$\Psi_{\vec{k}} = \phi_{\vec{k}} \left\{ 1 + \sum_{\vec{g}}{}' \frac{U_{\vec{g}} \cdot e^{i\vec{g}\cdot\vec{r}}}{\hbar^2 \left[\frac{\vec{k}^2-(\vec{k}+\vec{g})^2}{2m_0}\right]} \right\}$$

and secondly as:

$$\Psi_{\vec{k}} = \left\{ 1 + \sum_{\vec{g}}{}' \frac{U_{\vec{g}} \cdot e^{i\vec{g}.\vec{r}}}{\hbar^2 \left[\frac{\vec{k}^2-(\vec{k}+\vec{g})^2}{2m_0}\right]} \right\} e^{i\vec{k}\cdot\vec{r}}$$

Let:

$$u_{\vec{k}}(\vec{r}) = \left\{ 1 + \sum_{\vec{g}}{}' \frac{U_{\vec{g}} \cdot e^{i\vec{g}\cdot\vec{r}}}{\hbar^2 \left[\frac{\vec{k}^2-(\vec{k}+\vec{g})^2}{2m_0}\right]} \right\}$$

So, we can write the following identity:

$$\Psi_{\vec{k}}(\vec{r}) = u_{\vec{k}}(\vec{r}) \cdot e^{i\vec{k}\cdot\vec{r}}$$

known as the **Bloch theorem: the wave-functions of a periodic potential are given as the product of a plane wave $e^{i\vec{k}.\vec{r}}$ with a Bloch function $u_{\vec{k}}(\vec{r})$ possessing the lattice periodicity**.

3.1.5 *The Born-von Karman Cycling Conditions and the Concept of Spatial Folding*

We remark that both the potential $U(\vec{r})$ and the Bloch wave $u_{\vec{k}}(\vec{r})$ have the same periodicity, i.e. they can be expressed using a similar mathematical formalism: $f(\vec{r}) = f(\vec{r} + \vec{L})$ whereas the wave-function writes as a straightforward result of Bloch theorem:

$$\Psi_{\vec{k}}(\vec{r} + \vec{L}) = u_{\vec{k}}(\vec{r} + \vec{L})e^{i\vec{k}.(\vec{r}+\vec{L})} = e^{i\vec{k}\cdot\vec{L}} \cdot \Psi_{\vec{k}}(\vec{r})$$

The Born-von Karman approximation assumes that a given electron state wave-function has identical phases at both edges of a crystal.

This may be expressed as:

$$\Psi_{\vec{k}}(\vec{r} + N_1\vec{a_1} + N_2\vec{a_2} + N_3\vec{a_3}) = \Psi_{\vec{k}}(\vec{r})$$

where N_1, N_2, N_3 are the integer numbers of elementary cells, stacked along the three directions of the real space to form the macroscopic crystal. As a consequence:

$$e^{i\vec{k}\cdot(N_1\vec{a_1}+N_2\vec{a_2}+N_3\vec{a_3})} = 1$$

Let us take a vector $\vec{k}$ of the reciprocal lattice:

$$\vec{k} = p_1\overrightarrow{a_1^*} + p_2\overrightarrow{a_2^*} + p_3\overrightarrow{a_3^*}$$

Then, we rewrite:

$$e^{i2\pi(p_1N_1+p_2N_2+p_3N_3)} = 1$$

This equality is fulfilled, provided that:

$$N_ip_i = n_i$$

with:

$$|n_i| \in \mathbb{N} \quad \forall i \in \{1, 2, 3\}$$

It is worthwhile noticing the length of the unit cell vectors is $\approx 10^{-9}$m while the dimensions of the macroscopic crystal are typically millimeters (10^{-3}m). The values of the N_i's are large numbers, due to the size of the crystal, compared to the size of the elementary crystal cell.

The equation above also indicates that **the p_i's form a pseudo-continuum** of real numbers with step $1/N_i$, obtained when increasing the n_i's by one.

Thanks to the symmetry properties of the energy dispersion relations, we gain access to all the needed information by representating them in a restricted portion of the reciprocal space. This will impact the value of the relevant numbers among n_i's. Before that, it is necessary to study in detail how expanding a non-interacting electron gas in a periodic potential impacts its wave-functions.

3.1.6 *The Effect of a Periodic Lattice: The Formation of Energy Gaps at the Edges of the Brillouin Zone*

In Sect. 3.1.4, we have demonstrated that the wave-functions of an electron satisfy the Bloch theorem and may be expressed as an expansion of plane waves, describing a non interacting free electron gas. Now, let's focus on its eigen-values or energies.

The second-order perturbation theory describes the energy spectrum as:

$$E_{\vec{k}} = \hbar^2 \frac{\vec{k}^2}{2m_0} + U_0 + \sum_{\vec{g}}{}' \frac{|U_{\vec{g}}|^2}{\hbar^2 \left[\frac{\vec{k}^2 - (\vec{k}+\vec{g})^2}{2m_0}\right]}$$

U_0 being a correction for the mean potential energy of the electrons in the lattice. This equation indicates a departure from a parabolic behavior. This is not surprising: the potential energy is different compared to the free-electron problem.

The question is: how shall we solve the divergence of the energy given by the above equation when

$$\vec{k}^2 = (\vec{k} + \vec{g})^2$$

when two un-perturbated states of the same energy are linked by a vector of the reciprocal lattice, we should use the perturbation theory for degenerated systems. We solve the secular equation:

$$\begin{vmatrix} \frac{\hbar^2 \vec{k}^2}{2m_0} - E & U_{\vec{g}} \\ U_{\vec{g}} & \hbar^2 \frac{(\vec{k}+\vec{g})^2}{2m_0} - E \end{vmatrix}$$

Its solutions are:

$$E(\vec{k}) = \frac{1}{2}\left[\frac{\hbar^2 \vec{k}^2}{2m_0} + \frac{\hbar^2 (\vec{k}+\vec{g})^2}{2m_0} \pm \sqrt{\frac{\hbar^4}{4m_0^2}\left[\vec{k}^2 - (\vec{k}+\vec{g})^2\right]^2 + 4|U_{\vec{g}}|^2}\right]$$

The dispersion relations $E(\vec{k})$ clearly depart from the parabolic behavior, characteristic of the free electron model.

We are led to investigate: for which $\vec{k}$ occurs such a lifting of degeneracy?

From identity of the moduli, $\vec{k}^2 = (\vec{k} + \vec{g})^2$ we can also write:

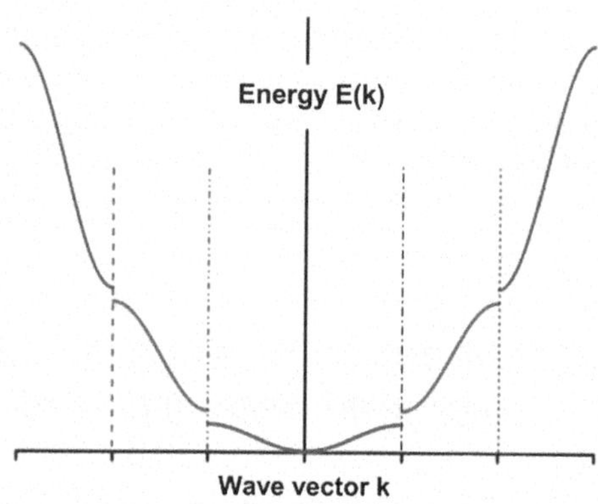

Fig. 3.3 Dispersion relations for the three first energy bands in the context of a periodic potential. When the wave-vector equals half a vector G of the reciprocal lattice, a band gap occurs with value $2|U_{\vec{g}}|$

$$\vec{k}^2 = \vec{k}^2 + \vec{g}^2 + 2\vec{k}\cdot\vec{g}$$

And, we obtain:

$$\vec{k} = -\frac{\vec{g}}{2}$$

The degeneracy is lifted. Relatively to the vectors of the reciprocal lattice, this writes:

$$\vec{k} = 2\pi\left[\frac{p_1}{2}\vec{a}_1^* + \frac{p_2}{2}\vec{a}_2^* + \frac{p_3}{2}\vec{a}_3^*\right]$$

For wave-vectors that equal half a vector of the reciprocal lattice, an energy gap is formed. Due to the formation of pseudo-continuum series of values for $\|\vec{k}\|$, the allowed energies form bands.

The dispersion relations of the three fondamental energy bands are plotted in Fig. 3.3, in the context of the Brillouin representation. This clearly illustrates the departure from the parabolic dispersion relations and we observe energy splittings. The latter constitute energy gaps and are strictly connected to the periodic structure of the potential energy.

$E(\vec{k})$ shows complementary properties. It is worthwhile noticing from the equation above that $E(\vec{k})$ equals $E(-\vec{k})$: the energy is an even function of $\vec{k}$. One can, in addition, easily demonstrate that $E(\vec{k}) = E(\vec{k}+\vec{g})$ when $\vec{g}$ is a vector of the reciprocal lattice. Figure 3.4 illustrates, in the context of the so-called periodic zone scheme, the influence of a periodic potential on the electronic states of the free electron gas.

Due to the periodicity of the wave-function in reciprocal space, one can restrict the set of p_i's, to $(p_1, p_2, p_3) \in [-1, 1]^3$, i.e. limit the $\vec{k}$ to those contained in the Wigner-Seitz cell, the first Brillouin zone. Figure 3.5 represents the dispersion relations of three energy bands when folded into the first Brillouin zone, according to the symmetry properties of $E(\vec{k})$.

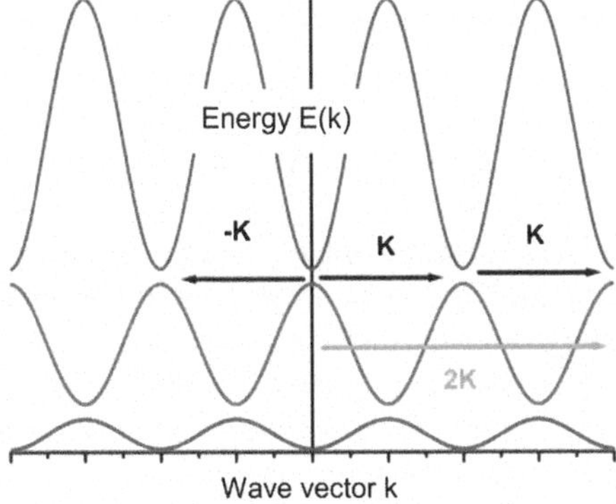

Fig. 3.4 Energy dispersion relations $E(\vec{k})$ for the fundamental bands of a spinless non-interacting electron gas in a periodic potential, in the context of the periodic zone scheme. Note the even parity of the dispersion relations, their periodicity $E(\vec{k}) = E(\vec{k} + p\vec{K})$ in reciprocal space ($p \in \mathbb{Z}$) and the occurrence of some specific energy splittings and energy gaps when wave-vectors equal half a wavevector of the reciprocal lattice

Fig. 3.5 The dispersion relations of three energy bands when folded into the first Brillouin zone

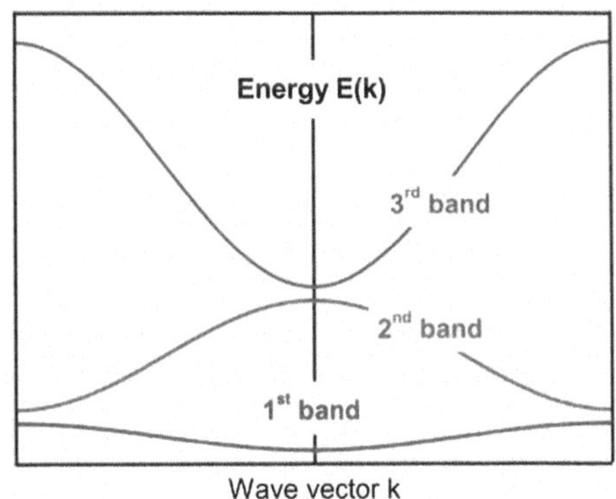

3.1.7 The Concept of the Effective Mass

If $\vec{k}$ represents the wave-vector of an electron possessing the energy $E(\vec{k})$, its frequency is defined by the ratio $\frac{E(\vec{k})}{\hbar}$. The group velocity is by definition:

$$v(\vec{k}) = \frac{1}{\hbar}\left(\frac{\partial(E(\vec{k})}{\partial\vec{k}}\right)$$

At the extremum of a quasi-free electron band, the energy may be written:

$$E(\vec{k}) = \frac{\hbar^2}{2}\frac{1}{m^*(\vec{k})}$$

where an effective mass is defined as:

$$\frac{1}{m^*(\vec{k})} = \frac{1}{\hbar^2}\left(\frac{\partial^2 E(\vec{k})}{\partial k^2}\right)$$

This effective mass differs from the one of the electron at rest; it is the bending of a dispersion relation, and it may be strongly non-parabolic versus $\vec{k}$.

We note, from the plot in Fig. 3.5, that at zone center the first (resp. second) dispersion relation or band is concave (resp. convex). On the contrary, at zone boundaries, the dispersion relation of the first band is convex but concave for the second band. This plot illustrates that, given a band, the sign of the effective mass changes when moving from the zone center to zone edges. In the most general way, there is no reason for the effective mass to be orientation-independent and we define it using this tensor expression:

$$\left[\frac{1}{m^*}\right]_{\alpha\beta} = \frac{1}{\hbar^2}\left[\frac{\partial^2 E(\vec{k})}{\partial k_\alpha \partial k_\beta}\right]$$

One can demonstrate that, near an extremum $E_{0\vec{K}}$ of the band structure occuring at wave-vector $\vec{K}$,

$$N(E_{\vec{k}} - E_{0\vec{K}}) = V\frac{1}{4\pi^2}(\frac{2}{\hbar^2})^{3/2}(m_x^* m_y^* m_z^*)^{1/2}(E_{\vec{k}} - E_{0\vec{K}})^{1/2}$$

where the m_i^* are *the effective masses in the principal directions of the quadric surface that represents the effective masses near the band extremum* $E_{0\vec{K}}$ of the band structure, at wave-vector $\vec{K}$. A general result is that, the flatter the dispersion relation, the higher the density of states.

3.1.8 The Tight-Binding Method

• *General equations.* A large variety of different methods have been used to compute the band structure of crystals. A famous one is called the tight-binding approach and its essence differs from the preceding one. Instead of using expansions of the wave-functions through plane waves and the components of the Fourier-transform of the crystal potential, we now consider the atomic orbital $\phi_a(\vec{r}-\vec{L})$ of a free atom located at $\vec{L}$ from the origin. Then, we construct Bloch waves as a series of such strongly localized ϕ_as, with the help of a phase factor $e^{i\vec{k}.\vec{L}}$ as follows:

$$\Phi_{\vec{k}}(\vec{r}) = \sum_{\vec{L}} e^{i\vec{k}.\vec{L}}\phi_a(\vec{r}-\vec{L})$$

we write the expected energy for this wave-function as follows:

$$E(\vec{k}) = \frac{\int \Phi^*_{\vec{k}}(\vec{r}) \left[-\frac{\hbar^2}{2m} \nabla^2 + U(\vec{r})\right] \Phi_{\vec{k}}(\vec{r}) d^3r}{\int \Phi^*_{\vec{k}}(\vec{r}) \Phi_{\vec{k}}(\vec{r}) d^3r}$$

where ∇ is the gradient operator.

$$E(\vec{k}) = \frac{\sum_{\vec{L},\vec{L}'} \int \Phi^*_a(\vec{r} - \vec{L}') \left[-\frac{\hbar^2}{2m} \nabla^2 + U(\vec{r})\right] \Phi_a(\vec{r} - \vec{L}) d^3r}{\sum_{\vec{L},\vec{L}'} \int \Phi^*_a(\vec{r} - \vec{L}') \Phi_a(\vec{r} - \vec{L}) d^3r}$$

when the overlap of orbitals in neighboring cells is small, we can transform this equality into:

$$E(\vec{k}) = \sum_{\vec{L},\vec{L}'} e^{i\vec{k}(\vec{L}-\vec{L}')} \int \Phi^*_a(\vec{r} - \vec{L}') \left[-\frac{\hbar^2}{2m} \nabla^2 + U(\vec{r})\right] \Phi_a(\vec{r} - \vec{L}) d^3r$$

$$E(\vec{k}) = \sum_{\vec{H}} e^{i\vec{k}\vec{H}} \int \Phi^*_a(\vec{r} + \vec{H}) \left[-\frac{\hbar^2}{2m} \nabla^2 + U(\vec{r})\right] \Phi_a(\vec{r}) d^3r$$

giving:

$$I(\vec{k}) = \sum_{\vec{H}} e^{i\vec{k}\vec{H}} I_{\vec{H}}$$

where

$$I_{\vec{H}} = \int \Phi^*_a(\vec{r} + \vec{H}) \left[-\frac{\hbar^2}{2m} \nabla^2 + U(\vec{r})\right] \Phi_a(\vec{r}) d^3r$$

The various integrals $I_{\vec{H}}$ only depend on $\vec{H} = \vec{L} - \vec{L}'$.

Let us now represent as $v_a(\vec{r})$ the potential of an isolated atom at $\vec{r} = \vec{0}$ and suppose the eigenfunction $\phi_a(\vec{r})$ satisfies the following Schrödinger equation:

$$\left[-\frac{\hbar^2}{2m} \nabla^2 + U(\vec{r})\right] \Phi_a(\vec{r}) = E_a \Phi_a(\vec{r})$$

Obviously, the integral $I_{\vec{0}} = E_a$
and

$$I_{\vec{H}} = \int \Phi^*_a(\vec{r} + \vec{H}) \left[U(\vec{r}) - v_a(\vec{r})\right] \Phi_a(\vec{r}) d^3r$$

In the context of the empirical tight-binding approach, these quantities $I_{\vec{H}}$ are fitting parameters. The whole band structure, and thus the dispersion relations, depend on them.

• *Linear mono-atomic lattice with first neighbor interactions*. We consider a N-atom linear mono-atomic lattice, of period a; each atom being centered at $n\vec{a}$ from the origin. The wave-functions centered on atoms at positions $n\vec{a}$ from the origin are called Wannier functions, noted $|n)$. In the atomic basis, the eigenstates of the crystal are obtained as the solutions of the (N × N) matrix below:

$$\begin{bmatrix} |1) & |2) & |3) & . & . & |N) \\ E & t & . & . & . & t \\ t & E & t & . & . & . \\ . & t & . & . & . & . \\ . & . & . & . & . & t \\ t & . & . & . & t & E \end{bmatrix}$$

where $E_0 = (n|H|n')\delta_{n,n'}$; $t = (n|H|n')\delta_{n,n\pm1}$ and $t = (1|H|N)$ to fulfill the Born-von Karman cyclic conditions. Let us define the integer $j = \frac{N-1}{2}$.

Then, in the basis of Bloch states $|k\rangle = \Sigma_{-j}^{j} e^{ikna}|n\rangle$, the energy expresses as a dispersion relation:

$$\mathrm{E(k)} = E_0 + t\cos(ka)$$

We note that $\mathrm{E} = E_0 + t$ at zone center and $E_0 - t$ at zone boundaries. The sign of t rules the shape of the dispersion relation. Expanding this model to include second-neighbor interaction, we introduce the parameter $u = (n|H|n')\delta_{n,n\pm2}$

$$\mathrm{E(k)} = \mathrm{E}_0 + t\cos(ka) + u\cos(2ka)$$

In general $|t| < |u|$ but their sign may be different.

• *Simple three-dimensional case*. In the case of a cubic mono-atomic lattice where atoms—with a single orbital—are located in first neighbor positions with respect to the origin sit at $\vec{h} = (a, 0, 0)$; $(0, a, 0)$; $(0, 0, a)$. The summation over the $\vec{H}$s gives the following dispersion relation:

$$E(\vec{k}) = E_a + 2I_{100}\left[\cos(ak_x) + \cos(ak_y) + \cos(ak_z)\right]$$

To describe accurately the band structure with many bands and correctly account for the chemical bonds, it is necessary to hybridize many orbitals. In general, one uses the s and p orbitals of the last atomic shell, but including d states of the inner shell (if any) as well as empty states (atomic shell with high quantum numbers) is often required.

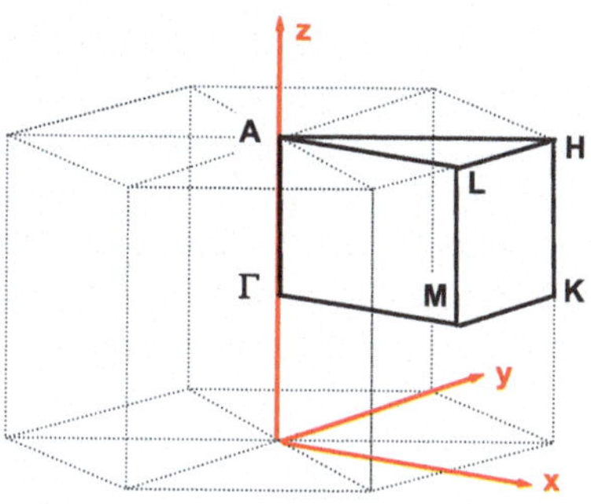

Fig. 3.6 The first Brillouin zone of the wurtzite crystal

3.1.9 Band Structure of Wurtzite Semi-conductors in the Context of a Spinless Tight-Binding Description

We have introduced in the first chapter, the concept of the reciprocal lattice. Here, its–starred–its vectors, as defined in solid-state physics, are:

$$\vec{a_1^*} = 2\pi(2/(a\sqrt{3}), 0, 0)$$

$$\vec{a_2^*} = 2\pi(1/(a\sqrt{3}), 1/a, 0)$$

$$\vec{c^*} = 2\pi(0, 0, 1/c)$$

versus those of the hexagonal lattice. They differ from the definition of crystallography by a factor 2π.
The first Brillouin zone of the wurtzite structure is constructed as shown in Fig. 3.6.

One can demonstrate that the coordinates of the relevant points away from zone center $\Gamma = (0, 0, 0)$ are: $A = (2\pi/c)(0, 0, 1/2)$, $L = (2\pi/a)(1/\sqrt{3}, 0, a/2c)$, $M = (2\pi/a)(1/\sqrt{3}, 0, 0)$, $H = (2\pi/a)(1/\sqrt{3}, 1/3, a/2c)$, $K = (2\pi/a)(1/\sqrt{3}, 1/3, 0)$.

In the preceding chapter, we have learned that the wurtzite unit cell contains four atoms, two anions and two cations.

In orthogonal coordinates, the crystallographic vectors of wurtzite are $\vec{a}_1 = a(\sqrt{3}/2, -1/2, 0)$, $\vec{a}_2 = a(0, 1, 0)$ and $\vec{c} = a(0, 0, c/a)$.

Then, anions sit at (0,0,0), and $a(1/\sqrt{3}, 0, c/2a)$ while cations positions are $a(1/\sqrt{3}, 0, (1/2 - u)c/a)$, and $(0, 0, (1 - u)c)$.

Lets us consider the perfect hexagonal close-packed stacking so that $u = 3/8$. Then, in the same cordinate system, anions locate at $(0, 0, 0)$, and $(a/\sqrt{3}, 0, c/2)$ while cations positions are $(a/\sqrt{3}, 0, c/8)$, and $(0, 0, 5c/8)$.

If we assume an sp^3 basis—one s and three p orbitals—centered at each of the four atoms sites, called Wannier functions here after, the eigenstates of a N-cell problem would be obtained after diagonalizing the $(16N \times 16N)$ matrix.

An alternative consists in constructing Block waves from these Wannier states.

Let $\vec{R}$ represent the position of a given unit cell in the crystal and let $\vec{t_1}$, $\vec{t_2}$, $\vec{t_3}$ and $\vec{t_4}$ be the translations internal to the cell $\vec{R}$ that give the atomic positions. In the perfect hexagonal close-packed stacking $\vec{t_1} = (0, 0, 0)$, $\vec{t_2} = (a/\sqrt{3}, 0, c/2)$, $\vec{t_3} = (a/\sqrt{3}, 0, c/8)$, and $t_4 = (0, 0, 5c/8)$.

Let $|n, \vec{t_i}, \vec{R})$ be a Wannier (localized) wave-function centered at a site of cell $\vec{R}$, with n running from 1 to 4 for s, p_x, p_y and p_z states. By construction, these states obey orthogonalisation properties:

$$(n, \vec{t_j}, \vec{R}||m, \vec{t_i}, \vec{R'}) = \delta_{mn}\delta_{ij}\delta_{\vec{R}\vec{R'}}$$

For each wave-vector $\vec{k}$, we construct the Bloch state as a linear combination of the Wannier functions as follows:

$$|n, \vec{t_i}, \vec{k}\rangle = \frac{1}{\sqrt{N}}\sum_{\vec{R}} e^{i\vec{k}.\vec{R}}|n, \vec{t_i}, \vec{R})$$

The above equation indicates that the Bloch state is the Fourier-transform of the Wannier function and vice-versa.

The crystal eigen-states may be written:

$$|\lambda, \vec{k}\rangle = \frac{1}{\sqrt{N}}\sum_{n, \vec{t_i}} |n, \vec{t_i}, \vec{k}\rangle\langle n, \vec{t_i}, \vec{k}||\lambda, \vec{k}\rangle$$

so that Schrödinger equation expresses:

$$\sum_{m,\vec{t_j}} \langle n, \vec{t_j}, \vec{k}|\mathbf{H}|m, \vec{t_j'}, \vec{k}\rangle - \epsilon\langle m, \vec{t_j'}, \vec{k}|\lambda, \vec{k}\rangle$$

where

$$\langle n, \vec{t_j}, \vec{k}|\mathbf{H}|m, \vec{t_j'}, \vec{k}\rangle = \sum_{\vec{R}} e^{i\vec{k}(\vec{R}-\vec{t_j}-\vec{t_j'})}(n, \vec{t_j}, \vec{0}|\mathbf{H}|m, \vec{t_j'}, \vec{R})$$

Now, we can assume that the only non-zero matrix elements of the Hamiltonian concern orbitals centered on the same, on nearest-neighbor or on second-nearest neighbor atoms.

The degree of sophistication of the band structure calculation depends on these approximations which, in the above equation, impact the summation over $\vec{t_j'}$ and $\vec{R}$.

If we assume nearest–neighbor interactions as shown in Fig. 3.7, the atoms of the cell $\vec{R}$ located at $\vec{t_1}$ (anion) is coupled to atom at $\vec{t_3}$ (cation) and to atom at $\vec{t_4}$(cation) but not coupled to atom at $\vec{t_2}$ (anion). Similarly, the atom at $\vec{t_3}$(cation) is not coupled to atom at t_4 (cation).

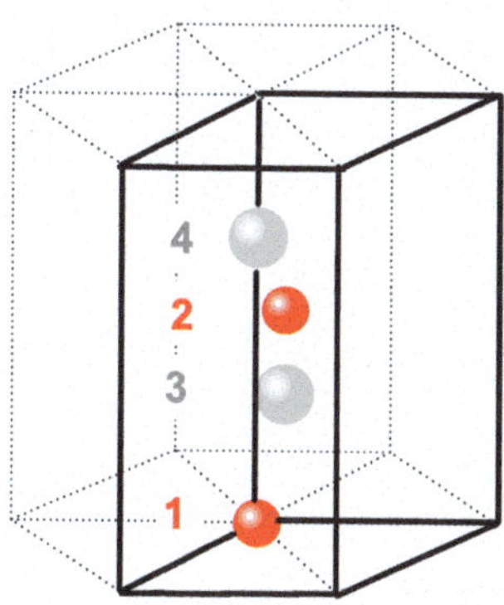

Fig. 3.7 The relative positions of the four basis atoms in the wurtzite cell

A small crystal field splitting (absent for cubic crystals) is introduced, so that the p_z orbitals deviate from the p_x and p_y ones. Of course, a smarter description would require to add interactions with more distant neighbors.

In the Bloch states $|m, \vec{t_i}, \vec{k}\rangle$ basis, this simple description requires to solve eigen-energies of the (16×16) matrix:

$$\begin{bmatrix} & 1_a & 2_a & 3_c & 4_c \\ 1_a & H_a & 0 & H_{1a,3c} & H_{1a,4c} \\ 2_a & 0 & V_a & H_{1a,4c} & H_{2a,4c} \\ 3_c & H^*_{1a,3c} & H^*_{1a,4c} & H_c & 0 \\ 4_c & H^*_{1a,4c} & H^*_{2a,4c} & 0 & H_c \end{bmatrix}$$

Each of these matrix elements is itself a (4×4) matrix.

The on-site matrix for anions or cations is a (4×4) matrix of the kind:

$$\begin{bmatrix} & |s,\alpha_i) & |p_z,\alpha_i) & |p_x,\alpha_i) & |p_y,\alpha_i) \\ (s,\alpha_i| & V_{s,i} & V_{sp_z,i} & 0 & 0 \\ |p_z,\alpha_i| & V_{sp_z,i} & V_{p_z,i} & 0 & 0 \\ (p_x,\alpha_i| & 0 & 0 & V_{p_x,i} & 0 \\ (p_y,\alpha_i| & 0 & 0 & 0 & V_{p_x,i} \end{bmatrix}$$

where α indicates the position of the atom within the cell and i runs for a and c for anions and cations respectively. Please note that i value is 1 or 2 for anions (a) and 3 or 4 for cations (c) and the modifications of the energies and splittings by the wurtzite crystal-field are included. The matrix elements above are empirical parameters.

Let us now define the wave-vector $\vec{k} = \frac{1}{2\pi}(k_1\vec{a}_1^* + k_2\vec{a}_2^* + k_3\vec{a}_3^*)$.

The off-site matrix elements may be written:

$$H_{1,4} = g_3(\vec{k})M_{1,4} = e^{-i\frac{3}{8}k_3} \cdot M_{1,4}$$

indicating a dispersion relation along k_3 via term $g_3(\vec{k})$ with:

$$M_{1,4} = \begin{bmatrix} & |s,4) & |p_z,4) & |p_x,4) & |p_y,4) \\ (s,1| & V_{s,s} & V_{s,z} & 0 & 0 \\ (p_z,1| & V_{s,z} & V_{z,z} & 0 & 0 \\ (p_x,1| & 0 & 0 & V_{x,x} & 0 \\ (p_y,1| & 0 & 0 & 0 & V_{x,x} \end{bmatrix}$$

Similarly:

$$H_{2,4} = g_2(\vec{k})M_{2,4} = e^{i\left[\frac{1}{3}(k_1-k_2)+\frac{3}{8}k_3\right]} \cdot M_{2,4}$$

where:

$$M_{2,4} = \begin{bmatrix} & |s,4) & |p_z,4) & |p_x,4) & |p_y,4) \\ (s,2| & g_0^* V'_{s,s} & g_0^* V'_{s,z} & -g_1^* V'_{s,x} & -\frac{\sqrt{3}}{2} g_-^* V'_{s,x} \\ (p_z,2| & g_0^* V'_{s,z} & g_0^* V'_{z,z} & -g_1^* V'_{z,x} & -\frac{\sqrt{3}}{2} g_-^* V'_{z,x} \\ (p_x,2| & -g_1^* V'_{s,x} & -g_1^* V'_{x,z} & g_1^* V'_{x,x} + \frac{3}{4} g_+^* (V'_{x,x} + V'_{y,y}) & -\frac{\sqrt{3}}{4} g_-^* (V'_{x,x} - V'_{y,y}) \\ (p_y,2| & -\frac{\sqrt{3}}{2} g_-^* V'_{s,x} & -\frac{\sqrt{3}}{2} g_-^* V'_{x,z} & -\frac{\sqrt{3}}{4} g_-^* (V'_{x,x} - V'_{y,y}) & g_1^* V'_{y,y} + \frac{3}{4} g_+^* (V'_{x,x} + V'_{y,y}) \end{bmatrix}$$

and

$$g_0(\vec{k}) = 1 + e^{i(k_1-k_2)}$$

$$g_1(\vec{k}) = \frac{2e^{ik_1} - e^{-ik_2} - 1}{2}$$

$$g_\pm(\vec{k}) = 1 \pm e^{-ik_2}$$

At last,

$$H_{1,3} = g_1(\vec{k})M_{1,3} = e^{i\left[\frac{1}{3}(k_2-k_1)+\frac{3}{8}k_3\right]} \cdot M_{1,3}$$

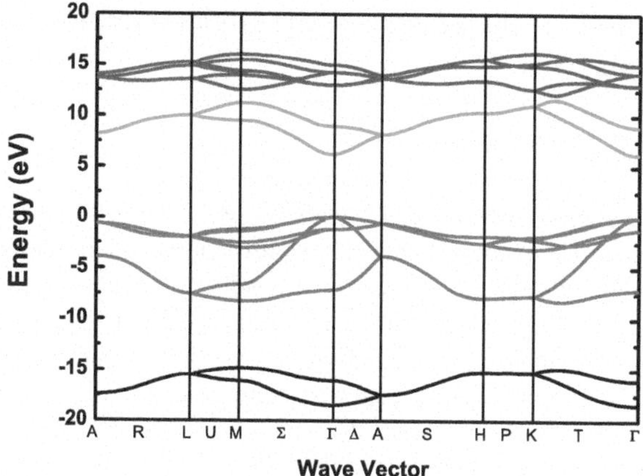

Fig. 3.8 The band structure of wurtzitic aluminum nitride computed using the spinless empirical tight-binding approach

where:

$$M_{1,3} = \begin{bmatrix} & |s,3) & |p_z,3) & |p_x,3) & |p_y,3) \\ (s,1| & g_0 V'_{s,s} & g_0 V'_{s,z} & g_1 V'_{s,x} & \frac{\sqrt{3}}{2} g_- V'_{s,x} \\ (p_z,1| & g_0 V'_{s,z} & g_0 V'_{z,z} & g_1 V'_{z,x} & \frac{\sqrt{3}}{2} g_- V'_{z,x} \\ (p_x,1| & g_1 V'_{s,x} & g_1 V'_{x,z} & g_1 V'_{x,x} + \frac{3}{4} g_+ (V'_{x,x} + V'_{y,y}) & -\frac{\sqrt{3}}{4} g_- (V'_{x,x} - V'_{y,y}) \\ (p_y,1| & \frac{\sqrt{3}}{2} g_- V'_{s,x} & \frac{\sqrt{3}}{2} g_- V'_{x,z} & -\frac{\sqrt{3}}{4} g_- (V'_{x,x} - V'_{y,y}) & g_1 V'_{y,y} + \frac{3}{4} g_+ (V'_{x,x} + V'_{y,y}) \end{bmatrix}$$

These tedious calculations indicate that there are very complicated dispersion relations, even for a simple calculation neglecting relativistic effects like spin-orbit interaction. We report here as Fig. 3.8 an example of band structure obtained for aluminum nitride with having a 6 eV band-gap value. The calculation of the band structure is performed by empirically fitting all parameters to match with optical and transport properties; it indicates that the maximum (resp. minimum) energy of the valence (resp. conduction) band occur at Brillouin zone center when $\vec{k} = \vec{0}$.

3.2 The Semi-classical Theory of the Dielectric Function in Crystals

3.2.1 Intuitive Description

In the preceding section, we have given an examples of band structure for aluminum nitride. Considering a transition between the state of the top filled valence band having energy E_v and the bottom state of the lowest conduction band at energy E_c, we can establish from group theory arguments that its matrix element is not vanishing when $\hbar\omega \geq E_v - E_c = E_{g0}$ and we call it an allowed transition. A photon with the specific energy $\hbar\omega \geq E_{g0}$ does not violate a spectroscopic selection rule in a transition that is vertical in the reciprocal space since the lattice parameter equals about 1 nm whilst the photon wavelength is 1 micrometer, thus the wave-vector of the photon can be neglected.

Further assuming the matrix element to slightly vary with the transition energy, the absorption coefficient $\alpha(E_{g0})$ in the three-dimensional semi-conductor can be naturally and intuitively written as proportional to the three-dimensional joint density of states $D\sqrt{\hbar\omega - E_{g0}}$, in extenso, $\alpha(E_{g0}) = A\sqrt{\hbar\omega - E_{g0}}$.

The free electron state missing in the valence band is analogous to a positive charge, named a valence-band hole by Wilson in 1930 who established the symmetry properties between the valence missing electron and the concept of valence hole. It corresponds to a free electron in excess in the conduction band, both states share the same value of $\vec{k}$ but are separated in energy by $\hbar\omega - (E_v - E_c)$.

Let us assume, keeping in mind that the physics will be kept, that the valence band is not degenerate. We write m_e and m_h electron and hole effectives masses, extracted respectively from the curvature of the bottom conduction and top valence bands in the neighborhood of Γ. Then, the joint density of states' effective mass writes:

$$m_r = \frac{m_e m_h}{m_e + m_h}$$

We want to outline that the joint density of states $D\sqrt{\hbar\omega - E_{g0}}$ is proportional to $m_r^{3/2}$, according to the model of the free particle in a three-dimensional box. Thus, the absorption coefficient for any direct transition in the reciprocal space can be written:

$$\alpha_i(\hbar\omega) = A_i m_{ri}^{3/2}\sqrt{\hbar\omega - E_{gi}}$$

The index i indicates that, to each transition E_{gi}, we attribute a given matrix element A_i and a given joint density of states:

$$m_{ri}^{3/2}\sqrt{\hbar\omega - E_{gi}}$$

The next important point to outline here is the expression of the joint density of state effective mass:

$$m_{ri} = \frac{m_{ei} m_{hi}}{m_{ei} + m_{hi}}$$

that rules the amplitude of the absorption. In particular, these m_{ri}s vary with band index i and wave-vector; thus, one can again intuitively imagine that these conditions may lead to specific situations featuring the absorption coefficient we write:

$$\sum_i \alpha_i(\hbar\omega)$$

or

$$\sum_{\vec{k}} \sum_{i=n_c-n_v} \alpha_i(\hbar\omega)$$

The summation is extended to all vectors $\vec{k}$ in the Brillouin zone and to all the couples of valence (n_v) and conduction bands (n_c). Obviously, there is a need for developing a more mathematical treatment.

3.2.2 Microscopic Theory of the Dielectric Constant

An excellent tutorial description of the semi-classical theory of the dielectric constant can be found in *Fundamentals of semiconductors*, by Peter Y. Yu and Manuel Cardona, Springer-Verlag, partly reproduced below.

When an electromagnetic radiation shines a solid, it splits in four parts corresponding to reflection, transmission, absorption and inelastically or elastically scattering. The solid may also emit some fluorescence. In this section, we will limit ourselves to the three first processes, directly connected with the semi-classical theory of the dielectric constant.

We know, from quantum mechanics, that a vector potential $A(\vec{r}, t)$ and a scalar potential $\phi(\vec{r}, t)$ have to be added into the classical one-electron Hamiltonian to account for the influence of the radiation field on the electronic structure of the solid. In the context of the Coulomb gauge we do not wish to detail here, the electron-radiation interaction Hamiltonian is written as:

$$H_{er} = \frac{e}{mc} \vec{A} \cdot \vec{p}$$

versus the impulsion operator $\vec{p}$, when second-order non-linear effects—proportional to $\vec{A}^2$—are neglected.

Applying time-dependent perturbation theory, the transition probability per unit volume R for an electron in the initial state $|i\rangle$ with energy E_i and wave-vector $\vec{k_i}$ towards the final state $|f\rangle$ with energy E_f and wave-vector k_f requires calculating the matrix element:

$$|\langle i|H_{er}|f\rangle|^2 = \frac{e^2}{m^2c^2}|\langle i|\vec{A}\cdot\vec{p}|f\rangle|^2 = \frac{\vec{A}^2e^2}{m^2c^2}|\langle i|\vec{u}\cdot\vec{p}|f\rangle|^2$$

We define $\vec{u}$ as the unit vector in the direction of $\vec{A}$. The Bloch states of the crystal are written:

$$|\vec{\beta}\rangle = u_{\vec{\beta},\vec{k}_\beta(\vec{r})} exp(i\vec{k}_\beta\cdot\vec{r})$$

for both initial and final states. Further introducing the amplitude of the electric field $E(\vec{q},\omega)$, $\vec{A}$ expresses as the sum of two complex conjugate quantities:

$$\vec{A} = \frac{\vec{E}}{2\|\vec{q}\|}[e^{i(\vec{q}\cdot\vec{r}-\omega t)} + e^{-i(\vec{q}\cdot\vec{r}-\omega t)}]$$

Time integration of $\langle i|A.p|f\rangle$ gives two terms: $\delta(E_i(\vec{k_i})-E_f(\vec{k_f})-\hbar\omega)$ and $\delta(E_i(\vec{k_i})-E_f(\vec{k_f})+\hbar\omega)$ for the two previous quantities. They respectively correspond to absorption of the photon and excitation of the electron from initial state $|i\rangle$ with energy E_i towards final state $|f\rangle$ with energy E_f, and to stimulated emission of a photon from energy relaxation from state $|f\rangle$ to $|i\rangle$. The second term is ignored hereafter since we disregard emission processes. Assuming wave-vector conservation $\vec{q}=\vec{k_i}-\vec{k_f}$ and rewriting the matrix element $|\langle i|\vec{u}.\vec{p}|f\rangle|^2$ as $|P_{if}|^2$, some lengthy mathematical manipulations lead to the transition probability for photon absorption per unit of time:

$$R = \frac{2\pi}{\hbar}\Sigma_{\vec{k_i},\vec{k_f}}|\langle i|H_{er}|f\rangle|^2\delta(E_i(\vec{k_i}) - E_f(\vec{k_f}) - \hbar\omega)$$

as:

$$R = \frac{2\pi}{\hbar}\frac{e^2}{m^2c^2}\left|\frac{E(\omega)}{2}\right|^2 \Sigma_{\vec{k}}|P_{if}|^2\delta(E_i(\vec{k_i}) - E_f(\vec{k_f}) - \hbar\omega)$$

where the $\vec{k}$-summation is restricted to the $\vec{k}$'s allowed per unit volume of the crystal.

The power lost per unit volume by the incident beam I due to absorption is $-\frac{dI}{dt}$ or $R\hbar\omega$. Introducing the refractive index n, the imaginary part ε_2 of the dielectric constant, the absorption coefficient α, one writes:

$$R = -\frac{dI}{dt} = \frac{\varepsilon_2}{n^2}c\omega\frac{n^2}{8\pi}|\vec{E}(\omega)|^2$$

which straight-forwardly furnishes us the expression of the imaginary part of the dielectric constant:

$$\varepsilon_2 = \left|\frac{2\pi e}{m\omega}\right|^2 \sum_{\vec{k}} |P_{if}|^2 \delta(E_i(\vec{k_i}) - E_f(\vec{k_f}) - \hbar\omega)$$

Straight-forward application of Kramers-Kronig analysis gives the expression of the real part:

$$\varepsilon_1 = 1 + \frac{4\pi^2 e^2}{m} \sum_{\vec{k}} \left(\frac{2}{m[E_i(\vec{k}) - E_f(\vec{k})]}\right) \frac{|P_{if}|^2}{[E_i(\vec{k}) - E_f(\vec{k})]^2 - \hbar^2\omega^2}$$

In the equation above,

$$\left(\frac{2}{m[E_i(\vec{k}) - E_f(\vec{k})]}\right)$$

is the number of oscillators with energy $E_i(\vec{k}) - E_f(\vec{k})$ and is called oscillator strength and noted f_{if}.

Most of the dispersion in ε_2 comes from the summation over the delta function. This summation can be replaced by an integral over the energy by defining a joint density of state that writes as follows for the doubly-degenerated (spin is included) conduction and valence bands:

$$D_j(E_{if}) = \frac{1}{4\pi^3} \int \frac{dS_{\vec{k}}}{|\nabla \vec{k} E_{if}(\vec{k})|}$$

where $E_{if}(\vec{k}) = E_i(\vec{k}) - E_f(\vec{k})$ and $S_{\vec{k}}$ is the energy surface where $E_{if}(\vec{k})$ is constant.

Then, $\sum_{\vec{k}}$ is replaced by $\int D_i(E_{if}) dE_{if}$.

It has been shown that the density of states displays singularities when $\nabla_{\vec{k}} E_{if}(\vec{k})$ vanishes, in general at high-symmetry points of the Brillouin zone. These singularities are the so-called Van Hove singularities. The imaginary part of the dielectric constant exhibits a series of resonances peaking at these specific energies which correspond to transitions between valence band and conduction band extrema, their intensities are energy-modulated by the spectral variation of the joint density of states. In wurtzitic semi-conductors, at the fundamental band-gap transition energy, instead of being represented by a broadened Dirac peak, the more complete theory predicts, for the imaginary part of the dielectric function, a sudden variation from zero to a square root energy-dependence:

For $E < E_g$, $\epsilon_2 = 0$.

For $E > E_g$, $\epsilon_2 \sim \sqrt{E - E_g}$.

This behavior is specific to the fundamental band-gap, at what is called an M_0 van-Hove singularity.

From the equations above, giving the real and imaginary parts of the dielectric function, one can now derive the absorption and the reflection using well-known formulae. The equations intuitively established in the preceding section are recovered.

The absorption coefficient of a normal incident beam from air on the surface of a semi-conductor is expressed as:

$$\alpha = \frac{\omega \epsilon_2}{c} = \frac{\sqrt{2}}{\sqrt{\epsilon_1 + (\epsilon_1^2 + \epsilon_2^2)}} \sim \sqrt{E - E_g}$$

Regarding the reflection coefficient, in normal incidence conditions, from air to semi-conductor, it writes:

$$R = \frac{\sqrt{\epsilon_1^2 + \epsilon_2^2} - \sqrt{2}[\sqrt{\epsilon_1^2 + \epsilon_2^2} + \epsilon_1]^{1/2} + 1}{\sqrt{\epsilon_1^2 + \epsilon_2^2} + \sqrt{2}[\sqrt{\epsilon_1^2 + \epsilon_2^2} + \epsilon_1]^{1/2} + 1}.$$

3.2.3 Experimental Values of the Spectral Dependence of the Dielectric Constants of Nitrides

Figure 3.9 represents the real (a) and imaginary parts (b) of the dielectric constants of aluminum nitride: black plots are for polarization of the electric field orthogonal to the z- or c-axis or [0001] direction, red plots are for $\vec{E}$ along z direction. The ordinary (resp. extraordinary) component of the dielectric constant is obtained when the photon propagates along (resp. orthogonal to) the c axis. In plot (c) are represented the ordinary (black) and extraordinary (red) imaginary parts of the dielectric function of wurtzite GaN. We remark that the lowest energy feature labelled E_0 is measured for the $\vec{E}//\vec{z}$ polarization for AlN at some 200 meV lower than the feature reported in the orthogonal polarization, while the situation is reversed for GaN. The latter shows a weaker15 meV splitting, the transition being clearly at higher energy for the feature measured in the $\vec{E}//\vec{z}$ polarization condition. By examining the relative amplitudes and energies of the different features that contribute to the dielectric constant, it is theoretically possible to self-consistently determine the fitting parameters of the tight-binding calculations of the band structure so that one fits the whole spectral shapes of the dielectric constants. We wish to outline that both ellipsometry and band structure calculations indicate that the top valence band of AlN at zone center ($\vec{k} = \vec{0}$) is dominantly built from the p_z state of the anion while the p_x and p_y states contribute to the 200 meV deeper valence band states. Regarding the conduction band, it is dominantly built from the *s* states of the anion (aluminum here).

This situation is more or less common in the case of wurtzitic semi-conductors: the top valence band is *p*-type at zone center while the conduction band is *s*-type.

There are however some deviations with respect to this statement: in indium nitride, the 5*d* states of indium strongly interact with the 2*p* states of nitrogen, which produces a quenching of the band-gap at zone center. In zinc oxide, the 3*d* states of zinc interact with the 2*p* states of oxygen, possibly changing the sign of the spin-

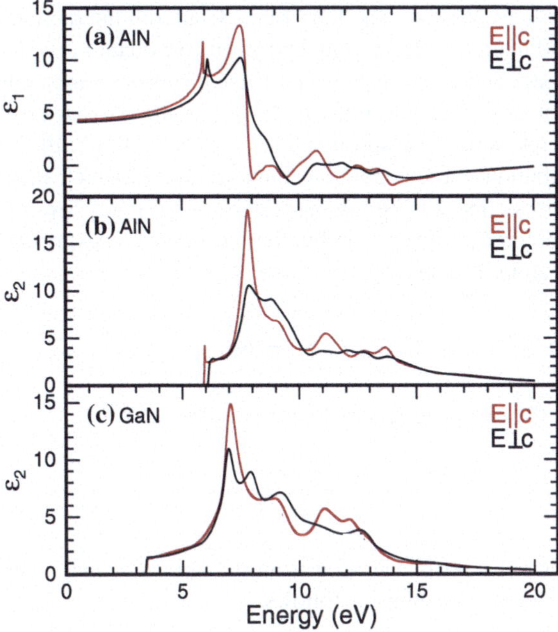

Fig. 3.9 The real (**a**) and imaginary (**b**) parts of the dielectric constants of wurtzitic aluminum nitride, and the imaginary parts of the dielectric constant of gallium nitride (**c**). Courtesy Rüdiger Goldhahn and Matthias Feneberg

orbit interaction from positive to negative. These elements indicate that those optical properties of semi-conductors are linked with atomic bonds, themselves being fumed by the quantum mechanics of atoms that constitute the crystal.

3.2.4 Excitonic Contributions to the Dielectric Constant

The equations above neither account for the experimental behavior in terms of the **exact shape** for the measured dielectric constant, nor for the absorption coefficient, nor the reflectance properties. In particular, the absorption coefficient is predicted to be zero at energies below the band-gap and to increase with the square root of the energy, measured relatively to the band-gap. The experimental situation is very different, as measured as early as 1940s in the case of ZnO by Mollwo. He observed the absorption versus photon energy displaying a rapid onset, saturation, then decrease and increase again, forming some kind of plateau at about 2×10^5 cm^{-1}. Excitonic effects, invented by Frenkel and Peierls, described more quantitatively later by Wannier and Mott, are responsible for this behavior. In the simplest picture, they can be viewed as the quantum of electrostatic interaction between the photo-created conduction electron—a negative charge—and the valence-missing electron, a positive charge named hole by Wilson. The long-range Coulomb interaction between these two particles lead them to behave as an hydrogen-like quantum system, with bound and un-bound states and electron-hole short-range spin-exchange interaction (analogous to the contact interaction in the hydrogen atom). The absorption coefficient consists of a discrete series of absorption lines, split according to the hydrogen series, and to the contribution of the continuous unbound states, which superimpose to the band to band process. The absorption onset starts at an energy lower than the energy gap E_g by an amount E_R, the exciton binding energy. E_R is expressed as a function of the hydrogen atom Rydberg R_y as:

$$E_R = \frac{\mu}{\varepsilon^2} R_y$$

In this equation, the excitonic reduced mass μ is defined as:

$$\frac{1}{\mu} = \frac{1}{m_e} + \frac{1}{m_v}$$

and the dielectric constant is averaged from the low-frequency $\varepsilon(0)$ and high frequency $\varepsilon(\infty)$ dielectric constants as follows:

$$\frac{1}{\varepsilon} \approx \frac{11}{16\varepsilon(0)} + \frac{5}{16\varepsilon(\infty)}$$

In the strictest sense, the amplitudes of the absorption of the excited excitonic states vary like ν^{-3} where ν is the principal quantum number, relatively to the absorption at the energy of the ground state exciton. It is important to remind that:

$$\sum_{\nu=1}^{\infty} \nu^{-3} = \zeta(3) \simeq 1{,}2020569$$

The value of the zeta function for three indicates that about 80 % of the total oscillator strength of excitonic bound states is concentrated on the ground state $\nu = 1$.

Excitonic levels have a finite life-time. The excitonic lines have a finite line-width and one generally does not observe the whole excitonic series, except in high quality zinc-blende, in wide band-gap or in chalcopyrite semi-conductors. The reason for this is the large value of the exciton binding energy, in the latter case as indicated in Fig. 3.10.

The spectral dependence of the absorption coefficient according to the band to band model is never observed, including at high temperatures as evidenced in Fig. 3.11.

From the mathematical point of view, the absorption coefficient including excitonic effect contains two excitonic contributions: the contributions of bound excitonic hydrogen-like and un-bound states. The contribution of bound states consists in a series of Lorentzian-shaped peaks, centered at the energies of the hydro-

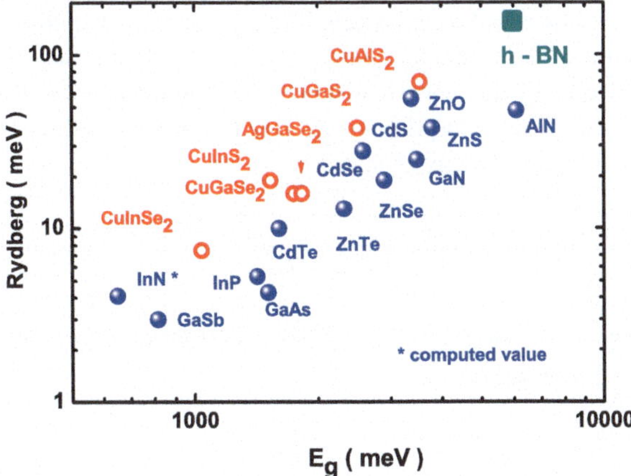

Fig. 3.10 Exciton binding energies versus energy gap for the most common semi-conductors

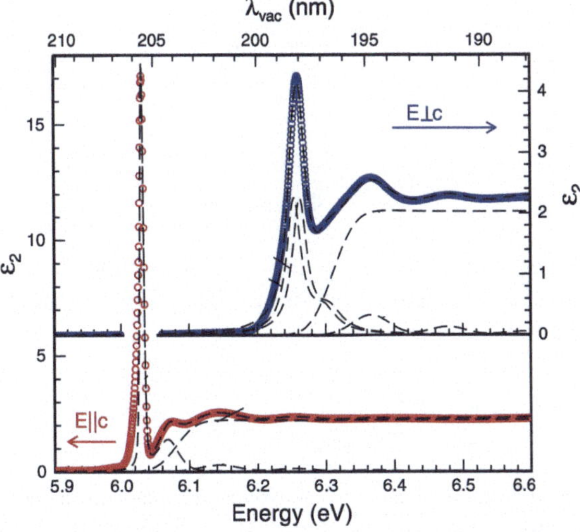

Fig. 3.11 Absorption coefficient for both polarizations in high-quality aluminum nitride. Courtesy Rüdiger Goldhahn and Matthias Feneberg

gen series (E_R/ν^2), with amplitudes varying accordingly to (ν^{-3}). For large values of the principal quantum number, the density of excitonic states increases, they merge and amalgamate to form a pseudo-continuum of density:

$$D_\nu(E) = \frac{\nu^3}{2E_R}$$

below the energy of the band to band gap. The excitonic contribution to the absorption for high values of the principal quantum number is the product of the oscillator strength ($\sim\nu^{-3}$) timed by the density of states of the pseudo-continuum ($\sim\nu^{3}$), compensating each other. The absorption coefficient has at finite value. Now, regarding un-bound states, the joint density of states is:

$$D_{3D} = \frac{1}{\sqrt{2}h^3}(\hbar\omega - E_g)\mu^{3/2}\sqrt{\hbar\omega - E_g}$$

which is rewritten versus the excitonic Rydberg, its Bohr radius a_B, and as a function of a dimensionless quantity:

$$\gamma = \sqrt{\frac{E_R}{\hbar\omega - E_g}}$$

as follows:

$$D_{3D} = \frac{1}{4\pi^2 a_B^3 R_y \gamma}$$

and:

$$\alpha_{3D}(\hbar\omega) = \frac{1}{4\pi^2 a_B^3 R_y} \frac{e^{\pi\gamma}}{\sinh \pi\gamma}$$

Beyond the scope of shifting the energy onset of absorption, Coulomb interaction and excitons introduce an enhancement of the absorption compared to the band to band process. This enhancement factor:

$$S(\hbar\omega - E_g) = \frac{K_{continuum}(\hbar\omega - E_g)}{K_{interbands}(\hbar\omega - E_g)} = \pi\gamma \frac{e^{\pi\gamma}}{\sinh \pi\gamma}$$

is called Sommerfeld factor. This ratio between the absorption including excitons and the band to band result: is always larger than 1 as shown in Fig. 3.11.

3.3 The Spin-Orbit Interaction

Including spin-orbit interaction, in order to improve the description of the electronic structure of semi-conductors, requires to handle Dirac relativistic equation rather than Schrödinger one. This complicated equation writes:

$$\left[c.\alpha.\mathbf{p} + \beta m_0 c^2 + V(\vec{r}) - W\right]\Psi = 0$$

where Ψ is a four-component spinor,

p is the operator $-i\hbar\nabla$;

$$\alpha = \begin{pmatrix} 0 & \sigma \\ \sigma & 0 \end{pmatrix}, \quad \text{where } \sigma \text{ indicates the Pauli operator matrices}$$

$$\beta = \begin{pmatrix} \tilde{1} & 0 \\ 0 & \tilde{1} \end{pmatrix}, \quad \text{where } \tilde{1} \text{ is a } (2 \times 2) \text{ unity matrix,}$$

and W = E + m_0c^2 is the total energy including the electron mass at rest, m_0c^2.

The Pauli matrices are defined here as:

$$\sigma_x = \begin{pmatrix} 0 & 1 \\ 1 & 0 \end{pmatrix}, \sigma_y = \begin{pmatrix} 0 & -i \\ i & 0 \end{pmatrix} \text{ and } \sigma_z = \begin{pmatrix} 1 & 0 \\ 0 & -1 \end{pmatrix}$$

in the representation where σ_z is diagonal. The eigenstates of the operator σ_z are $\begin{bmatrix} 1 \\ 0 \end{bmatrix} \equiv \uparrow$ and $\begin{bmatrix} 0 \\ 1 \end{bmatrix} \equiv \downarrow$ respectively.

In spin space, the intrinsic angular momentum of the electron is $s = \frac{1}{2}\hbar\sigma$.

To discuss the relationship between Schrödinger $\left(\left[\frac{\mathbf{p}^2}{2m_0} + V(\vec{r}) - E\right]\Psi = 0\right)$ and Dirac equation, and to get rid of a sophisticated, but exact mathematical approach, one generally expresses the kinetic energy in power of **p**:

$$\sqrt{c^2\mathbf{p}^2 + m_0^2c^4} = m_0c^2 + \frac{\mathbf{p}^2}{2m_0} - \frac{\mathbf{p}^4}{8m_0^3c^2} + \cdots$$

At low orders of approximation, the wave equation for the upper two-component spinor becomes:

$$\left[\frac{\mathbf{p}^2}{2m_0} + V(\vec{r}) - \frac{\mathbf{p}^4}{8m_0^3c^2} - \frac{\hbar^2}{4m_0^2c^2}\nabla V(\vec{r})\cdot\nabla + \frac{\hbar}{4m_0^2c^2}\sigma\cdot(\nabla V(\vec{r})\times\mathbf{p})\right]\Psi = E\Psi$$

The first correction term comes from the expansion of the relativistic correction in powers of **p**:

$$\mathbf{H}_v = -\frac{\mathbf{p}^4}{8m_0^3c^2}$$

It is a correction to the kinetic energy and it gives a smooth deformation of the band structure.

The second relativistic correction is called the Darwin correcting term of the potential.

$$\mathbf{H}_D = -\frac{\hbar^2}{4m_0^2c^2}\nabla V(\vec{r})\cdot\nabla$$

The third relativistic term is the spin-orbit coupling interaction correcting term

$$\mathbf{H}_{so} = \frac{\hbar}{4m_0^2c^2}\sigma \cdot (\nabla V(\vec{r}) \times \mathbf{p})$$

It mixes the two components of the Pauli spinor. Its origin is the interaction of the electron-spin moment with the magnetic field "seen" by the electron. A classical analysis gives twice the correct value. This term is spin-dependent and removes degeneracies of non-relativistic results.

In quantum mechanics text-books, this term is sometimes written phenomenologically:

$$\mathbf{H}_{so} = \Delta_{so}\vec{L}.\vec{\sigma} = \Delta_{so}(L_x\sigma_x + L_y\sigma_y + L_z\sigma_z) = \Delta_{so}\frac{(\vec{L}+\vec{\sigma})^2 - L^2 - \sigma^2}{2}$$

versus orbital angular $\vec{L}$ and spin $\vec{\sigma}$ momenta.

Note: *We remind that the eigen-value of angular momentum A^2 is $A(A+1)$.*

Let us consider a spin-less s state with orbital angular momentum $\vec{L} = \vec{0}$. When including the spin, the total angular momentum $\vec{J}$ becomes $\vec{J} = \vec{L} + \vec{\sigma} = \vec{\frac{1}{2}}$. No spin-induced splitting occurs.

Let us now construct the eigenstates of the valence band by coupling a $\vec{\sigma} = \vec{\frac{1}{2}}$ spin to a set of three $\vec{L} = \vec{1}$ p-type states.

Lets represent the three p states in terms of angular momentum $\vec{L}$, and its projection along the z axis as $|L, m_L\rangle$.

$$|1,1\rangle = -\frac{1}{\sqrt{2}}(p_x + ip_y);\ |1,0\rangle = p_z;\ |1,-1\rangle = \frac{1}{\sqrt{2}}(p_x - ip_y);$$

We remind here that the degeneracy of angular momentum $\vec{L}$ is 2L + 1.

The $\vec{J} = \vec{\frac{3}{2}}$ states are obtained as:

$$\left|\frac{3}{2},\frac{3}{2}\right\rangle = |1,1\rangle \uparrow$$
$$\left|\frac{3}{2},\frac{1}{2}\right\rangle = \sqrt{\frac{1}{3}}|1,1\rangle \downarrow + \sqrt{\frac{2}{3}}|1,0\rangle \uparrow$$
$$\left|\frac{3}{2},\frac{1}{2}\right\rangle = \sqrt{\frac{2}{3}}|1,0\rangle \downarrow + \sqrt{\frac{1}{3}}|1,-1\rangle \uparrow$$
$$\left|\frac{3}{2},-\frac{3}{2}\right\rangle = |1,-1\rangle \downarrow$$

The $\vec{J} = \frac{1}{2}$ states are obtained as:

$$\left|\frac{1}{2}, \frac{1}{2}\right\rangle = \sqrt{\frac{2}{3}}|1, 1\rangle \downarrow - \sqrt{\frac{1}{3}}|1, 0\rangle \uparrow$$
$$\left|\frac{1}{2}, -\frac{1}{2}\right\rangle = \sqrt{\frac{1}{3}}|1, 0\rangle \downarrow - \sqrt{\frac{2}{3}}|1, -1\rangle \uparrow$$

In the (6 × 6) basis, the spin-orbit interaction writes:

$$\begin{bmatrix} & |\frac{3}{2},\frac{3}{2}\rangle & |\frac{3}{2},\frac{1}{2}\rangle & |\frac{3}{2},-\frac{1}{2}\rangle & |\frac{3}{2},-\frac{3}{2}\rangle & |\frac{1}{2},\frac{1}{2}\rangle & |\frac{1}{2},-\frac{1}{2}\rangle \\ \langle\frac{3}{2},\frac{3}{2}| & \Delta_{so} & 0 & 0 & 0 & 0 & 0 \\ \langle\frac{3}{2},\frac{1}{2}| & 0 & \Delta_{so} & 0 & 0 & 0 & 0 \\ \langle\frac{3}{2},-\frac{1}{2}| & 0 & 0 & \Delta_{so} & 0 & 0 & 0 \\ \langle\frac{3}{2},-\frac{3}{2}| & 0 & 0 & 0 & \Delta_{so} & 0 & 0 \\ \langle\frac{1}{2},\frac{1}{2}| & 0 & 0 & 0 & 0 & -2\Delta_{so} & 0 \\ \langle\frac{1}{2},\frac{1}{2}| & 0 & 0 & 0 & 0 & 0 & -2\Delta_{so} \end{bmatrix}$$

when considering atomic sodium, it evidences a yellow doublet split by 2.25 meV. Starting from the same ground $3s_{1/2}$, there are two possible transitions: the first one to the $3p_{1/2}$ state gives the 589.59 nm radiation whereas the second transition leads to $3p_{3/2}$ state—588.99 nm radiation. It corresponds to a spin-orbit interaction of about 1.5 milli-electron-volt for the $3p$ state.

In a wurtzite crystal, this interaction would be written as:

$$\mathbf{H}_{so}^{wurtzite} = \Delta_2(L_x\sigma_x + L_y\sigma_y) + \Delta_3 L_z\sigma_z$$

in order to introduced anisotropy between the z direction and the isotropic plane (x, y).

Instead of working in the context of a six-fold representation of the $\left\{\vec{\frac{3}{2}}, \vec{\frac{1}{2}}\right\}$ representation, very convenient for treating the energy spectrum of p-type states with spin in spherical symmetry, one uses the following representation that matches exactly with the symmetry of a p electron in a wurtzite environnement:

$$\{|1, 1\rangle \uparrow; |1, -1\rangle \downarrow; |1, 1\rangle \downarrow; |1, 0\rangle \uparrow; |1, -1\rangle \uparrow; |1, 0\rangle \downarrow\}$$

In this representation, the spin-orbit interaction produces a splitting of the four-fold $\vec{\frac{3}{2}}$ and two-fold $\vec{\frac{1}{2}}$ states as represented below:

$$\begin{bmatrix} |1,1\rangle\uparrow & |1,-1\rangle\downarrow & |1,1\rangle\downarrow & |1,0\rangle\uparrow & |1,-1\rangle\uparrow & |1,0\rangle\downarrow \\ \Delta_2 & 0 & 0 & 0 & 0 & 0 \\ 0 & \Delta_2 & 0 & 0 & 0 & 0 \\ 0 & 0 & -\Delta_2 & \sqrt{2}\Delta_3 & 0 & 0 \\ 0 & 0 & \sqrt{2}\Delta_3 & 0 & 0 & 0 \\ 0 & 0 & 0 & 0 & -\Delta_2 & \sqrt{2}\Delta_3 \\ 0 & 0 & 0 & 0 & \sqrt{2}\Delta_3 & 0 \end{bmatrix}$$

There are now three eigen-states, which eigen-values are:

$$\left\{ \Delta_2; -\frac{\Delta_2 + \sqrt{\Delta_2^2 + 8\Delta_3^2}}{2}; -\frac{\Delta_2 - \sqrt{\Delta_2^2 + 8\Delta_3^2}}{2} \right\}$$

The state $\{|1,1\rangle \uparrow; |1,-1\rangle \downarrow\}$ with eigen-value Δ_2 is two-fold degenerated; it corresponds to the states $|\frac{3}{2},\frac{3}{2}\rangle$ and $|\frac{3}{2},-\frac{3}{2}\rangle$ in spherical symmetry. In the language of group theory that we shall see later, and in terms of the group representation theory, we attribute them the Γ_9 symmetry. The four other states are also two-fold degenerated. The eigenstates are linear combinations of $\{|1,1\rangle \downarrow, |1,0\rangle \uparrow\}$ on one side and of $\{|1,-1\rangle \uparrow |1,0\rangle \downarrow\}$ on the other. In relation with group theory, states $\{|1,1\rangle \downarrow, |1,0\rangle \uparrow\}$ and $\{|1,-1\rangle \uparrow, |1,0\rangle \downarrow\}$ are basis vectors for two different representations of Γ_7 symmetry. There are thus two two-fold eigenvalues corresponding to wave-functions having the Γ_7 symmetry.

Let us consider eigen-value

$$-\frac{\Delta_2 + \sqrt{\Delta_2^2 + 8\Delta_3^2}}{2}$$

The eigen-vectors are:

$$\left\{ -\frac{\left[\Delta_2 + \sqrt{\Delta_2^2 + 8\Delta_3^2}\right]}{\sqrt{\left[\Delta_2 + \sqrt{\Delta_2^2 + 8\Delta_3^2}\right]^2 + 8\Delta_3^2}} |1,-1\rangle \uparrow + \frac{2\sqrt{2}\Delta_3}{\sqrt{\left[\Delta_2 + \sqrt{\Delta_2^2 + 8\Delta_3^2}\right]^2 + 8\Delta_3^2}} |1,0\rangle \downarrow \right\}$$

and

$$\left\{ -\frac{\left[\Delta_2 + \sqrt{\Delta_2^2 + 8\Delta_3^2}\right]}{\sqrt{\left[\Delta_2 + \sqrt{\Delta_2^2 + 8\Delta_3^2}\right]^2 + 8\Delta_3^2}} |1,1\rangle \downarrow + \frac{2\sqrt{2}\Delta_3}{\sqrt{\left[\Delta_2 + \sqrt{\Delta_2^2 + 8\Delta_3^2}\right]^2 + 8\Delta_3^2}} |1,0\rangle \uparrow \right\}$$

Now, regarding the eigen-value

$$-\frac{\Delta_2 - \sqrt{\Delta_2^2 + 8\Delta_3^2}}{2}$$

its eigen-vectors are:

$$\left\{-\frac{\left[\Delta_2 - \sqrt{\Delta_2^2 + 8\Delta_3^2}\right]}{\sqrt{\left[\Delta_2 - \sqrt{\Delta_2^2 + 8\Delta_3^2}\right]^2 + 8\Delta_3^2}}|1,-1\rangle \uparrow + \frac{2\sqrt{2}\Delta_3}{\sqrt{\left[\Delta_2 - \sqrt{\Delta_2^2 + 8\Delta_3^2}\right]^2 + 8\Delta_3^2}}|1,0\rangle \downarrow\right\}$$

and

$$\left\{-\frac{\left[\Delta_2 - \sqrt{\Delta_2^2 + 8\Delta_3^2}\right]}{\sqrt{\left[\Delta_2 - \sqrt{\Delta_2^2 + 8\Delta_3^2}\right]^2 + 8\Delta_3^2}}|1,1\rangle \downarrow + \frac{2\sqrt{2}\Delta_3}{\sqrt{\left[\Delta_2 - \sqrt{\Delta_2^2 + 8\Delta_3^2}\right]^2 + 8\Delta_3^2}}|1,0\rangle \uparrow\right\}$$

These simple calculations shows that the components of the wave-functions change rapidly versus the difference between Δ_2 and Δ_3. To get a more accurate description of the energy spectrum for the valence band states, one generally introduces a phenomenological parameter, the crystal-field splitting parameter that permits to distinguish the p_z state from the p_x and p_y ones. Related to the representation in terms of wave-functions of the angular momentum, the corresponding Hamiltonian requires to introduce a complementary matrix element Δ_1 and writes:

$$\mathbf{H}_{crist} = \Delta_1 L_z^2.$$

The full valence band spectrum becomes, at zone center.

$$\begin{bmatrix} |1,1\rangle\uparrow & |1,-1\rangle\downarrow & |1,1\rangle\downarrow & |1,0\rangle\uparrow & |1,-1\rangle\uparrow & |1,0\rangle\downarrow \\ \Delta_1+\Delta_2 & 0 & 0 & 0 & 0 & 0 \\ 0 & \Delta_1+\Delta_2 & 0 & 0 & 0 & 0 \\ 0 & 0 & \Delta_1-\Delta_2 & \sqrt{2}\Delta_3 & 0 & 0 \\ 0 & 0 & \sqrt{2}\Delta_3 & 0 & 0 & 0 \\ 0 & 0 & 0 & 0 & \Delta_1-\Delta_2 & \sqrt{2}\Delta_3 \\ 0 & 0 & 0 & 0 & \sqrt{2}\Delta_3 & 0 \end{bmatrix}$$

In terms of eigen-values, the energy of the two-fold levels of of Γ_9 symmetry is shifted by Δ_1. Regarding the Γ_7, their energies are obtained in the above equations, by replacing $-\Delta_2$ by $\Delta_1 - \Delta_2$. The same procedure permits to obtain the expansions of the eigen-vectors.

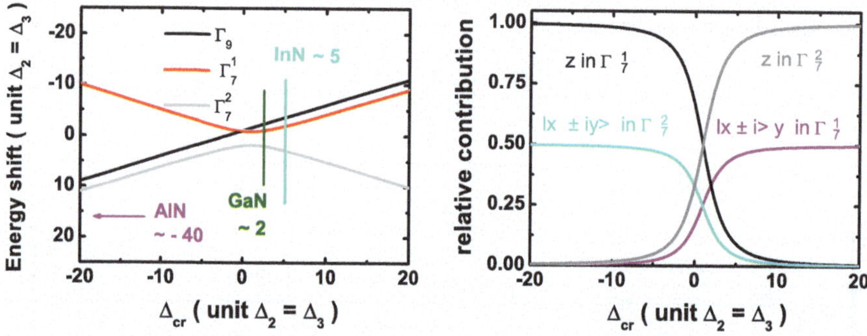

Fig. 3.12 (*left*) Evolution of the valence-band splittings versus crystal-field splitting. All energies are expressed versus the spin-orbit interaction parameter in the context of the quasi-cubic approximation (here $\Delta_1 = \Delta_{cr}$ and $\Delta_2 = \Delta_3 = \Delta_{so}$). The average positions for *AlN*, *GaN* and *InN* are indicated. (*right*) Relative contributions of z, x, and y states in the wave-function of valence bands having Γ_7 symmetry

Table 3.1 Values of the ratio c/a for various wurtzitic semi-conductors

Material	BN	GaN	InN	AlN	ZnS	CdS	CdSe	ZnO
c/a	1.64	1.62	1.61	1.6	1.64	1.62	1.63	1.60
Δ_1 (meV)	–	10.1	25	220	58	28	40	30.5
Δ_2 (meV)	–	5.7	6	6	27	27	138	4.2
Δ_2 (meV)	–	5.9	6	6.5	27	30	138	11.5

The number of digits is compatible with the determination method

It is important to indicate that the quasi-cubic approximation assumes $\Delta_2 = \Delta_3$; it is commonly adopted although it is now established that most of the properties of wurtzitic materials clearly depart from it. In Fig. 3.12 are plotted the evolutions of the splittings versus Δ_1 in units of $\Delta_2 = \Delta_3$ (left) and the evolutions of the expansions of the wave-vectors in the same conditions (right).

The relative values of the c/a ratio (as determined optically), of Δ_1, Δ_2 and Δ_3 are gathered in Table 3.1 for various semi-conductors, nitrides and chalcogenides. There is no real trend except at the scale of the spin-orbit interaction parameters which are, in general, increasing with the atomic number of the anion and the dominant contribution to the valence band states at its top most maximum. ZnO and InN are two exceptions: the 3*d* states of zinc interact with the *p* states of oxygen and the 5*d* states of indium interact with the 2 states of nitrogen, strongly reducing the band-gap energy. The crystal-field splitting parameter is a complex function of both the c/a ratio and internal displacement parameter. At the time of writing, the author has not found any unambiguous investigation in the literature.

At this stage, we would like to drop a few lines regarding the difficulty one faces when computing band structures. In the simple case of the tight-binding calculation, there are a lot of fitting parameters required. To get them, many experimental investigations are required such as fluorescence studies and ellipsometry measurements.

The ab-initio methods are more recent, more sophisticated with less empirical inputs. However, they need more and more computation time or power. Given a material, when its band structure is more or less established from both experimental and theoretical sides, experimentalists develop adapted, flexible approaches such as the $\vec{k} \cdot \vec{p}$ method. From now on, the author of this book, an experimentalist, will be extensively use this method.

3.4 The $\vec{k} \cdot \vec{p}$ Method and the Description of Band Dispersion at Zone Center in Wurtzitic Semi-conductors

This method is a very suitable phenomenological approach that permits to describe the band structure of semi-conductors at a given point—here Γ—of the Brillouin zone. It is extensively based on group theory arguments. An exhaustive description of this approach has been published by Kikuo Cho in 1974, in a paper referenced in the bibliography section at the end of this chapter.

3.4.1 *The Simplest Spinless Description of Conduction and Valence Bands Dispersions at Zone Center in Wurtzitic Semi-conductors*

We first consider, in the context of a spin-less description, the three valence band states $\{x(\vec{r}), y(\vec{r}), z(\vec{r})\}$ and the conduction state $s(\vec{r})$. Related to this basis, the theory of invariants indicates that the dispersion relations are derived, as follows, as functions of the three components of the wave-vector $\vec{k}$: k_x, k_y and k_z.

$$\begin{bmatrix} & s(\vec{r}) & x(\vec{r}) & y(\vec{r}) & z(\vec{r}) \\ s(\vec{r}) & \frac{\hbar^2}{2m_0}\alpha k_z^2 + \frac{\hbar^2}{2m_0}\beta(k_x^2 + k_y^2) & 0 & 0 & 0 \\ x(\vec{r}) & 0 & Ak_x^2 + Bk_y^2 + Ck_z^2 & Fk_xk_y & Gk_xk_z \\ y(\vec{r}) & 0 & Fk_xk_y & Ak_y^2 + Bk_x^2 + Ck_z^2 & Gk_yk_z \\ z(\vec{r}) & 0 & Gk_xk_z & Gk_yk_z & Dk_z^2 + E(k_x^2 + k_y^2) \end{bmatrix}$$

We want to make a point here: this representation requires 9 parameters even though theory necessitates 8 at the strictest sense. This is because we have not fully taken into account the predictions of group theory. Actually, some of them are bound: F = A − B

$$\begin{bmatrix} & s(\vec{r}) & x(\vec{r}) & y(\vec{r}) & z(\vec{r}) \\ s(\vec{r}) & \frac{\hbar^2}{2m_0}\alpha k_z^2 + \frac{\hbar^2}{2m_0}\beta(k_x^2 + k_y^2) & 0 & 0 & 0 \\ x(\vec{r}) & 0 & Ak_x^2 + Bk_y^2 + Ck_z^2 & (A-B)k_xk_y & Gk_xk_z \\ y(\vec{r}) & 0 & (A-B)k_xk_y & Ak_y^2 + Bk_x^2 + Ck_z^2 & Gk_yk_z \\ z(\vec{r}) & 0 & Gk_xk_z & Gk_yk_z & Dk_z^2 + E(k_x^2 + k_y^2) \end{bmatrix}$$

This (4×4) representation reveals the intrinsic symmetry of this problem, in the case of a wurtzite crystal: both $x(\vec{r})$ and $y(\vec{r})$ are treated on a same footing, differently from $z(\vec{r})$. The conduction band state $s(\vec{r})$ is not coupled with the valence band states at this degree of sophistication: the matrix elements that represent the dispers ion relations in the valence band are quadratic functions of the components of $\vec{k}$.

- Let $k_x = k_y = 0$
  The dispersion relations are:

$$\begin{bmatrix} & s(\vec{r}) & x(\vec{r}) & y(\vec{r}) & z(\vec{r}) \\ s(\vec{r}) & \frac{\hbar^2}{2m_0}\alpha k_z^2 & 0 & 0 & 0 \\ x(\vec{r}) & 0 & Ck_z^2 & 0 & 0 \\ y(\vec{r}) & 0 & 0 & Ck_z^2 & 0 \\ z(\vec{r}) & 0 & 0 & 0 & Dk_z^2 \end{bmatrix}$$

  The conduction band is characterized by a dispersion relation of the kind: $E_c(k_z) = E_g + \frac{\hbar^2}{2m_0}\alpha k_z^2$. The electron effective mass is $\frac{1}{\alpha}$ in units of the electron mass at rest.
- Let $k_z = k_y = 0$
  The dispersion relations are:

$$\begin{bmatrix} & s(\vec{r}) & x(\vec{r}) & y(\vec{r}) & z(\vec{r}) \\ s(\vec{r}) & \frac{\hbar^2}{2m_0}\beta(k_x^2 & 0 & 0 & 0 \\ x(\vec{r}) & 0 & Ak_x^2 & 0 & 0 \\ y(\vec{r}) & 0 & 0 & Bk_x^2 & 0 \\ z(\vec{r}) & 0 & 0 & 0 & Ek_x^2 \end{bmatrix}$$

  The conduction band is characterized by a dispersion relation of the kind: $E_c(k_z) = E_g + \frac{\hbar^2}{2m_0}\beta k_x^2$. The electron effective mass is $\frac{1}{\beta}$ in units of the electron mass at rest.

3.4.2 The Simplest (6×6) $\vec{k} \cdot \vec{p}$ Description for Valence Band Dispersions at Zone Center in Wurtzitic Semi-conductors

To have a more realistic description, one has to include the spin-orbit interaction and the crystal-field splitting.

$$\begin{bmatrix} |1,1\rangle\uparrow & |1,-1\rangle\downarrow & |1,1\rangle\downarrow & |1,0\rangle\uparrow & |1,-1\rangle\uparrow & |1,0\rangle\downarrow \\ H_{11} & 0 & 0 & H_{14} & H_{15}^* & 0 \\ 0 & H_{11} & H_{15}^* & 0 & 0 & H_{14}^* \\ 0 & H_{15} & H_{22} & \sqrt{2}\Delta_3 & 0 & H_{14} \\ H_{14}^* & 0 & \sqrt{2}\Delta_3 & H_{33} & H_{14} & 0 \\ H_{15} & 0 & 0 & 0 & H_{22} & \sqrt{2}\Delta_3 \\ 0 & H_{14} & H_{14}^* & 0 & \sqrt{2}\Delta_3 & H_{33} \end{bmatrix}$$

$$H_{11} = \Delta_1 + \Delta_2 + Ck_z^2 + (A+B)k_x^2$$
$$H_{22} = \Delta_1 - \Delta_2 + Ck_z^2 + (A+B)k_x^2$$
$$H_{33} = Dk_z^2 + E(k_x^2 + k_y^2)$$
$$H_{14} = \frac{1}{\sqrt{2}} Gk_z(k_x - ik_y)$$
$$H_{15} = \frac{1}{2}(A-B)(k_x^2 - k_y^2 + 2ik_xk_y)$$

- Let $k_x = k_y = 0$.

Related to the valence band functions expressed in terms of angular momenta description, one obtains the following Hamiltonian:

$$\begin{bmatrix} |1,1\rangle\uparrow & |1,-1\rangle\downarrow & |1,1\rangle\downarrow & |1,0\rangle\uparrow & |1,-1\rangle\uparrow & |1,0\rangle\downarrow \\ \Delta_1+\Delta_2+Ck_z^2 & 0 & 0 & 0 & 0 & 0 \\ 0 & \Delta_1+\Delta_2+Ck_z^2 & 0 & 0 & 0 & 0 \\ 0 & 0 & \Delta_1-\Delta_2+Ck_z^2 & \sqrt{2}\Delta_3 & 0 & 0 \\ 0 & 0 & \sqrt{2}\Delta_3 & Dk_z^2 & 0 & 0 \\ 0 & 0 & 0 & 0 & \Delta_1-\Delta_2+Ck_z^2 & \sqrt{2}\Delta_3 \\ 0 & 0 & 0 & 0 & \sqrt{2}\Delta_3 & Dk_z^2 \end{bmatrix}$$

The dispersion relations for the Γ_9 valence band are $E_v(k) = E_v(0) + Ck_z^2$. The hole effective-mass is $\frac{\hbar^2}{2}\frac{1}{C}$ in units of the electron mass at rest. Concerning the Γ_7 valence band, one has to solve a (2×2) matrix and to perform two derivations. The expressions are tedious; we, therefore, refrain to detail them here.

- Let $k_z = k_y = 0$.

The 6×6 valence band Hamiltonian is obtained as two similar 3×3 hamiltonians which we write as follows:

$$\begin{bmatrix} |1,\pm1\rangle\begin{pmatrix}\uparrow\\\downarrow\end{pmatrix} & |1,\mp1\rangle\begin{pmatrix}\uparrow\\\downarrow\end{pmatrix} & |1,0\rangle\begin{pmatrix}\uparrow\\\downarrow\end{pmatrix} \\ \Delta_1+\Delta_2+\frac{\hbar^2}{2m_0}(A+B)k_x^2 & -\frac{\hbar^2}{2m_0}\frac{1}{2}(A-B)k_x^2 & 0 \\ -\frac{\hbar^2}{2m_0}\frac{1}{2}(A-B)k_x^2 & \Delta_1-\Delta_2+\frac{\hbar^2}{2m_0}(A+B)k_x^2 & \sqrt{2}\Delta_3 \\ 0 & \sqrt{2}\Delta_3 & \frac{\hbar^2}{2m_0}Ek_x^2 \end{bmatrix}$$

The dispersion relations for the valence band are not simple analytical equations. For dispersion relations away from these two simple orientations of the wave-vector, the situation is much more tricky.

3.4.3 Including Strain Field to the $\vec{k} \cdot \vec{p}$ Description of Band Dispersion in Wurtzitic Semi-conductors

Again, symmetry arguments permit to represent the impact of a strain Hamiltonian on the energy spectrum using a Hamiltonian H_{strain} represented in a very simple manner:

$$\begin{bmatrix} s(\vec{r}) & x(\vec{r}) & y(\vec{r}) & z(\vec{r}) \\ a_1 e_{zz} + a_2(e_{xx} + e_{yy}) & 0 & 0 & 0 \\ 0 & l_1 e_{xx} + m_1 e_{yy} + m_2 e_{zz} & (l_1 - m_1) e_{xy} & n_2 e_{xz} \\ 0 & (l_1 - m_1) e_{xy} & l_1 e_{yy} + m_1 e_{xx} + m_2 e_{zz} & n_2 e_{yz} \\ 0 & n_2 e_{xz} & n_2 e_{yz} & l_2 e_{zz} + m_3(e_{xx} + e_{yy}) \end{bmatrix}$$

These new proportionality constants are matrix elements that are called the deformation potentials.

We remark the similarity between the strain Hamiltonian and the dispersion relations. Both conduction and valence band Hamiltonians of strained wurtzitic semi-conductors can be obtained by superposition of both contributions, giving the equations above:

$$E_c = E_g + \alpha(k_z^2 + \beta(k_x^2 + k_y^2) + a_1 e_{zz} + a_2(e_{xx} + e_{yy})$$

for the conduction band. On the valence band side, the matrix elements H_{ij} change:

$$H_{11} = \Delta_1 + \Delta_2 + Ck_z^2 + (A + B)k_x^2 + (l_1 + m_1)(e_{xx} + e_{yy}) + m_2 e_{zz}$$
$$H_{22} = \Delta_1 - \Delta_2 + Ck_z^2 + (A + B)k_x^2 + (l_1 + m_1)(e_{xx} + e_{yy}) + m_2 e_{zz}$$
$$H_{33} = Dk_z^2 + E(k_x^2 + k_y^2) + l_2 e_{zz} + m_3(e_{xx} + e_{yy})$$
$$H_{14} = \frac{1}{\sqrt{2}} Gk_z(k_x - ik_y) + n_2(e_{xz} - ie_{zy})$$
$$H_{15} = \frac{1}{2}(A - B)(k_x^2 - k_y^2 + 2ik_xk_y) + (l_1 - m_1)(e_{xx} - e_{yy} + 2ie_{xy})$$

At this stage, we have ignored any coupling between the conduction and valence bands. Let us introduce it now.

3.4.4 The Simplest (8 × 8) $\vec{k} \cdot \vec{p}$ Description of Valence Band Dispersions at Zone Center in Wurtzitic Semi-conductors

We work in the context of the 8-dimension basis which contains the two conduction and the six valence band states described above:$\{|s\rangle \uparrow; |s\rangle \downarrow; |1,1\rangle \downarrow; |1,-1\rangle \downarrow; |1,1\rangle \downarrow; |1,0\rangle \uparrow; |1,-1\rangle \uparrow, |1,0\rangle \downarrow\}$. Then, the Hamiltonian has the following form:

$$\mathbf{H}(\vec{k}) = \begin{pmatrix} H_{cc} & H_{cv} \\ H_{cv}^{\dagger} & E_{vv} \end{pmatrix}$$

In the $\{|s\rangle \uparrow; |s\rangle \downarrow\}$ basis of the conduction states, H_{cc} writes:

$$\mathbf{H}_{cc} = \begin{pmatrix} E_c & 0 \\ 0 & E_c \end{pmatrix}$$

The valence band Hamiltonian H_{vv} is the (6 × 6) one given previously.

The interaction Hamiltonien H_{cv} couples the conduction and valence band states; we write it as:

$$H_{cv} = \begin{pmatrix} Q & 0 & 0 & R & Q^* & 0 \\ 0 & Q^* & Q & 0 & 0 & R \end{pmatrix}$$

with

$$Q = \frac{1}{\sqrt{2}} \frac{\hbar}{m_0} \langle s|p_x|x(\vec{r})\rangle (k_x + ik_y) = \frac{1}{\sqrt{2}} P_{\perp}(k_x + ik_y)$$

$$R = \frac{\hbar}{m_0} \langle s|p_z|z(\vec{r})\rangle k_z = P_{//} k_z$$

$P_{//}$ and $P_{\perp}$ are matrix elements would be identical in cubic symmetry. However, the dispersion relations would also express differently. So ... the comparison between zinc blende and wurtzite is dangerous.

$$P_{//}^2 = \frac{\hbar^2}{2m_0} \left[\frac{1}{\alpha} - \frac{1}{\beta} \right] \frac{(E_g + \Delta_1 + \Delta_2)(Eg + 2\Delta_2) - 2\Delta_3^2}{E_g + 2\Delta_2}$$

$$P_{\perp}^2 = \frac{\hbar^2}{2m_0} \left[\frac{1}{\beta} - \frac{1}{\alpha} \right] \frac{\{(E_g + \Delta_1 + \Delta_2)(Eg + 2\Delta_2) - 2\Delta_3^2\}E_g}{(E_g + \Delta_2)(E_g + \Delta_1 + \Delta_2) - 2\Delta_3^2}$$

At this stage, once the inputs are known, the dispersion relations of the coupled bands are obtained numerically. It is possible to go further than we did here but the complexities necessitated by the treatment are beyond the scope of this book. In Table **??** is listed the full set of parameters useful to perform a numerical calculation

Table 3.2 Typical values

Material	GaN	InN	AlN
Eg (eV)	3.5	0.6	6
Δ_1 (meV)	10.1	25	−220
Δ_2 (meV)	5.7	6	6
Δ_3 (meV)	5.9	6	6.5
$1/\alpha$	0.19	0.065	0.32
$1/\beta$	0.20	0.068	0.33
$A_1(\frac{2m0}{\hbar^2})$	−5.95	−15.8	−3.99
$A_2(\frac{2m0}{\hbar^2})$	−0.53	−0.50	−0.31
$A_3(\frac{2m0}{\hbar^2})$	5.41	15.25	3.67
$A_4(\frac{2m0}{\hbar^2})$	−2.51	−7.15	−1.15
$A_5(\frac{2m0}{\hbar^2})$	−2.51	−7.06	−1.33
$A_6(\frac{2m0}{\hbar^2})$	−3.2	−10.08	−1.95
C_1 (eV)	−6.5	−3.62	−4.3
C_2 (eV)	−11.2	−4.60	−11.5
C_3 (eV)	4.9	2.68	6.8
C_4 (eV)	−5.0	1.74	−3.6
C_5 (eV)	−2.8	2.07	−2.80
C_6 (eV)	−3.1	8.55	−4.5

of the band structure of nitride semi-conductors in the presence of any strain field. There are several conventions in the literature, regarding the labels of the different quantities that are introduced. In the table, we have used a convention different from one used above to obtain the hole masses. The relationships are A_1 = D; A_2 = E; A_3 = C − D; A_4 = A + B − E; A_5 = (A − B)/2andA_6 = G. Regarding the deformation potentials, $C_1 = a_1 - l_2$; $C_2 = a_2 - m_3$; $C_3 = m_2 - l_3$; $C_4 = l_1 + m_1 - m_3$; $C_5 = l_1 - m_1$; $C_6 = n_2$. A quasi-cubic approximation has been invoked to connect the quantities having similar meaning. In terms of deformation potentials, it writes: $C_1 + C_3 = C_2$; $C_3 = -C_4$ and $C_3 + 4C_5 = \sqrt{2}C_6$. These approximations were motivated by the lack of knowledge of the band structure parameters, a situation now clarified.

3.5 Phonons in Wurtzitic Semi-conductors

The crystal is not frozen: even at 0 K, atoms oscillate from their equilibrium position as it will be shown in the section about quantum description of phonons.

3.5.1 Longitudinal and Transverse Waves in Continuous Media

We consider here large wavelengths when compared to the interatomic spacing, so that the crystal may be considered as a continuous medium.

A wave that propagates in a continuous and isotropic medium can be decomposed into a longitudinal wave and two transversal ones. When the particle displacement ξ occurs in the direction of the wave propagation x, it is called longitudinal. Such a wave is described by its motion—or propagation—equation:

$$\frac{\partial^2}{\partial x^2}\xi = \frac{1}{c^2}\frac{\partial^2}{\partial t^2}\xi \quad \text{where} \quad c^2 = \frac{\lambda}{\rho}$$

In the propagation equation, λ is the constant of elasticity, ρ the density and c the velocity of all the waves.

The general solution of this one-dimensional equation is $A(x - ct) + B(x + ct)$ where A and B are arbitrary twice differentiable functions of x and t.

Specific solutions are exp $(i(\omega t \pm kx))$ with $c = \omega/k$.

The transverse waves have the same equations but their velocity is reduced, compared to the longitudinal one.

3.5.2 The Classical Model and the Concept of Normal Coordinates

We have shown, at the beginning of this chapter, that the Born-Oppenheimer approximation permits to express the Schrödinger equation for the nuclei motion in the crystal under the form:

$$\left[\mathbf{T}_K + \mathbf{V}(\{\overrightarrow{R_J}\})\right]\chi(\{\overrightarrow{R_J}\})s\chi(\{\overrightarrow{R_J}\})$$

The kinetic energy is indexed using K, because we write the Hamiltonian for the single atom of the crystal having the mass M_K. Note that this index is a capital K while a small k is used to represent the wave-vector.

We write $\{\vec{R_J}^0\}$ a the whole set of atomic positions at the equilibrium, where the potential function has a minimum.

We represent by $\{\vec{\zeta_J}\}$ the whole set of small atomic displacements from this equilibrium position. Suppose that v at a given time t, atoms are located at positions

$$\{\vec{R_J}\} = \{\vec{R_J}^0\} + \{\vec{\zeta_J}\}$$

In the framework of the harmonic approximation, the potential is a quadratic function of the displacements $\{\vec{\zeta_J}\}$.

We write this:

$$V = \frac{1}{2}\sum_{\alpha=1}^{3N}\sum_{\beta=1}^{3N} V_{\alpha\beta}\zeta_\alpha\zeta_\beta \quad \text{with} \quad V_{\alpha\beta} = \left[\frac{\partial^2 V}{\partial R_\alpha \partial R_\beta}\right]_{\vec{R_\alpha}^0}$$

We assign the kinetic energy part of the Hamiltonien to be the half-product of the mass by the square of the velocity:

$$T = \frac{1}{2}\sum_{\alpha=1}^{3N} M_\alpha \dot{\zeta}_\alpha^2 \quad \text{with} \quad \dot{\zeta}_\alpha = \frac{d}{dt}\zeta_\alpha$$

That further gives:

$$T = \frac{1}{2}\sum_{\alpha=1}^{3N} \dot{\eta}_\alpha^2 \quad \text{with} \quad \eta_\alpha = \sqrt{M_\alpha}\zeta_\alpha \quad \text{and} \quad \dot{\eta}_\alpha = \frac{d}{dt}\eta_\alpha$$

$$V = \frac{1}{2}\sum_{\alpha=1}^{3N}\sum_{\beta=1}^{3N} F_{\alpha\beta}\eta_\alpha\eta_\beta \quad \text{with} \quad F_{\alpha\beta} = V_{\alpha\beta}/\sqrt{M_\alpha M_\beta}$$

The quantities $\vec{\zeta}_\alpha$'s (respectively the $\vec{\eta}_\alpha$'s) are the components of a 3N-component uni-column matrix $[\zeta]$ (respectively $[\eta]$) which transpose form is an uni-row 3N-column matrix $[\zeta]^\dagger$ (resp. $[\eta]^\dagger$). The $V_{\alpha\beta}$'s and $F_{\alpha\beta}$'s are components of (3N × 3N) rank-two symmetric tensors $\vec{\vec{V}}$ and $\vec{\vec{F}}$.

Using the tensor notation, the kinetic energy writes: $T = \frac{1}{2}[\dot{\eta}]^\dagger[\dot{\eta}]$ and the potential energy term writes: $V = \frac{1}{2}[\eta]^\dagger \vec{\vec{F}}[\eta]$.

Let us define a new set of coordinates by an orthogonal transformation U.

$$\eta_\alpha = \sum_\beta U_{\alpha\beta}Q_\beta \quad \text{that may be written} \quad [\eta] = [U][Q]$$

After some algebraic manipulations, one can easily demonstrate that:

$$\mathrm{T} = \frac{1}{2}[\dot{\eta}]^\dagger[\dot{\eta}] = \frac{1}{2}\left[U\dot{Q}\right]^\dagger\left[U\dot{Q}\right] = \frac{1}{2}\left[\dot{Q}^\dagger U^\dagger U\dot{Q}\right] = \frac{1}{2}\left[\dot{Q}\right]^\dagger\left[\dot{Q}\right]$$

where $\left[\dot{Q}\right]^\dagger$ is the transpose matrix of the column matrix $\left[\dot{Q}\right]$.

Similarly,

$$\mathrm{V} = \frac{1}{2}[UQ]^\dagger \vec{\vec{F}}[UQ] = \frac{1}{2}Q^\dagger\left[U^\dagger\vec{\vec{F}}U\right]Q$$

The transformation can always be chosen so that $\left[U^\dagger\vec{\vec{F}}U\right]$ is a diagonal matrix:

$$\left[U^\dagger FU\right]_{\alpha\beta} = \omega_\alpha^2\delta_{\alpha\beta}$$

From Hamilton equations, the momentumP_α is defined as:

$$P_\alpha = \frac{\partial}{\partial \dot{Q}_\alpha} T = \dot{Q}_\alpha$$

then,

$$H = \frac{1}{2} \sum_{\alpha=1}^{3N} (P_\alpha^2 + \omega_\alpha^2 Q_\alpha^2) \text{ and } \dot{P}_\alpha = \frac{d}{dt} P_\alpha = \ddot{Q}_\alpha$$

From another Hamilton equation, $\dot{P_\alpha} = -\frac{\partial}{\partial Q_\alpha} H = -\omega_\alpha^2 Q_\alpha$
After a derivation of the Hamiltonian with respect to Q_α, one obtains:

$$\ddot{Q}_\alpha = -\omega_\alpha^2 Q_\alpha \text{for all} \alpha\text{'s}$$

They are 3N classical harmonic oscillators with solutions:

$$Q_\alpha = Q_\alpha^0 \sin(\omega_\alpha t + \varphi_\alpha)$$

The Q_α's are the normal coordinates.

Since the particle displacements are real, the normal coordinates must also be real; atoms oscillate harmonically in time with pulsations ω_α's. There are 3N coordinates and 3N pulsations corresponding to the 3N degrees of freedom of the crystal.

The quantities $\eta_\alpha = \sum_\beta U_{\alpha\beta} Q_\beta$ *and* $\zeta_\alpha = \sum_\beta \sqrt{M_\alpha} U_{\alpha\beta} Q_\beta$, *actually the real atomic displacements, are linear combinations of the normal coordinates.*

Back to the expression of the Hamiltonian in terms of mass-weighted coordinates η_α and still using Hamilton's equations, we can write:

$$\ddot{\eta_\alpha} = \sum_\beta F_{\alpha\beta} \eta_\beta \quad \text{for } \alpha = 1, 2, \ldots, 3N.$$

When $\eta_\beta = U_{\beta\gamma} Q_\gamma$, we can further write:

$$\sum_\alpha \left[F_{\alpha\beta} - \omega_\gamma^2 \delta_{\beta\alpha} \right] U_{\alpha\gamma} = 0$$

A non-trivial solution is obtained for U when the determinant of the coefficients is zero.

$$|F_{\alpha\beta} - \omega^2 \delta_{\alpha\beta}| = 0$$

This is a polynomial of degree 3N in ω^2, giving 3N values for ω.

In fact, six values of ω are zero (five for a linear molecule). They correspond to translational or rotational modes of the crystal. Clearly, if each atom is given the same translation, there is no change in the potential of the crystal. There are thus 3N-6 internal coordinates with non-zero and real frequencies.

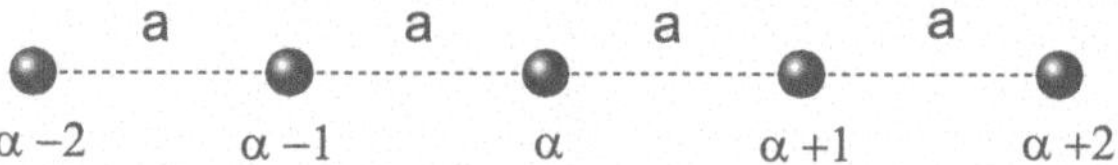

Fig. 3.13 Sketch of the linear mono-atomic lattice

3.5.3 Group Theory and Normal Modes

A normal mode can be described by the 3N coordinates $\eta^0_\alpha = U_{\alpha\beta} Q^0_\beta$, giving the maximum displacements to atoms from their equilibrium positions. The crystal symmetry may demand that different modes possess the same frequency (pulsation). Then, they are degenerate modes. For a nondegenerate frequency ω_α, the normal coordinate must transform according to a real one dimensional representation of the symmetry group of the molecule or crystal under study:

$$RQ_\alpha = \pm Q_\alpha$$

where R is any symmetry element of the group. For a n-fold degenerate frequency, the n normal coordinates $Q_1, \ldots Q_n$, must transform according to a real n-fold representation of the group:

$$RQ_\alpha = \sum_{\beta=1}^{\beta=n} C_{\beta\alpha} Q_\alpha$$

Apart from accidental degeneracy, this representation will be irreducible. In the simple group, only cubic symmetries allow three-dimensional degeneracies. Uniaxial symmetries allow two-fold symmetries. Abelian groups only allow uni-dimensional irreducible representations, therefore non-phonon frequencies are degenerated in that case.

3.5.4 The Linear Mono-atomic Lattice

The infinite chain of identical atoms

We consider an infinite linear chain of identical atoms of mass μ separated by a distance a at rest, as shown in Fig. 3.13. If these atoms experience a perturbation that moves them, in the context of the nearest neighbor approximation, the mutual force between two atoms is *proportional to their relative displacement* from their equilibrium position. Let x_α be the displacement of atom α. The resultant force on the α^{th}atom is the resultant of the forces acting on its right and left sides.

$$F_\alpha = \kappa(x_{\alpha+1} - x_\alpha) - \kappa(x_\alpha - x_{\alpha-1})$$

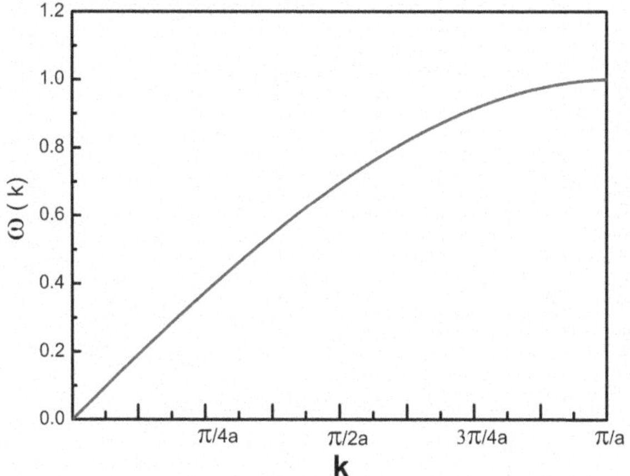

Fig. 3.14 Dispersion relation $\omega(k)$ for the linear mono-atomic lattice in units of $\sqrt{\frac{4\kappa}{\mu}}$

The proportionality constant κ is an elastic constant. The equation of motion is:

$$\mu \ddot{x_\alpha} = \kappa(x_{\alpha+1} - x_\alpha) - \kappa(x_\alpha - x_{\alpha-1}) \quad \text{where } \ddot{x_\alpha} = \frac{d^2}{dt^2} x_\alpha$$

From Bloch theorem, we can affirm that there are traveling waves and atoms vibrating at the same frequency $\omega(k)$.

The solutions will be of the kind $x = Ae^{i(\omega(k)t+kx)}$ where k is a **real and continuous** wave-vector. Thanks to the periodicity of the linear chain that is infinite here, $x = \alpha a$, with $\alpha \in \mathbb{Z}$. The amplitude A is a complex constant.

Then, one obtains, from the resolution of the equation of motion, the dispersion relation:

$$\omega(k) = \sqrt{\frac{4\kappa}{\mu}} \left| \sin \frac{ka}{2} \right|$$

And the eigen-states of the problem:

$$x_\alpha(k,t) = A.exp \left\{ i \left(\sqrt{\frac{4\kappa}{\mu}} \cdot \left| \sin \frac{ka}{2} \right| t + k_{\alpha a} \right) \right\}$$

A plot of the dispersion relation $\omega(k)$ is reported in Fig. 3.14.

Due to the periodicity of the dispersion relation versus k since ω is a periodic function with period $2\pi/a$, the plot is restricted to values of k belonging to the first Brillouin zone of the linear lattice. We remark the parity property of the dispersion relation in k-space. For each value of ω, there are two values of k that satisfy the dispersion relation: k and $-k$, which correspond to waves travelling in opposite

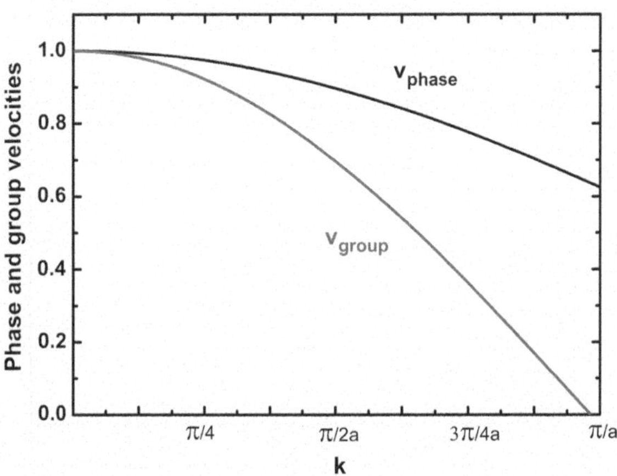

Fig. 3.15 Phase and group velocities for the linear mono-atomic lattice in units of $a\sqrt{\frac{\kappa}{\mu}}$

directions. This two-fold degeneracy would be also found when studying two- and three-dimensional mono-atomic lattices.

Note: when k is a complex number, some solutions are larger than the maximum frequency $\omega_m = \sqrt{\frac{4\kappa}{\mu}}$. Then, $\left|\sin\frac{ka}{2}\right|\rangle 1$. The wave-vectors are:

$k = \frac{\pi}{a} + i\varpi$, ϖ being a real number. The quantity ϖ is called attenuation coefficient. It rapidly increases with ω, according to equation $\left[\cos\frac{\varpi a}{2}\right]^2 = \frac{\omega^2\mu}{4\kappa}$.

At this stage, we would like to remind the reader the definitions of two quantities of paramount importance:

The phase velocity of the wave is defined by:

$$v_{phase} = \frac{\omega}{k}$$

The group velocity at which the energy or information is transferred by the wave is defined as:

$$v_{group} = \frac{\partial\omega}{\partial k}$$

For the mono-atomic linear lattice:

$$v_{phase} = a\sqrt{\frac{\kappa}{\mu}\frac{2}{ka}\sin|k\frac{a}{2}|} \quad \text{and} \quad v_{group} = a\sqrt{\frac{\kappa}{\mu}}\cos(k\frac{a}{2})$$

The evolutions of v_{phase} and v_{group} are plotted in Fig. 3.15.

The finite length chain of identical atoms

If we consider a linear chain of N + 1 atoms, both ends atoms being fixed, for a frequency ω the general solution is:

$$x_n = e^{i\omega t}(Ae^{ikna} + Be^{-ikna})$$

Using the boundary conditions $x_n = 0$ for n=0 and N, one can easily show that:

$$\mathrm{A} = -\mathrm{B} \quad \text{and} \quad \sin(kNa) = 0.$$

Then, the values of k are no longer continuous, at the strictest sense:

$$k = p\pi Na \quad \text{with } \mathrm{p} \in \mathbb{N} \quad \text{and} \quad x_n = Ce^{i\omega t} \sin \frac{p\pi}{N}$$

Since particles n=0 and N are fixed, there are **N − 1 allowed standing waves** corresponding to the $N - 1$ particles free to vibrate at frequencies:

$$\omega(p) = \sqrt{\frac{4\kappa}{\mu}} | \sin \frac{p\pi}{2N} | \quad \text{with } p \in \{1, 2, \ldots, N-2, N-1\}$$

The periodic chain of identical atoms

The boundary conditions for a periodic chain of N atoms give traveling wave solutions.

These boundary solutions are written $x_n = x_{n+N}$ for all n.

$$\mathrm{k} = 2p\pi Na \quad \text{with } \mathrm{p} \in \mathbb{Z}$$

and $p \in \{\pm 1, \pm 2, \ldots, \pm \frac{N-1}{2}\}$ if N is odd and $p \in \{\pm 1, \pm 2, \ldots, \pm \frac{N}{2}\}$ if N is even.

The allowed values of k are twice spaced compared to the preceding situation.

In the latter case, states $\pm \frac{N}{2}$ are obviously equivalent.

There are N − 1 traveling modes and, in addition, a non-zero solution: a translation of the whole crystal without change of the potential energy when $p = 0$.

The frequency distribution

A frequency distribution $g(\omega)$ may be also defined so that $g(\omega)d\omega$ is the number of allowed frequencies, between ω and $\omega + d\omega$.

From equation $k = p\pi Na$, or alternatively from equation $k = 2p\pi Na$, one can calculate the number of states between $|k|$ and $|k + dk|$: $dp = \frac{kNa}{\pi} = dk$.

Then, $g(\omega)d\omega = \frac{Na}{\pi}\frac{dk}{d\omega}d\omega$

since $k = \frac{2}{a} \arcsin \frac{\omega}{\omega_m}$.

One can demonstrate that:

$$g(\omega) = \frac{2N}{\pi\omega_m} \frac{1}{(\omega_m^2 - \omega^2)^{1/2}}$$

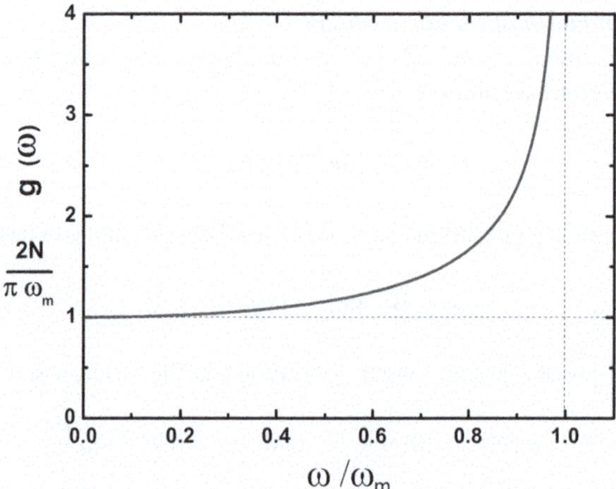

Fig. 3.16 Frequency distribution $g(\omega)$ for the mono-atomic linear lattice

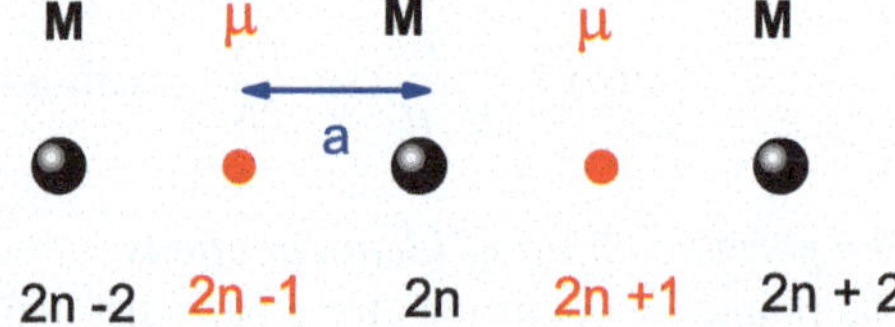

Fig. 3.17 Representation of a linear lattice with two different atoms

for a N-atom linear chain to which are applicated Born-von Karman cyclic conditions. From this distribution function, we evidence that the majority of atoms vibrate at a frequency close to the the cut-off as shown in Fig. 3.16.

3.5.5 The Linear Lattice with Two Different Atoms: Acoustic and Optical Branches

We consider here a line of two alternating atoms equally spaced at a distance a from each other. Alternate atoms have mass M and μ, we suppose here that M$\gtrsim \mu$. The primitive cell has a length $2a$ and contains 2 atoms. The heavy atoms are labelled $2n$, with $n \in \mathbb{Z}$ while light atoms are labelled $2n + 1$ as sketched in Fig. 3.17. The equations of motion write:

$$M\ddot{x}_{2n} = \kappa(x_{2n+1} + x_{2n-1} - 2x_{2n})$$

$$\mu\ddot{x}_{2n+1} = \kappa(x_{2n+2} + x_{2n} - 2x_{2n+1})$$

We look for solutions of the following kind, satisfying the Bloch theorem:

$$x_{2n} = A_1 e^{i[\omega t + k2na]}$$

$$x_{2n+1} = A_2 e^{i[\omega t + k(2n+1)a]}$$

The equations of motion are:

$$\begin{pmatrix} \omega^2 M - 2\kappa & 2\kappa\cos(ka) \\ 2\kappa\cos(ka) & \omega^2\mu - 2\kappa \end{pmatrix} \begin{pmatrix} A_1 \\ A_2 \end{pmatrix} = \begin{pmatrix} 0 \\ 0 \end{pmatrix}$$

These two equations have non-trivial solutions when the determinant of coefficients is zero. The solutions are quadratic in ω^2:

$$\omega^2 = \kappa\left(\frac{1}{M} + \frac{1}{\mu}\right) \pm \kappa\left[\left(\frac{1}{M} + \frac{1}{\mu}\right)^2 - \frac{4}{M\mu}\sin^2(ka)\right]^{\frac{1}{2}}$$

For long waves, in extenso for small k's, at zone center, the two roots are:

$$\omega_+^2 = 2\kappa\left(\frac{1}{M} + \frac{1}{\mu}\right) \quad \text{and} \quad \omega_-^2 = \frac{2\kappa}{M+\mu}k^2a^2$$

for the optical branch and for the acoustical branches respectively.

At the edges of the Brillouin zone, when $k = \pm\frac{\pi}{2a}$, the two roots become:

$$\omega_+^2 = \frac{2\kappa}{\mu} \quad \text{and} \quad \omega_-^2 = \frac{2\kappa}{M}$$

In the limit of small k's, it can be shown that $\frac{A_1}{A_2} = 1$ for the acoustical branch and $\frac{A_1}{A_2} = -1$ for the optical branch.

In Fig. 3.18 are plotted these dispersion relations in the first Brillouin zone. The low-energy branch, $\omega = 0$ at zone center, is called the acoustic phonon branch. The high-energy branch is the optical phonon branch. Note the asymptotic values at the zone center and at zone edges, as well as the formation of an energy gap, being straightforward consequences of the periodicity of the linear crystal and atomic mass mismatch.

3.5.6 Quantum Theory of Lattice Vibrations

Previously in this section, we have treated the lattice vibrations in a context of classical physics. Although vector k is pseudo-continuous, no quantum effect was

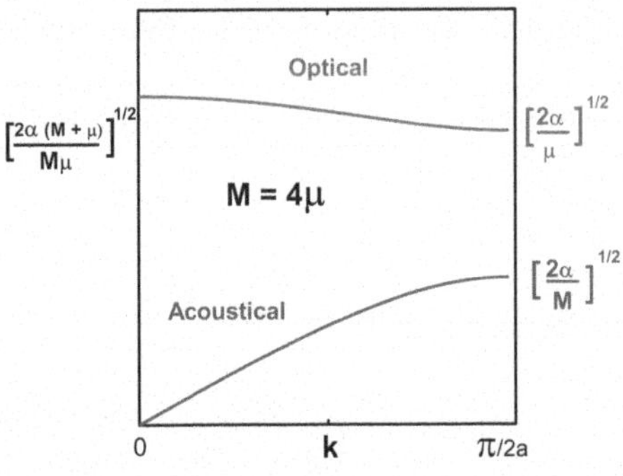

Fig. 3.18 Dispersion relations for a linear lattice with two different atoms

evidenced. Let us consider the total kinetic and potential energies of the linear mono-atomic lattice (N patterns).

$$T = \frac{\mu}{2}\sum_{n=0}^{N-1} \dot{x}_n^2 \quad \text{and} \quad V = \frac{\kappa}{2}\sum_{n=0}^{N-1} x_n^2$$

In this problem, we take k as: $k = \frac{2\pi}{Na}l$ with $l = 0, \pm 1, \ldots \pm \frac{N-1}{2}$; N odd.
We then construct complex Bloch states from the $\{x_n\}$ as follows:

$$R_k = \frac{1}{\sqrt{N}}\sum_{n=0}^{N-1} e^{-ikna} x_n \quad \text{and} \quad R_l = \frac{1}{\sqrt{N}}\sum_{n=0}^{N-1} e^{-i\frac{2\pi}{N}ln} x_n$$

After some algebraic manipulations, that require to keep in mind that:

$$\sum_l = -\frac{N-1}{2}^{l=\frac{N-1}{2}} e^{i\frac{2\pi}{N}l(n-n')} = N\delta_{nn'}$$

and

$$x_n = \sum_{l=-\frac{N-1}{2}}^{l=\frac{N-1}{2}} e^{i\frac{2\pi}{N}ln} R_l$$

One gets: $\ddot{R}_l + \omega_l^2 R_l = 0$
The general solution of this equation is

$$R_l = A_l^- e^{-i\omega t} + A_l^+ e^{+i\omega t}$$

with $R_l = R_{-l}^*$ so that atomic displacements are real quantities.

$$T = \frac{\mu}{2} \sum_{l=-\frac{N-1}{2}}^{l=\frac{N-1}{2}} \dot{R}_l \dot{R}_l^* \quad \text{and} \quad V = \frac{\mu}{2} \sum_{l=-\frac{N-1}{2}}^{l=\frac{N-1}{2}} \omega_l^2 R_l R_l^*$$

We write the Hamiltonian

$$\mathrm{H} = \frac{\mu}{2} \sum_{l=-\frac{N-1}{2}}^{l=\frac{N-1}{2}} (\dot{R}_l \dot{R}_l^* + \omega_l^2 R_l R_l^*)$$

N real coordinates are defined: $Q_{1l} = \frac{1}{2}(R_l^* + R_l)$ and $Q_{2l} = \frac{1}{2}(R_l^* - R_l)$

$$\mathrm{H} = \frac{\mu}{2} \sum_{l=1}^{l=\frac{N-1}{2}} \sum_{i=1}^{i=2} (\dot{Q}_{il}^2 + \omega_l^2 Q_{il}^2)$$

The sum is over the positive half of the Brillouin Zone; the $l = 0$ mode gives the translational energy.

We note P_{il} the conjugate to Q_{il}.

We know from Hamilton's equations that:

$$P_{il} = \frac{\partial T}{\partial \dot{Q}_{il}} = \mu \dot{Q}_{il}.$$

The classical Hamiltonian writes:

$$\mathrm{H} = \sum_{l=1}^{l=\frac{N-1}{2}} \sum_{i=1}^{i=2} \left(\frac{P_{il}^2}{2\mu} + \frac{\mu}{2} \omega_l^2 Q_{il}^2 \right)$$

The Schrödinger Hamiltonian writes:

$$\mathrm{H} = \sum_{l=1}^{l=\frac{N-1}{2}} \sum_{i=1}^{i=2} \left(\frac{\hbar^2}{2\mu} \frac{\partial^2}{\partial Q_{il}^2} + \frac{\mu}{2} \omega_l^2 Q_{il}^2 \right)$$

This Schrödinger equation is separable into N − 1 single particle wavefunctions, all of them satisfying:

$$\left(\frac{\hbar^2}{2\mu} \frac{\partial^2}{\partial Q_{il}^2} + \frac{\mu}{2} \omega_l^2 Q_{il}^2 \right) \Psi_{il} = E \Psi_{il}$$

These are Schrödinger equations of a uni-dimensional harmonic oscillator which eigenvalues are:

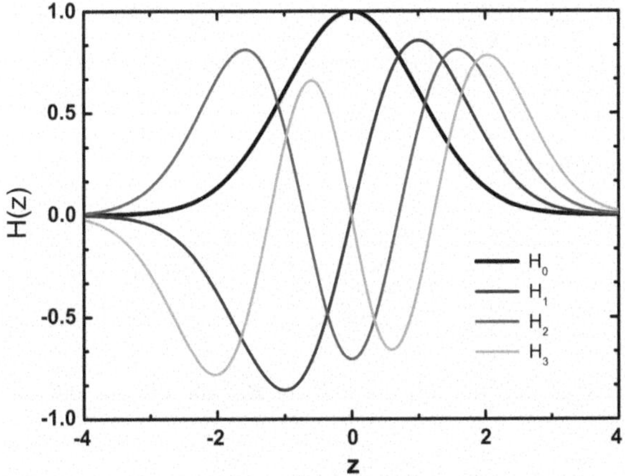

Fig. 3.19 Hermite functions H(z) of the four lowest quantum numbers plotted as a function of z

$E_{il} = \left(n_{il} + \frac{1}{2}\right) \hbar\omega_l$ with n_{il} being a quantum number and $n_{il} \in \mathbb{N}$.

The quantum description prescripts the ground state energy to be $\frac{1}{2}\hbar\omega$; which indicates that the crystal is not frozen at zero kelvin. This is an expression of the uncertainty principle.

The crystal is considered as a non-interacting gas of quanta of excitations of the normal coordinates. These quanta are called phonons and obey Bose-Einstein statistics.

The wave-function of the crystal is a product of Hermite functions of the normal coordinates. The Hermite functions are the solutions of the harmonic oscillator in quantum mechanics.

In Fig. 3.19 are plotted the four lowest energy Hermite functions H(z) as a function of an abstract parameter z.

The total energy, obtained by summing over the N − 1 vibrational modes, is:

$$E = \sum_{l \neq 0} \hbar\omega_l(n_l + 1/2)$$

The summation is not extended to the $l = 0$ (translational) mode of the crystal.

3.5.7 Phonons in Wurtzitic Crystals

• *Normal modes and symmetries of phonon frequencies in wurtzite semi-conductors*
To have an idea of the possible phonon modes and of their symmetry, it is necessary to consider the number of atoms in the unit cell. Then, since each atom A is allowed to move, its coordinates may be expressed within the unit cell, using a local cartesian

or orthogonal basis (A, x, y, z). If there are N atoms in the unit cell, there are 3N − 3 optical phonon modes and 3 acoustical phonon modes. The wurzite unit cell contains there are four atoms. Anions sit at $A_1 = (0, 0, 0)$, and $A_2 = a(1/\sqrt{3}, 0, c/(2a))$ in orthogonal coordinates while cations positions are $C_3 = a(1/\sqrt{3}, 0, c/(8a))$, and $C_4 = a(0, 0, 5c/(8a))$. We consider as a representation of the vibrations, the 12-fold basis $\{x_1, x_2, x_3, x_4, y_1, y_2, y_3, y_4, z_1, z_2, z_3, z_4\}$ to which we apply the symmetry elements of C_{6v}. Then, we calculate the character of the reducible representation and decompose it into the irreducible representations of C_{6v}.

The character table of $C_{6v}$is the following one:

Character table and basis functions for C_{6v}

C_{6v}	E	$\bar{E}$	$(C_2, \bar{C_2})$	$2C_3$	$2\bar{C_3}$	$2C_6$	$2\bar{C}_6$	$(3\sigma_d, 3\bar{\sigma}_d)$	$(3\sigma_v, 3\bar{\sigma}_v)$	basis functions
Γ_1	1	1	1	1	1	1	1	1	1	z
Γ_2	1	1	1	1	1	1	1	−1	−1	
Γ_3	1	1	−1	1	1	−1	−1	1	−1	$x^3 - 3xy^2$
Γ_4	1	1	−1	1	1	−1	−1	−1	1	$y^3 - 3yx^2$
Γ_5	2	2	−2	−1	−1	1	1	0	0	(x, y)
Γ_6	2	2	2	−1	−1	−1	−1	0	0	$\Gamma_3 \bigotimes \Gamma_5$
Γ_7	2	−2	0	1	−1	$\sqrt{3}$	$-\sqrt{3}$	0	0	$\|1/2, \pm 1/2\rangle$
Γ_8	2	−2	0	1	−1	$-\sqrt{3}$	$\sqrt{3}$	0	0	$\Gamma_3 \bigotimes \Gamma_7$
Γ_9	2	−2	0	−2	2	0	0	0	0	$\|3/2, \pm 3/2\rangle$

Group theory predicts that there are:

• Three acoustical modes that transform at Brillouin zone center according to the A_1 and E_1 phonon modes. We remind that the energies of these acoustical modes vanish at zone center.

• Nine optical modes that transform according to A_1, two B_1 , one E_1, and two E_2 phonons modes. The energies of such modes is not vanishing at the Brillouin zone center.

The normal coordinates of optical phonons vibration modes are given in Fig. 3.20 where they are plotted versus coordinates z and x (degeneracy of the E_1, E_2, and $E_2^{'}$ phonon modes is found by replacing x by y).

To measure the energies of the phonon modes, one generally perform two kinds of experiments:

• Infrared absorption or reflectivity experiments permit to probe phonon modes which produce a local modification of the crystal dipole proportional to the atomic displacement. Then, a dipolar-interaction may occur with the electromagnetic field at the exact phonon frequency. A photon of adapted energy may be absorbed by the crystal, producing a measurable variation of light attenuation or reflectance. One generally uses Fourier-Transform spectrometers to work at the wavelengths these processes occur. The electric field $\vec{E}$ produces a variation of the dipolar momentum $\vec{\mu}$ that is proportional to the atomic displacement. The Hamiltonian used to model this process has the symmetry of a polar vector. It transforms according to z or (x, y) for wurtzite; it is an odd parity interaction Hamiltonian in real space.

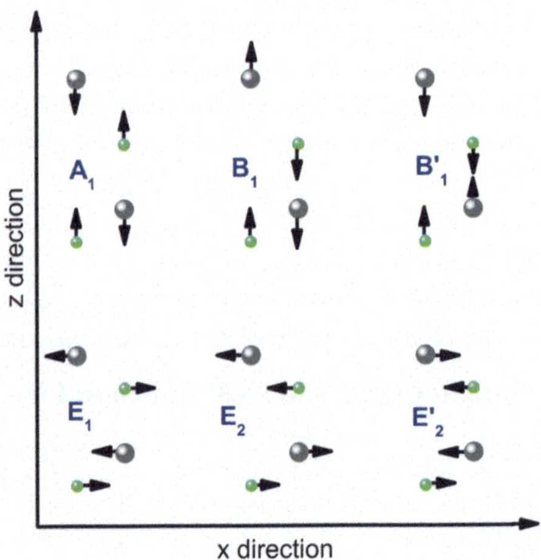

Fig. 3.20 Optical phonon modes of the wurtzite structure

• An inelastic scattering process, through a Raman experiment, can be produced by light-matter interaction in the crystal. Such experiments probe the aptitude of the electronic cloud to deform when submitted to the electric field of an incident photon. If so, part of the photon energy is transferred to the crystal (a phonon is created) and the rest of the electric energy is scattered out of the crystal, a photon with an energy smaller than the incident one can be detected.

The electric field $\vec{E}$ produces a variation of the dipolar momentum $\vec{\mu}$ that is proportional to the second-rank tenso $[\alpha]$:

$$\begin{bmatrix} \mu_x \\ \mu_y \\ \mu_z \end{bmatrix} = \begin{bmatrix} \alpha_{xx} & \alpha_{xy} & \alpha_{xz} \\ \alpha_{yx} & \alpha_{yy} & \alpha_{yz} \\ \alpha_{xz} & \alpha_{yz} & \alpha_{zz} \end{bmatrix} \cdot \begin{bmatrix} E_x \\ E_y \\ E_z \end{bmatrix}$$

Here, the interaction occurs via the second-rank tensor$[\alpha]$ and the possible symmetries of the interaction Hamiltonian are x^2, y^2, z^2, xy, yz, zx. This is an even parity Hamiltonian in real space.

At this stage, let us remind that the selection rules between an initial state $|i\rangle$ of symmetry Γ_i that may be reducible towards a final state $|f\rangle$ of symmetry Γ_f via an interaction Hamiltonian H_{int} of symmetry Γ_{int}, that may also be reducible, indicates that interaction is not vanishing under the condition that the product of the representations:

$$\Gamma_i * \bigotimes \Gamma_{int} \bigotimes \Gamma_f \subset \Gamma_1$$

where Γ_1 represents the fully-symmetric representation. It is noted preferentially A_1 by people studying the phonon frequencies.

Although infrared absorption (or reflectance experiments) couple initial and final states of different spatial parities, regarding a Raman experiment, there are many

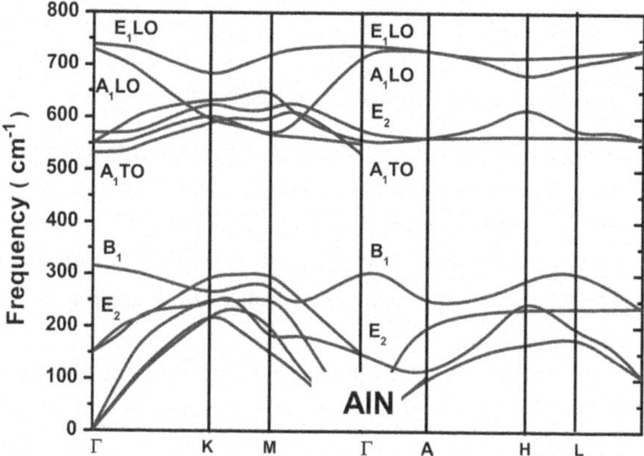

Fig. 3.21 Dispersion relations for the phonon modes of the wurtzite structure, after [2]

more possibilities: depending of the orientation of the electric field, the modification of the dipolar momentum may be forbidden of if allowed, may have an intensity depending of the values of the components of the second rank tensor $[\alpha]$. These values vary from a wurtzite semi-conductor to another.

The A_1 (polarized in the z direction) and E_1 (polarized in the (x,y) plane) modes are both Raman-active and infrared-active. The two E_2 modes are Raman-active only and the B_1 modes are silent.

Calculation of the phonon modes is very complicated and requires to solve the dynamical matrix in a three-dimensional space. Then, one obtains phonon three-dimensional dispersion relations. The dispersion relations for the phone modes in gallium nitride are plotted in Fig. 3.21.

3.5.8 Contribution of Phonons to the Dielectric Constant in Bulk Wurtzitic Semi-conductors

• Electromagnetic waves in solids

In vacuum, light propagates with vector $\vec{k}$ as a transverse electromagnetic wave, solution of the Maxwell equation $\vec{\nabla} \cdot \vec{E} = 0$, where $\vec{E}$ is the electric field and $\vec{E} \perp \vec{k}$.

In a real medium with a dispersive dielectric constant $\epsilon(\omega)$, the solution for the propagation becomes $\nabla \cdot \epsilon_0 \epsilon(\omega) \vec{E} = 0$. Apart from the transverse solution, there is a new solution that does not exist in vacuum namely $\epsilon(\omega) = 0$. Such solutions may be found at the frequencies ω_L where $\epsilon(\omega)$ vanishes.

$$\epsilon(\omega_L) = 0 \quad \text{and} \quad \vec{E} // \vec{k}.$$

Global consideration of Maxwell equations indicates that such longitudinal waves are not electromagnetic: $\vec{D} = \vec{0}$, $\vec{H} = \vec{0}$, $B = \vec{0}$, and $\vec{P} = -\epsilon_0 \vec{E}$. They are pure polarization waves with $\vec{E}$ and $\vec{P}$ opposite to each other.

- **Contribution of transverse waves to the dielectric contribution**

To account for the contributions of phonons in the framework of the classical description—Lorentz or Selmaier models—of the dielectric constant, one uses the classical equation of motion for a damped oscillator under a radiation field.

Let $\vec{E} = (E_x, 0, 0)e^{i(kz-\omega t)}$ this radiation field which propagates along z and is polarized along the x direction.

In this elementary model, the oscillations of the atom occur in the x direction, the atom mass is M, the phenomenological damping constant is γ and the restoring force constant is β. Let also $\omega_0^2 = \frac{\beta}{M}$ used to represent the phonon frequency squared.

We write the equation of motion of the damped oscillator under the electromagnetic field as:

$$M\ddot{x} - \gamma M\dot{x} + \beta x = eE_x e^{-i\omega t} \text{ which gives:}$$
$$(-M\omega^2 - i\gamma\omega M + \beta)x = eE_x \quad \text{and}$$
$$x = \frac{E_x}{M} e(\omega_0^2 - i\gamma\omega - \omega^2)^{-1}$$

The charge of the oscillator creates a dipole moment ex. If the density of oscillators is N_x, then the total polarization P_x of the crystal is $N_x ex$

$$P_x = \frac{N_x}{M} E_x e^2 (\omega_0^2 - i\gamma\omega - \omega^2)^{-1}$$

The dielectric displacement is $D_x = \epsilon_0 E_x + P_x = \epsilon_x(\omega) E_x$

$$D_x = \epsilon_0 [1 + \frac{N_x}{M} e^2 (\omega_0^2 - i\gamma\omega - \omega^2)^{-1}] E_x$$
$$\epsilon_x(\omega) = [1 + \frac{N_x}{M} e^2 (\omega_0^2 - i\gamma\omega - \omega^2)^{-1}]$$

In a three-dimensional situation, this model writes by analogy: $\vec{D} = \vec{\vec{\epsilon}}\vec{E}$ where tensor $\vec{\vec{\epsilon}}$ is the three-dimensional dielectric constant. In that case, the volume density of oscillator N replaces N_x.

One introduces the susceptibility tensor $\vec{\vec{\chi}} : \vec{\vec{\epsilon}} = 1 + \vec{\vec{\chi}}$.

Quantity $\frac{N}{M} e^2$ represents the magnitude of the coupling strength of the electromagnetic field to the oscillators. It is a quantity analogous to the oscillator strength in quantum mechanics. It is possible to write it $f_0 \omega_0^2$ by introducing a dimension-less quantity f_0 which simplifies the expression of $\epsilon(\omega)$:

$$\epsilon(\omega) = [1 + f_0 \omega_0^2 / (\omega_0^2 - i\gamma\omega - \omega^2)]$$

• Contribution of longitudinal modes to the dielectric contribution
The condition $\epsilon(\omega_L) = 0$ leads to equation:

$$1 + f_0\omega_0^2/(\omega_0^2 - i\gamma\omega_L - \omega_L^2) = 0$$

The solutions ω_L are the solutions of the second order algebraic equation:

$$\omega_L^2 + i\gamma\omega_L - \omega_0^2 - f_0\omega_0^2 = 0$$

In the asymptotic situation, when $\gamma \rightarrow 0$, convenient for handling calculations that indicate the physics behind equations:

$$\epsilon(\omega = \omega_T = \omega_0) \rightarrow \infty$$
$$\epsilon(\omega = \omega_L) = 0 \rightarrow \omega_L = \omega_0\sqrt{1 + f_0}$$

The splitting between the longitudinal and transverse frequencies is proportional to the oscillator strength:

$$\hbar(\omega_L - \omega_T) = \omega_0(\sqrt{1 + f_0} - 1)$$

The quantity $\hbar(\omega_L - \omega_T)$ is called Longitudinal-Transverse splitting energy Δ_{LT}.

The relationship between the oscillator strength and the energies of transverse and longitudinal modes is often expressed as:

$$\omega_L^2 - \omega_T^2 = f_0\omega_0^2$$

In the real case, there are several resonances and the dielectric function will write as a summation through the whole series of resonances as:

$$\epsilon(\omega) = \left[1 + \sum_j f_{0,j}\omega_{0,j}^2/(\omega_{0,j}^2 - i\gamma_j\omega - \omega^2)\right]$$

With the definitions of the parameters being obvious for each phonon.

In the vicinity of one resonance $\omega_{0,j}$, it is possible to neglect the contributions from all lower resonances and the constant contributions of all higher energy resonances can be summed in a so-called background dielectric constant ϵ_b.

In the spectral region around $\omega_{0,j}$, one can write:

$$\epsilon(\omega) = [\epsilon_b + f_{0,j}\omega_{0,j}^2/(\omega_{0,j}^2 - i\gamma_j\omega - \omega^2)]$$

Rescaling $f_{0,j}$ to $\phi_{0,j}$, one gets:

$$\epsilon(\omega) = \epsilon_b[1 + \phi_{0,j}\omega_{0,j}^2/(\omega_{0,j}^2 - i\gamma_j\omega - \omega^2)]$$

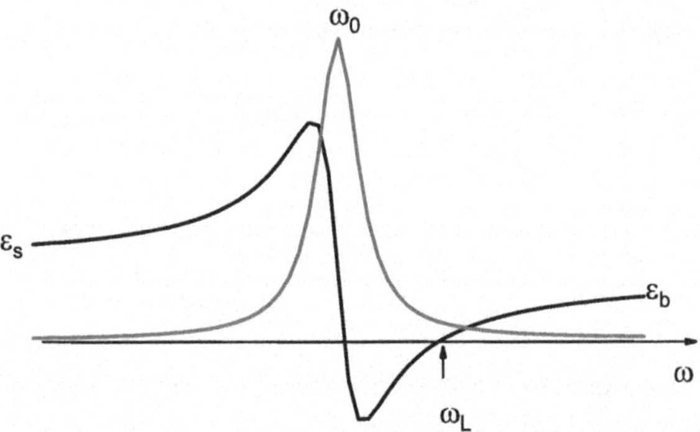

Fig. 3.22 Real and imaginary parts of the dispersive dielectric constant as given by the Drude-Lorentz model

The static constant ϵ_s is defined as the almost constant value of $\epsilon(\omega)$ below ω_T.

$$\epsilon_s = \epsilon_b(1 + \phi_0/\omega_0^2)$$

It leads us to formulate the so-called Lyddane-Sachs-Teller relation:

$$\epsilon_s/\epsilon_b = \omega_L^2/\omega_0^2$$

There are several conventions, and ϵ_s and ϵ_b are sometimes noted ϵ_0 and ϵ_∞ respectively in the literature. Then, the dielectric constant is given as:

$$\epsilon(\omega) = \epsilon_\infty[1 + (\omega_L^2 - \omega_T^2)/(\omega_T^2 - i\gamma\omega - \omega^2)]$$

and the Lyddane-Sachs-Teller relation rewrites:

$$\epsilon_0/\epsilon_\infty = \omega_L^2/\omega_T^2.$$

In Fig. 3.22, we have plotted the spectral dependence of the real and imaginary parts of the dielectric function near a resonance in a specific case when $\epsilon_s = \epsilon_s = 3$, $f_0 = 1$, and $\gamma = 0.5$. We have also indicated the longitudinal frequency ω_L when the dielectric constant passes from a negative value through zero back to positive values and the asymptotic values ϵ_0 and ϵ_∞ for the real part. At an angular frequency slightly higher than ω_L, the dielectric constant passes through unity. Then, the reflectivity is very close to zero. This frequency is called the Reststrahlen frequency ω_R.

$$\left(\frac{\omega_R}{\omega_T}\right)^2 = \frac{\epsilon_0 - 1}{\epsilon_\infty - 1}$$

• Selection rules

In quantum mechanics, the oscillator strength is given by the matrix element squared $|\langle i|H_{dip}|f\rangle|^2$ between an initial state $|i\rangle$ and a final state $|f\rangle$ through the action of the dipolar interaction, represented by Hamiltonian H_{dip} here.

The initial state is the ground—vacuum—state of the crystal. It transforms like A_1 (or Γ_1, depending on the notations), as indicated in the character table of C_{6v}.

The final sate $|f\rangle$ has the Γ_f symmetry of the phonon susceptible to be created under the action of the electromagnetic field which symmetry is noted $\Gamma_{H_{dip}}$.

We have shown, when discussing the contribution of electrons to the dielectric constant, that this quantity has the symmetry of a vector $\vec{r}$. In wurtzite, this symmetry $\Gamma_{\vec{r}}$ is reducible into $\Gamma_1 \bigoplus \Gamma_5$ (or $A_1 \bigoplus E_1$ depending of the notations). Let us now consider the selection rules. Equation $\Gamma_i * \bigotimes \Gamma_{int} \bigotimes \Gamma_f \subset \Gamma_1$ may be replaced by its equivalent:

$$\Gamma_i * \bigotimes \Gamma_{int} \subset \Gamma_f$$

For a photon polarized along the z direction, the symmetry Γ_{int} is A_1. The multiplication table of C_{6v} indicates that the only final state giving non-vanishing matrix element is A_1. The dielectric function writes:

$$\epsilon_{zz}(\omega) = [1 + f_{0,A_1}\omega_{0,A_1}^2/(\omega_{0,A_1}^2 - i\gamma_{A_1}\omega - \omega^2)]$$

For a photon polarized along the x direction, the symmetry Γ_{int} is E_1. The multiplication table of C_{6v} indicates that the only final state giving non vanishing matrix element is E_1. We remark the irreducible representation E_1 is two-fold (x and y are degenerated). The dielectric function writes:

$$\epsilon_{xx}(\omega) = \epsilon_{yy}(\omega) = [1 + f_{0,E_1}\omega_{0,E_1}^2/(\omega_{0,E_1}^2 - i\gamma_{E_1}\omega - \omega^2)].$$

3.5.9 Phonon Energies in Bulk Wurtzitic Semi-conductors

In Table 3.3 are summarized the phonon energies for some of the most studied wurtzitic semi-conductors. We emphasize that A_1 and E_1 modes correspond to atomic displacements where the pairs of group III- and group V- atoms move together in the same direction, each pair moving in the opposite direction with respect to the other one (as can be seen from Fig. 3.20). These modes are called polar modes and each one splits into propagation-parallel longitudinal (LO) and a propagation-perpendicular transverse (TO), due to the long-range macroscopic electric field. In general, the A_1 (*TO*) and E_1 (TO) modes and the E_1(LO) modes are active, propagating perpendicular to the c-axis, while A_1(LO) and E_1(*TO*) are active for the phonon traveling along the c-axis.

Table 3.3 Typical values of the phonon energies at zone center for nitride semi-conductors

Energy (cm^{-1})	GaN	InN	AlN
E_2 low	144	87	248
A_1 (TO)	533	447	614
E_1(TO)	561	472	673
E_2 high	569	488	660
A_1 (LO)	737	588	894
E_1 (LO)	743	593	917
B_1 low	334	225	547
B_1 high	690	576	719

Table 3.4 Typical values of the phonon deformation potentials in cm^{-1}

Phonon	a_{GaN}	b_{GaN}	b_{AlN}	b_{AlN}
E_2 low	115	−80	150	−227
A_1(TO)	−630	−1,290	−784	−390
E_1(TO)	−820	−680	−832	−746
E_2 high	−850	−920	−877	−911
A_1 (LO)	−685	−997	−743	−735
E_1 (LO)	−776	−704	−864	−809

3.5.10 Phonons in Strained Wurtzitic Semi-conductors

A strain field changes the positions of the atomic nuclei. In the linear strain limit, the phonon frequencies shift and split when two-fold degeneracies are lifted as:

$$\Delta\omega(A_1) = a(A_1)(e_{xx} + e_{yy}) + b(A_1)e_{zz}$$
$$\Delta\omega(B_1) = a(B_1)(e_{xx} + e_{yy}) + b(B_1)e_{zz}$$
$$\Delta\omega(E_{1,2}) = a(E_{1,2})(e_{xx} + e_{yy}) + b(E_{1,2})e_{zz} \pm c(E_{1,2})\left[(e_{xx} - e_{yy})^2 + 4e_{xy}^2\right]^{1/2}$$

The coefficients a(j), b(j) and c(j) are the corresponding phonon deformation potentials per unit strain and the $e_{\alpha\beta}$ are the components of the strain tensor.

The linear combinations $2a(j) + b(j)$ and $a(j) - b(j)$ directly give the effects of hydrostatic strain or axial strain parallel to the *c*-axis.

In the case of a biaxial strain, $e_{xx} = e_{yy}$ and $e_{xy} = 0$, then $\Delta\omega(j) = 2a(j)e_{xx} + b(j)e_{zz}$.

The effect of a strain field, which reduces the wurtzite symmetry to a symmetry sub-group of C_{6v}, lifts the two-fold degeneracy of the E_1 and E_2 modes as expected from group theory because all the irreducible representations are uni-dimensional.

In Table 3.4 are listed the values of the phonon deformation potentials for GaN and AlN.

3.5.11 Interaction of Phonons with Plasmons in Doped Wurtzitic Semi-conductors

When a semi-conductor is doped, collective oscillations of the electron gas occur. Plasmons are the quanta of oscillation of this electron gas. These oscillations can be described classically in the framework of a classical theory, similarly to phonons, as a function of their density n, but the electrons do not experience a restoring force and their mass is the effective mass m_{eff}. They oscillate at a plasma frequency:

$$\omega_P = \sqrt{\frac{ne^2}{\epsilon_0 m_{eff}}}$$

They create a polarization field and contribute to the effective dielectric constant as follows:

$$\epsilon_{eff} = 1 - \frac{\omega_P^2}{\omega(\omega + i\gamma_P)}$$

The electromagnetic field excites both electrons and phonons, and eventually the dielectric constant writes:

$$\epsilon(\omega) = \left[1 + \sum_j f_{0,j}\omega_{0,j}^2 / (\omega_{0,j}^2 - i\gamma_j\omega - \omega^2) - \frac{\omega_P^2}{\omega(\omega + i\gamma_P)} \right]$$

For the sake of a simplified calculation, we consider one phonon and neglect all damping parameters. The dielectric constant then writes:

$$\epsilon(\omega) = \epsilon_b \left[1 + \frac{\omega_L^2 - \omega_T^2}{\omega_T^2 - \omega^2} - \frac{\omega_P^2}{\omega^2} \right]$$

When solving $\epsilon(\omega) = 0$, one can couple the longitudinal phonon and longitudinal plasmon waves. This gives two new longitudinal waves with eigen frequencies ω_{L-} and ω_{L+} that are the following ones:

$$\omega_\pm^2 = \frac{\omega_L^2 + \omega_P^2}{2} \pm \left[\left(\frac{\omega_L^2 + \omega_P^2}{2} \right)^2 - \omega_L^4 \right]^{1/2}$$

they are clearly depart from the phonon and plasmon original energies with a non-linear dependence on the plasmon frequency, that is to say with the doping n.

Concerning the transverse waves, the dielectric constant is simply the superposition of the phonon and plasmon frequencies.

$$\epsilon(\omega) = \epsilon_b \left[1 + \frac{\omega_L^2 - \omega_T^2}{\omega_T^2 - \omega^2} - \frac{\omega_P^2}{\omega^2} \right]$$

The experimental investigation of the plasmon-LO phonon coupling is an interesting approach for determination of the residual doping. Not discussed here, but also of importance, is the evolution of the line shapes with doping. The real description requires, of course, to include dampings and their variations with doping as well as doping dependence of the oscillator strengths. Finally, the dependences of both phonon and plasmon modes with wave-vectors have to be included. All these implementations are obviously out of the scope of this book.

We would like to catch the opportunity to mention that such electron-phonon interaction has been predicted quite un-usually in graphene. The longitudinal plasmons couple only to transverse (TO) phonons while transverse plasmons couple only to LO phonons. The coupling between the longitudinal plasmon and the TO phonon is, for graphene, larger than the coupling between the transverse plasmon and the LO phonon [1].

Phonons are also interesting probes of strain fields. However, the sensivity of the LO phonon frequencies to both strain and doping does not make them the accurate strain probe. This argument also holds regarding the TO modes of the polar phonons: the oscillator strength (the LT splitting) being also doping-dependent, it is a second-order effect. Thus, people use as a strain probe, the frequency of the Raman active phonon mode E_2 and, therefore, a Raman scattering experiment to probe the strain.

Infrared reflectivity and Raman scattering measurements are thus very complementary characterization experiments.

References

1. M. Jablan, M. Soljacic, H. Buljan, Unconventional plasmon-phonon coupling in graphene. Phys. Rev. B **83**, 161409 (2011)
2. V. Yu Davydov et al., The wurtzite structure. Phys. Rev. B 58, 12899 (1998)
3. Energy Band Theory, by J. Callaway, Academic Press, New York, (1964).
4. *Electrons and Phonons*, by J. M. Ziman, Oxford University Press, (1979)
5. *Principles of the Theory of Solids*, by J. M. Ziman, Cambridge University Press, Cambridge, (1979), ISBN: 0521297338
6. Electronic States and Optical Transitions in Solids, by F. Bassani, G. Pastori Parravicini, Pergamon Press, Oxford, (1975).
7. Fundamentals of Semiconductors, *Physics and Materials Properties, by P* (Springer-Verlag, Y. Yu and M. Cardona, 1996). ISBN 3540614613
8. Solid State Physics, *by J* (Cambridge University Press, S. Blakemore, 1985). ISBN 0521313910
9. Optical Processes in Semiconductors, by J. I. Pankove, Dover Publications Inc., New York, (1975), ISBN: 9780486602752.
10. Quantum Theory of Solids, *by C* (John Wiley and Sons, Kittel, 1987). ISBN 0471624128
11. Elemental Theory of Angular Momentum, *by M* (John Wiley and Sons, E. Rose, 1957). ISBN : 0486684806
12. The Physics of Vibrations and Waves, *by H* (John Wiley and Sons, J. Pain, 1976). ISBN 0471994081

13. Symmetry and Strain-induced effects in semiconductors, *by G* (John Wiley and Sons, L. Bir and G. E. Pikus, 1974). ISBN 0470073217
14. Unified theory of symmetry-breaking effects on excitons in cubic and wurtzite structures, K. Cho, Phys. Rev. B 14, 4463, (1976).
15. Semi-empirical tight-binding band structures of wurtzite semiconductors, AlN, CdS, CdSe, ZnS, and ZnO, A. Kobayashi, O. F. Sankey, S. M. Volz, and J. D. Dow. Phys. Rev. B **28**, 935 (1983)
16. Consistent set of band parameters for the group, III nitrides AlN, GaN and InN, P. Rinke, M. Winkelnkemper, A. Qteish, D. Bimberg, J. Neugebauer, and M. D. Scheffler. Phys. Rev. B **77**, 075202 (2008)
17. All deformation potentials in GaN determined by reflectance spectroscopy under uniaxial stress: definite breakdown of the quasi cubic approximation, R. Ishii, A. Kaneta, M. Funato, and Y. Kawakami, Phys. Rev. B 81, 155202, (2010).
18. Complete set of deformation potentials for AlN determined by reflectance spectroscopy under uniaxial stress, R. Ishii, A. Kaneta, M. Funato, and Y. Kawakami, Phys. Rev. B 87, 235201, (2013).
19. Effective masses and valence-band splittings in GaN and, AlN, K. Kim, W. R. L. Lambrecht, B. Segall, and M. van Schilfgaarde. Phys. Rev. B **56**, 7363 (1997)
20. Elastic constants and related properties of tetrahedrally bonded, BN, AlN, GaN, and InN, K. Kim, W. R. L. Lambrecht, and B. Segall, Phys. Rev. B 53, 16310, (1996), Erratum: Elastic constants and related properties of tetrahedrally bonded BN, AlN, GaN, and InN by K. Kim, W. R. Lambrecht, and B. Segall. Phys. Rev. B **56**, 7018 (1997)
21. k.p method for strained wurtzite semiconductors, S. L. Chuang and C. S. Chang, Phys. Rev. B 54, 2491 (1996).
22. Analytical solutions of the block-diagonalized Hamiltonian for strained wurtzite semiconductors, M. Kumagai, S. L. Chuang, and H. Ando, Phys. Rev. B 57, 15303 (1998).
23. Crystal-orientation effects on the piezoelectric field and electronic properties of strained wurtzite semiconductors, S.-H. Park, and S. L. Chuang, Phys. Rev. B 59, 4725, (1999).
24. Phonon dispersion and Raman scattering in hexagonal GaN and AlN, V. Yu. Davydov, Yu. E. Kitaev, I. N. Goncharuk, A. N. Smirnov, J. Graul, O. Semchinova, D. Uffmann, M. B. Smirnov, A. P. Mirgorodsky, and R. A. Evarestov, Phys. Rev. B 58, 12899, (1998).
25. Piezospectroscopic studies of the Raman spectrum of Cadmium Sulfite, R. J. Briggs, and A. K. Ramdas, Phys. Rev. B 13, 12899, (1976)

Chapter 4
Optical Properties of Wurtzitic Semiconductors and Epilayers

We review the optical properties of bulk wurtzitic semiconductors. Reflectance and photoluminescence features are related to the crystal symmetry; selection rules for the optical process are introduced as well as the concepts of fee excitons and exciton-polaritons. The problem of the long-range Coulomb interaction in anisotropic uniaxial crystals is treated. Finally we discuss photoluminescence as an efficient tool for diagnosing doping in bulk semiconductors.

4.1 Pioneering Reflectivity Experiments on Cadmium Sulfite

4.1.1 The Valence Band Ordering in CdS

The optical properties of wurtzitic bulk samples were established a long time ago by means of reflectivity and using selection rules predicted by group theory. In Fig. 4.1 is plotted the 77 K reflectivity spectra collected in case of bulk CdS samples [2] for $\vec{E}//\vec{c}$ (left-hand side) and $\vec{E}\perp\vec{c}$ (right-hand side) polarizations of the incident photon. The experiment was performed in conditions of normal incidence of the incident beam ($\vec{k}_{photon}//\vec{c}$). The low energy feature is not observed for the $\vec{E}\perp\vec{c}$ condition. The energies of these features have been attributed to the signatures of optical transitions corresponding to the lowest band gap transitions at the zone center of the Brillouin zone.

By the help of the selection rules and using group theory arguments Thomas and Hopfield assigned to the three valence band of CdS a given symmetry and proposed relative orderings in terms of energy. It is established that the conduction band is a s-like state (even space parity) and that it has a Γ_7 symmetry. A photon polarized along the $\vec{c}$ direction has Γ_1 symmetry. A dipole-active transition may be produced by such photon from an even parity state of Γ_7 symmetry to an odd parity one of Γ_7 symmetry too. A photon polarized perpendicular to the $\vec{c}$ direction, with $\vec{k}_{photon}//\vec{c}$ has Γ_5 symmetry. A dipole-active transition may be produced by such photon from

B. Gil, *Physics of Wurtzite Nitrides and Oxides*, Springer Series in Materials Science 197, DOI: 10.1007/978-3-319-06805-3_4,

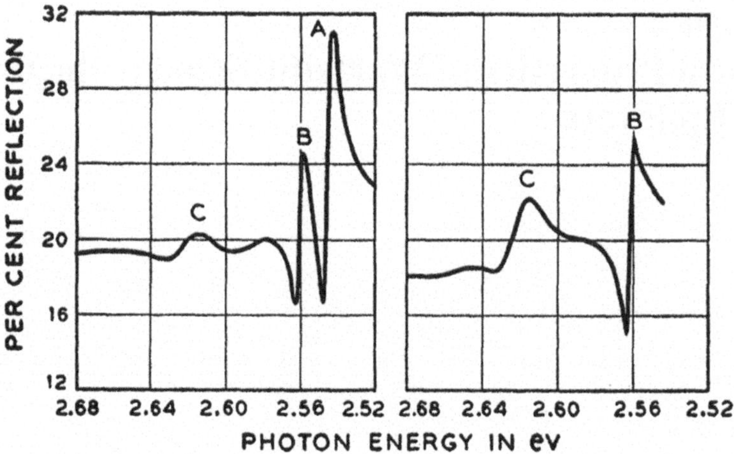

Fig. 4.1 The 77 K reflectivity spectra collected in case of bulk CdS samples [2] for $\vec{E}//\vec{c}$ (*left-hand side*) and $\vec{E}\perp\vec{c}$ (*right-hand side*) polarizations of the incident photon

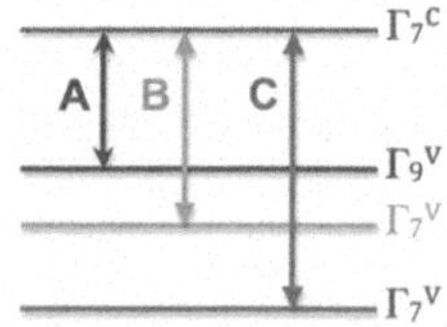

Fig. 4.2 Conduction and valence symmetries and their spectral ordering as proposed for bulk CdS samples [2]

an even parity state of Γ_7 symmetry to an odd parity one of Γ_9 symmetry. Therefore selection rules enable to discriminate between the three valence band and Thomas and Hopfield proposed the valence band ordering described in Fig. 4.2. They called the three transition A, B, and C in terms of increasing energy.

When reducing the crystal temperature to 4.2 K by immersion of the sample in liquid helium, the transition energies are slightly blue shifted, and the selection rules observed at 77 K are kept. The value of transition A is 2.554 eV. Values of transitions B and C are 2.570 and 2.632 eV respectively. The value of the A–B splitting is 16 meV, the value of the A–C splitting is 98 meV.

Some new transitions were found as seen in Fig. 4.3. These transitions follow the previously observed ones and are labeled A′, B′. They are split by 21 meV from their respective low energy associate, and are about a tenth in terms of relative intensities. These complementary transitions are the signatures of excited state excitonic transitions. An exciton as rapidly mentioned in Chap. 3, as we shall describe it with more details later, at the end of the present one, is an elementary excitation of the crystal that results of the long range Coulomb interaction between a photo-created free electron in the conduction band and the positive charge left (a hole) in the valence band. In the simplest picture it can be naively viewed as a pseudo hydrogen atom (described using an effective reduced mass μ) moving in the crystal (medium of dielectric constant ϵ)

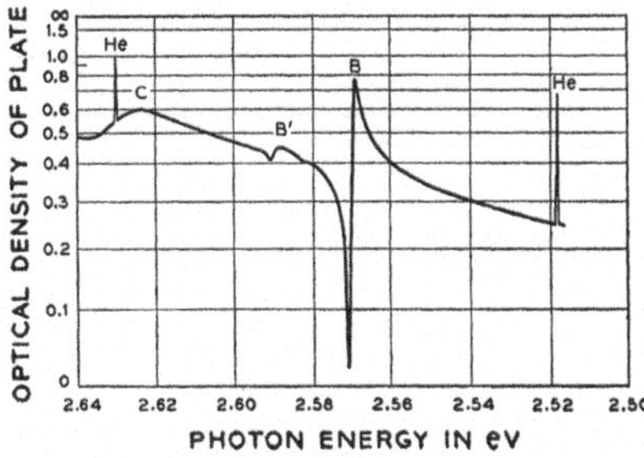

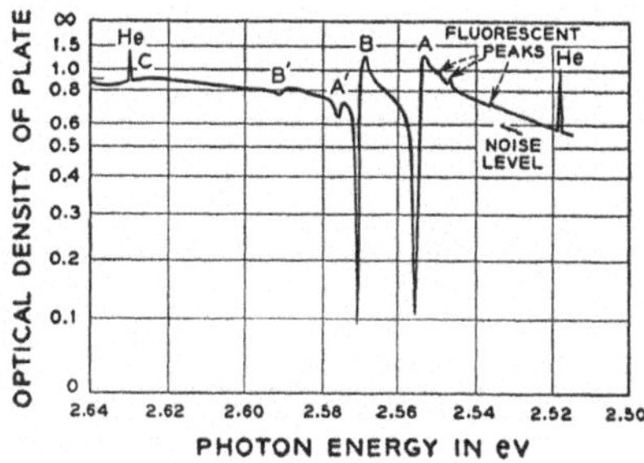

Fig. 4.3 The 4.2 K reflectivity spectra collected in case of bulk CdS samples [2] for $\vec{E}//\vec{c}$ (*left-hand side*) and $\vec{E}\perp\vec{c}$ (*right-hand side*) polarizations of the incident photon

with a translational mass M. The energy spectrum of this analogous of the hydrogen atom is described in terms of both bound states (a series of levels which spectrum is following the Rydberg series $E(n) \propto \nu^{-2}$ in the one hand, and intensities vary like ν^{-3}) and unbound states (a continuum of states). From the 21 meV splitting between A and A$'$ or B and B$'$, one can deduce an effective Rydberg energy of about 28 meV for both A-related excitons (Γ_9 valence band) and B-related (first Γ_7 valence band).

The measured value of the band to band energy gap of CdS is 2.554 eV at 4.2 K. One always measure an excitonic energy in undoped semiconductors of reasonable crystaline quality. The value of the band to band energy gap has to be enhanced by a quantity that equals the exciton binding energy. It is 2.582 eV at 4.2 K for CdS.

By modeling the dielectric constants using a series of resonances centered at the transition energies of the reflectivity structures, Thomas and Hopfield obtained the series of oscillator strengths for all these transitions in both polarizations that are expressed in Table 4.1 in terms of the intensity of line A.

The value of the splittings between the Γ_9 valence band and the Γ_7 valence bands are obtained by means of optical spectroscopy. They are deduced from measurements of *excitonic energies*. These excitonic energies are the values of a band to band energy gap reduced from an exciton binding energy. In CdS, the measured value of the *excitonic binding energy* deduced from splittings between reflectivity features of 1 and 2 s excitons is 28 meV. This value is larger than the experimental value of the A–B experimental splitting (16 meV) and it has obviously an influence concerning our knowledge of the splittings between the Γ_9 valence band Γ_7 valence bands. These splittings are defined out of the context of the long range excitonic interaction and are measured within its context. The valence band parameters that are the most commonly accepted values for bulk CdS have been given Table 2.1. They are $\Delta_1 = 28$ meV; $\Delta_2 = 22$ meV; $\Delta_3 = 28$ meV. We refer to most accepted values since the values of lattice parameters of materials can vary with doping or interaction with a foreign substrate, ... leading to different values of the bandgap.

Table 4.1 Values of the ratio c/a for various wurtzitic semiconductors

Transition	A (n = 1)	B (n = 1)	A′ (n = 2)	B′ (n = 2)	C (n = 1)
E ⊥ c	1	0.555	0.058	0.055	0.31
E // c	0	0.62	0	0.05	0.67

The number of digits is compatible with the determination method

4.1.2 Excitons and Polaritons in CdS

4.1.2.1 Spinless Excitons

The concept of exciton was invented by Peierls, Frenkel, Wannier and Mott in the early days of the 1930 s. It is a quantum of electronic excitation energy traveling with wave vector $\vec{K}$ through the crystal. The crystal being a periodic structure, the motion of the traveling exciton is characterized by a wave vector. Local deviations of the crystal structure from a perfectly periodic one may trap this traveling exciton. From free, it then becomes localized. The free exciton may be also scattered by repulsive potentials, like for instance the semiconductor to air interface.

The exciton is created by the long-range Coulomb interaction of an electron (a negative charge in the conduction band) and a hole (a missing electron in the valence band). The spinless Schrödinger equation permits at first order to describe these eigenstates. It may be written as follows:

$$\left[-\frac{\hbar^2}{2m_0M_h}\nabla_R^2-\frac{\hbar^2}{2m_0m_e}\nabla_r^2-\frac{e^2}{4\pi\epsilon\epsilon_0}\frac{e^2}{|\vec{r_e}-\vec{r_h}|}\right]\Phi_n(\vec{r_e},\vec{r_h})=E_n.\Phi_n(\vec{r_e},\vec{r_h})$$

One then defines a relative coordinate $\vec{r}$ to describe the relative motion of the electron and the hole and the coordinates $\vec{R}$ of the center of mass:

$$\vec{r}=\vec{r}_e-\vec{r}_h$$

and

$$\vec{R}=(m_e\vec{r}_e+m_h\vec{r}_h)/(m_e+m_h)$$

Further defining the translation mass $\mu^{-1}=m_e^{-1}+m_h^{-1}$ and the reduced mass μ as

$$\mu^{-1}=m_e^{-1}+m_h^{-1},\text{ n}$$

One can write the wave functions: $\Phi_n(\vec{r}_e,\vec{r}_h)=\Phi_\nu(\vec{r})\cdot\Psi_\beta(R)$ and two spinless Schrödinger equations:

$$\left[-\frac{\hbar^2}{2m_0M}\nabla_R^2\right]G_\beta=\epsilon_\beta G_\beta$$

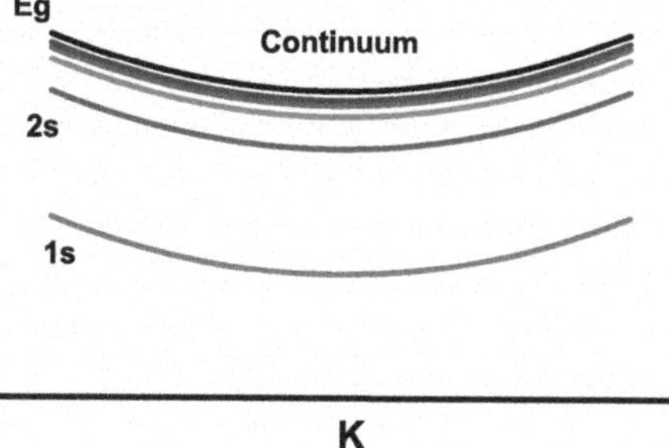

Fig. 4.4 Plot of the parabolic dispersion relations for the center of mass of the exciton for different excitonic levels

$$\left[-\frac{\hbar^2}{2m_0\mu}\nabla_r^2 - \frac{e^2}{4\pi\epsilon\epsilon_0}\frac{e^2}{|\vec{r}|}\right]\Psi_\nu(\vec{r}) = E_\nu \cdot \Psi_\nu(\vec{r})$$

The eigen values are in terms of the universal constants:

$$\epsilon_\beta = (\hbar^2/2m_0M)K^2$$

for the motion of the center of mass and

$$E_\nu = \frac{1}{\nu^2}\frac{m_0c^2\alpha^2}{2}\frac{\mu}{\epsilon^2}$$

for the bound state of the pseudo hydrogen atom.

The wave function $G_\beta(\vec{R})$ is a plane wave normalized at the scale of the size of crystal:

$G_\beta(\vec{R}) = Cte \cdot e^{i\vec{K}\vec{R}}$;

$\vec{K}$ is the quantum number ($\vec{K}$ is the sum of the electron and hole wave numbers $\vec{k_e}$ and $\vec{k_h}$).

The eigen states $\Psi_\nu(\vec{r})$ are the solutions of the classical spinless hydrogen atom. The total energy of the interacting spinless electron and hole pair is:

$$E_{\nu,\vec{K}} = \frac{1}{\nu^2}\frac{m_0c^2\alpha^2}{2}\frac{\mu}{\epsilon^2} + (\hbar^2/2m_0M)K^2$$

In Fig. 4.4 are plotted the dispersion relations of the center of mass of the electron-hole pair for several exciton states versus K.

Important note: Wave vectors of electron and holes $\vec{k_e}$ and $\vec{k_h}$ belong the to energy band structure scheme. Exciton center of mass motion is represented by a wave number $\vec{K}$ that does not belong to the energy band structure dispersion scheme.

The spectroscopy of the various excitonic levels roughly follows the classical model of the hydrogen atomic terms of energy distribution ($1/\nu^2$ law) and oscillator strength: the s states, the only states for which $|\Psi_\nu(\vec{0})|^2 \neq 0$ couple to the electro-magnetic field with an efficiency that varies like $|\Psi_\nu(\vec{0})|^2 \sim 1/\nu^3$. It is important to remark that the dispersion relations of the energy bands of the wurtzitic materials are anisotropic as well as the dielectric constant. This drastically complicates the calculation of the hydrogen levels, but as we shall see it in Sect. 4.6. Both anisotropies (band structure and dielectric constant) tend to compensate each other. The net result of axial symmetry is to lift the degeneracy of levels in terms of the value of angular moment in the one and to couple states of identical spatial parity and identical values of the squared projection of angular momentum. Before to consider this it is much more relevant to introduce spin effects.

4.1.2.2 Spin-Exchange Interaction and Internal Structure of the Exciton

We restrict ourselves here to the relative motion of the electron and hole pair. The electron wave functions expresses through conduction states, it is a s-type level, its symmetry is Γ_1 in the spinless description; Γ_7 when spin is included. The p-type hole wave function has Γ_{hole} symmetry that is reducible as $\Gamma_1 \oplus \Gamma_5$ in the frame work of a spin less description. It is reducible as $\Gamma_9 \oplus \Gamma_7 \oplus \Gamma_7$ when spin is included.

The symmetry of an exciton state is written in the most general way in wurtzitic crystals for which all irreducible representations have real characters:

$$\Gamma_{hole} \otimes \Gamma_{electron} \otimes \Gamma_{hydrogenic}$$

Let us consider the s-like hydrogenic (envelope) states which symmetries are Γ_1. The other ones are not couple with the electromagnetic field in the absence of symmetry breaking perturbation. They are therefore of little interest for us here. Multiplication table of C_{6v} indicates to us that the symmetry reduces to:

$$\Gamma_{hole} \otimes \Gamma_{electron} = \Gamma_6 \oplus 2\Gamma_2 \oplus 2\Gamma_1 \oplus 3\Gamma_5$$

The construction of the exciton states is sketched in Fig. 4.5.

The radiative transitions between these excited states of the crystal and the crystal ground state (vacuum state, state without exciton, state of Γ_1 symmetry) are allowed from the two Γ_1 excitons in π ($\vec{E}//\vec{c}$) polarization, while transitions from the three Γ_5 excitons to the ground state are allowed in σ polarization ($\vec{E}\perp\vec{c}$). Other excitonic levels are silent. These selection rules are consistent with the report of Thomas and Hopfield in their polarized reflectivity experiments. However, the transition energies measured in both polarizations should be energy splitted by a quantity that accounts for the existence of an electron spin and of a hole one. This quantity is the short-range exchange interaction.

To go further, one has to compute the excitonic wave functions from the electron and hole ones. Let σ_e and σ_h the electron and hole spin operators. We write the spin exchange term as the following operator:

$$H_{exch} = 1/2\gamma \cdot \vec{\sigma}_e \cdot \vec{\sigma}_h$$

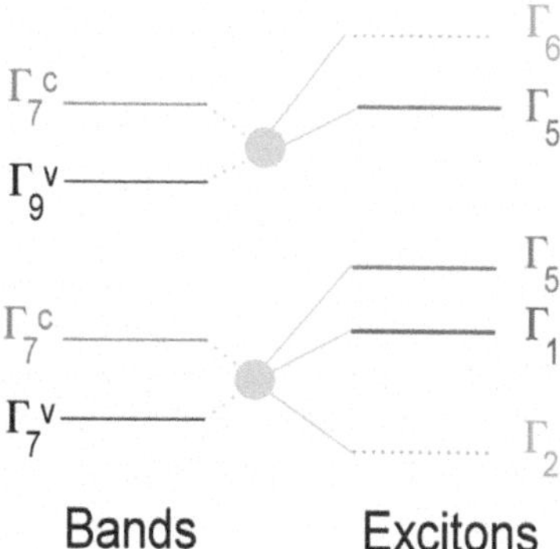

Fig. 4.5 Construction of the exciton states from the valence and conduction band

Table 4.2 Eigenvectors for excitons in wurtzite. p_+ represents $-(p_x + ip_y)/\sqrt{2}$ and p_- represents $(p_x - ip_y)/\sqrt{2}$

State	Basis functions
Γ_1	$(p_- \uparrow \beta + p_+ \downarrow \alpha)/\sqrt{2}$
Γ_1'	$p_z(\downarrow \beta + \uparrow \alpha)/\sqrt{2}$
Γ_2	$(p_- \uparrow \beta - p_+ \downarrow \alpha)/\sqrt{2}$
Γ_2'	$(\downarrow \beta - \uparrow \alpha)/\sqrt{2}$
Γ_6	$(p_+ \uparrow \beta, p_+ \downarrow \alpha)$
Γ_5	$(p_+ \uparrow \alpha, p_+ \downarrow \beta)$
Γ_5'	$(p_+ \downarrow \beta, p_+ \uparrow \alpha)$
Γ_5''	$(p_z \downarrow \alpha, p_z \uparrow \beta)$

The actions of operators $\vec{\sigma_e}$ and $\vec{\sigma_h}$ on spins have been detailed in Chap. 3 in the section relative to spin orbit interaction (Sect. 2.3). We emphasize the fact that the exchange interaction term has been expressed here using ONE phenomenological constant γ although the z direction should be distinguished from the (x, y) couple in a more complete description using two parameters.

$$H_{exch} = 1/2[\gamma_0 \cdot \sigma_{ez} \cdot \sigma_{hz} + \gamma_1(\sigma_{ex} \cdot \sigma_{hx} + \sigma_{ey} \cdot \sigma_{hy})]$$

In Table 4.2 are given the excitonic wave functions in terms of p-type states, electron spin and missing valence electron spin. The electron spin are expressed as $\uparrow$ and $\downarrow$ corresponding to electron spin eigenvalues 1/2 and $-1/2$ respectively. The states representing the valence missing electron are expressed using α and β corresponding to missing valence electron spin eigenvalues 1/2 and $-1/2$ respectively. The interest to us this concept is to construct exciton states in a representation for which the radiative exciton states are not spin flip states.The spin of the valence missing electron is obtain by application of operator $-i\sigma_y$ to the hole spin.

The matrix representation of the Γ_5 eigentates of the exciton writes:

$$\begin{matrix} |\Gamma_5> & |\Gamma_5'> & |\Gamma_5''> \\ \Delta_1+\Delta_2-\gamma/2 & -\gamma & 0 \\ -\gamma & \Delta_1-\Delta_2-\gamma/2 & \sqrt{2}\Delta_3 \\ 0 & \sqrt{2}\Delta_3 & \gamma/2 \end{matrix}$$

The exciton wave function writes: $\Psi(\Gamma_5) = \mathrm{a}|\Gamma_5 > +\mathrm{b}|\Gamma_5' > +c|\Gamma_5'' >$.

The oscillator strength (in σ polarization) is the square of the sum of the coefficients of the non spin-flip basis vector.

$$|\Omega(\Gamma_5)|^2 = (a+b)^2/2$$

The matrix representation of the Γ_1 eigentates of the exciton writes:

$$\begin{matrix} |\Gamma_1> & |\Gamma_1'> \\ \Delta_1-\Delta_2+\gamma/2 & \sqrt{2}\Delta_3 \\ \sqrt{2}\Delta_3 & -3\gamma/2 \end{matrix}$$

The exciton wave function writes: $\Psi(\Gamma_1) = \mathrm{a}'|\Gamma_1 > +\mathrm{b}'|\Gamma_1' >$.

The oscillator strength (in π polarization) is the coefficient of the non spin-flip basis vector.

$$|\Omega(\Gamma_1)|^2 = \mathrm{b}'^2$$

The matrix representation of the Γ_2 eigentates of the exciton writes:

$$\begin{matrix} |\Gamma_2> & |\Gamma_2'> \\ \Delta_1-\Delta_2+\gamma/2 & \sqrt{2}\Delta_3 \\ \sqrt{2}\Delta_3 & +\gamma/2 \end{matrix}$$

Eigenstates of the two fold Γ_6 exciton is $Delta_1 + \Delta_2 + \gamma/2$.

The oscillator strength of the excitons varies like varies like $|\Psi_\nu(\vec{0})|^2 \sim 1/\nu^3$. The exchange interaction varies, very roughly speaking like $|\Psi_\nu(\vec{0})|^4 \sim 1/\nu^6$.

This interaction has been measured from the splitting between Γ_1 and Γ_5 excitations in CdS. A value of 2.5 eV was found.

4.1.2.3 Polaritons

The free exciton travels through the crystal with wave vector $\vec{K}$. Its energy is then dispersive and parabolic in K. We write it:

$$E(\vec{K}) = \hbar\omega_{exc} + (\hbar^2/2m_0M)K^2$$

where $\hbar\omega_{exc} == E_g - E(\nu, \vec{0})$ is the energy of the excitonic resonance of quantum number ν. If one assimilate the exciton to a harmonic oscillator,by analogy with the situation of the contributions of the phonon resonances to the dielectric constant, the full series of s-like exitonic resonances will contribute to the dielectric constant, with selection rules. The Γ_5 excitons contribute in σ polarization, the Γ_1 excitons contribute in π polarization.

If one considers one resonance, for the transverse wave $(\vec{K} \cdot \vec{E} = 0)$:

$$K^2\vec{E} = \omega^2/c^2\vec{D}$$

and

$$\epsilon(\omega, \vec{K}) = c^2K^2/\omega^2$$

giving

$$\epsilon(\omega, \vec{K}) = \epsilon_\infty + 4\pi\alpha_0\omega_{exc}^2/\left[(\hbar\omega_{exc} + \hbar^2K^2/(2m_0M) - \omega)^2 - i\gamma(K)\omega\right]$$

where $4\pi\alpha_0(K)$ is the oscillator strength and $\gamma(K)$ is the damping constant (also called radiative homogeneous broadening parameter).

We suppose here the oscillator strength and homogeneous broadening to be K-independent and further choose $\gamma = 0$ so that we can get the analytical forms of the dispersion relations $\omega(K)$ for the two exciton polariton branches at the second order in K, that is to say neglecting the terms proportional to K^4. We obtain:

$$\omega_{1,2}^2(K) = \left[(\epsilon_\infty + 4\pi\epsilon_0)\omega_{exc}^2 + (\hbar\omega_{exc}/m_0M + c^2)K^2\right]/2\epsilon_\infty$$
$$\pm \left\{\left[(\epsilon_\infty + 4\pi\epsilon_0)\omega_{exc}^2 + (\hbar\omega_{exc}/m_0M - c^2)K^2\right]^2 + 16\pi\alpha_0\omega_{exc}^2c^2K^2\right\}^{1/2} /2\epsilon_\infty$$

These equations indicate that the propagations of the exciton and of the photon in the crystal are not independent but coupled in the exciton polariton scheme. the dispersion relations are not parabolic with $\vec{K}$.

For the longitudinal wave $(\vec{K} \times \vec{E} = \vec{0})$

$$\vec{D} = \vec{0}$$

and

$$\epsilon(\omega, \vec{K}) = 0$$

$$\omega(K) \approx \omega_{exc}\sqrt{(1 + 4\pi\alpha_0/\epsilon_\infty)} + \hbar/(2m_0M)K^2$$

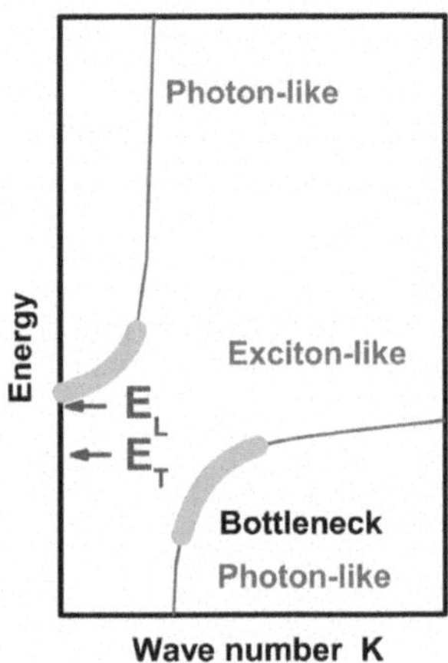

Fig. 4.6 Dispersion relation for the Transverse exciton polariton waves. The Transverse and Longitudinal energies are indicated as $E_T$and E_L respectively. The regions of maximum coupling (which will contribute to the photoluminescence are bolded)

The dispersion relation for the transverse exciton polariton states are sketched in Fig. 4.6.

When $M \to \infty$, and at K = 0, the two frequencies are given by:

$$\omega_1(0) = 0$$

and

$$\omega_2(0) = \omega_L(0) = \omega_{exc}\sqrt{(1 + 4\pi\alpha_0/\epsilon_\infty)}$$

The frequency of the transverse wave is:

$$\omega_T(0) = \omega_{exc}$$

The longitudinal-transverse splitting is proportional to the oscillator strength:

$$\omega_L(0) = \omega_T(0) = \omega_{LT} \simeq 2\pi\alpha_0\omega_{exc}/\epsilon_\infty$$

Alternatively one can express the oscillator strength versus the longitudinal-transverse splitting:

$$\alpha_0 \approx \epsilon_\infty\omega_{LT}/(2\pi\omega_{exc})$$

In a real situation one cannot assume that $\gamma \to 0$. Then the dielectric function has both real and imaginary parts. From a line shape fitting of the reflectivity near the exciton resinances, one can get the oscillator strength, the damping parameter, the value of the translational mass M, the value of ϵ_∞ (see for instance [3]). The value of the longitudinal transverse splitting for A exciton-polariton in CdS is 2.2 meV. regarding B exciton-polariton it is 1.73 meV for the Γ_1 exciton polariton and it is 1.39 meV for the Γ_5 exciton polariton as reported in [5]. Note that they took

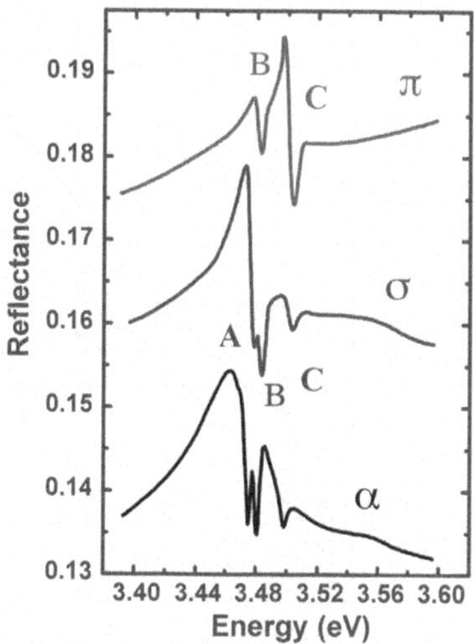

Fig. 4.7 Reflectance properties of bulk GaN at 2 K for various polarization conditions. After [6]

$\epsilon_\infty = 7.4$ in their paper. In fact, from the quality of the semiconductor surface results the average value of the reflectance and that $\epsilon\infty$ is often used as a fitting parameter which is chosen so that it fits the value of the reflectivity at energies below the resonance energy.

4.2 Optical Reflectivity in Gallium Nitride

4.2.1 The Pioneering Work

In Fig. 4.7 have been plotted some reflectivity spectra taken on GaN crystal at the beginning off of the 1970s. One clearly identifies the transition involving the Γ_9 valence band which is the ground state transition. One also remarks that the intensities of transitions B and C involving the Γ_7 valence bands are observed with different intensities in σ and π polarizations as an evidence of the different weights of p_x, p_y and p_z Bloch states in the composition of the wave functions of Γ_7 symmetry. These relative intensities permit to estimate the average values of the crystal field splitting parameter and of the spin-orbit interaction parameter as computed in Chap. 3, Sect. 2.3.

4.2.2 Strain-Fields in (0001) Epilayers

In Fig. 4.8a have been represented the evolutions of these energies of A, B, and C excitons measured at low temperature on a large variety of different substrates: Silicon, (0001)-oriented sapphire, (001)-oriented 6H-SiC, (0001)-oriented ZnO.

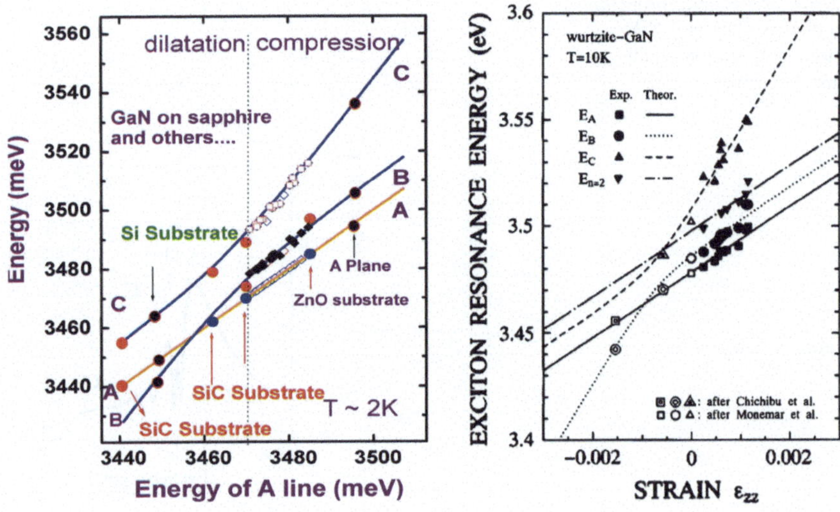

Fig. 4.8 *Left* plot of the transition energies of GaN epilayers grown on a large variety of foreign substrates versus energy of A-line. *Right* plot of the transition energies in GaN epilayers versus deformation in the c direction. After [7]

To demonstrate the correlation between these energies they have been plotted respectively to the energy of line A which involves a hole of Γ_9 symmetry is not coupled to the other valence bands. The A line shifts linearly under a (0001) strain-field. Anisotropy of the optical response in the growth plane was not reported as an indication of the conservation of the C_{6v} symmetry.

In Fig. 4.8b have been represented the evolutions of these energies of A, B, and C excitons measured at low temperature on a large variety of different substrates as a function of the deformation along the growth direction measured by x-diffraction. Both figures are compatible to each other as an indication of important strain fields in nitride epilayers grown on foreign substrates. For the (0001)-oriented growth, as indicated in Chap. 3, the Γ_9 valence band shifts linearly under a (0001) strain-field. The origin of the different strain fields is tricky. For hetero-epitaxies it includes the contribution of the interfacial region between the nitride layer and its foreign substrate, and more dominantly the contributions of the differences between the dilatation coefficients of the epilayer and those of the substrate. Last the thickness of the film is also impacting since interfacial strain relaxes progressively. These differences in dilatation coefficients partly explain why growth of SiC leads to a dilatation while growth on sapphire leads to biaxial compression.

In Fig. 4.9a and b have been plotted some typical low temperature reflectivity experiments recorded on GaN epilayers (0001)-grown on SiC and sapphire in case of $\vec{E}\perp\vec{c}$ polarizations of the incident photon in conditions of normal incidence of the incident beam ($\vec{k}_{photon}//\vec{c}$). As indicated in Fig. 4.9a, the excitonic energies are down shifted compared to bulk GaN when growth occurs on 6H-SiC. Simultaneously the

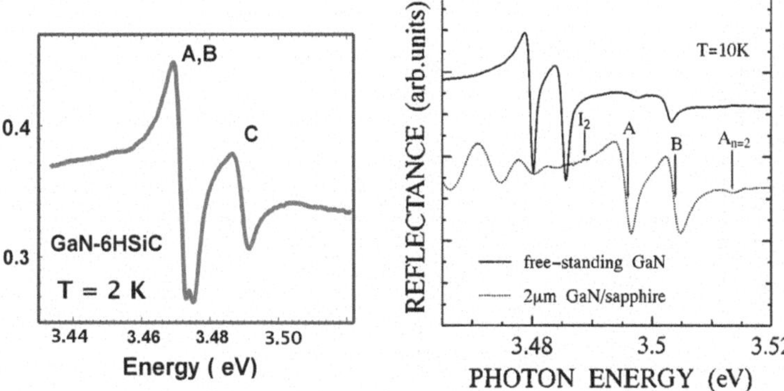

Fig. 4.9 *Left* low temperature reflectivity of a GaN epilayer grown on a SiC. The polarization of the photon is $\vec{E} \perp \vec{c}$ and $(\vec{k_{photon}} // \vec{c})$. Note the strong C line and the overlap of A and B related feature. *Right* reflectivity feature of a GaN epilayer grown on C-plane sapphire. The strong biaxial compression blue chift the overall spectrum with repeat to the one typical of bulk unstrained GaN. Note the weakness of the C line for this strain and this polarization of the photon. $\vec{E} \perp \vec{c}$ and $(\vec{k_{photon}} // \vec{c})$. Also note the observation of excited exciton states [8]

intensity of line C is large compared with Fig. 4.7 (middle spectrum). The dilatation reduces the splitting between (p_x, p_y) and p_z states (components e_{xx} and e_{yy} of the strain tensor are increased, component e_{zz} is decreased), dramatically increasing the amount of (p_x, p_y) Bloch states in the hole wave function of line C, thus drastically increasing its oscillator strength in the conditions of the experiment.

As indicated in Fig. 4.9b, the excitonic energies are blue-shifted compared to bulk GaN when growth occurs on sapphire. Simultaneously the intensity of line C is small compared with Fig. 4.7. The compression increases the splitting between (p_x, p_y) and p_z states (components e_{xx} and e_{yy} of the strain tensor are decreased, component e_{zz} is increased), dramatically decreasing the amount of (p_x, p_y) Bloch states in the hole wave function of line C, thus drastically decreasing its oscillator strength in the conditions of the experiment. In Fig. 4.10a are plotted the intensities of A, B and C lines versus position of the A line in GaN biaxially strained by growth in the (0001)-oriented plane. The plot is given in arbitrary units for both polarizations. In Fig. 4.10b is plotted the evolution of the crystal field splitting parameters city the ratio c/a of the lattice parameters. Clearly the sign of if varies with the ratio (or said alternatively with strain) which produces the reversal of symmetry of the ground state valence band from Γ_9 to Γ_7 when bulk GaN experiences a biaxial stretching in the (0001) plane.

The valence band parameters for bulk unstrained GaN are $\Delta_1 = 10$ meV; $\Delta_2 = 5.5$ meV; $\Delta_3 = 6.2$ meV.

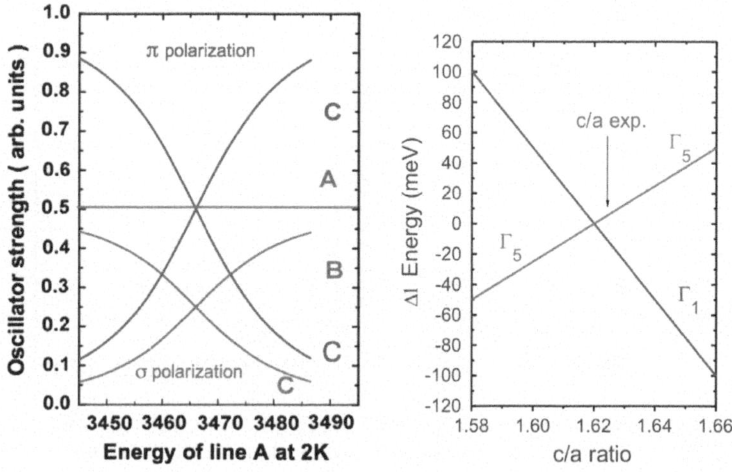

Fig. 4.10 *Left* evolution of the oscillator strength of the different transitions versus energy of A-line which is used to calibatre the biaxial strain. *Right* evolution of the crystal field splitting parameter in GaN versus c/a ratio

4.2.3 Longitudinal-Transverse Splitting and Exciton-Polaritons in GaN

In Fig. 4.11 are plotted the photoluminescence and reflectivity spectra of GaN taken at low temperature in case of a weakly strained (biaxial compression) GaN epilayer grown on C-plane sapphire. From the line shape fitting of the fluorescence line shape, the longitudinal-transverse splitting of the exciton-polariton is estimated in the 0.8 meV range for the Γ_5 A and B excitons in bulk unstrained GaN.The corresponding photoluminescence showing polariton accumulation in the low and high energy dispersion branches in the bottle neck region of the low polariton ranch and in the bottom of the high polariton branch (bolded parts of the dispersion relations in Fig. 4.6). The reasons for polariton accumulation here are multiple, they include in particular arguments based on the differences between group velocities ($\partial\omega/\partial K$) of the polariton and of acoustic phonon phonon branches (for scattering and energy relaxation processes) and consideration of the low density of states at energies below the polariton bottle neck spectral region.

4.2.4 Excitons in GaN Epilayers Grown with Strain on M-plane or A-plane Orientations

To measure the short-range interaction parameter, the method which was used first consisted in growing a (0001)-oriented GaN layer on A-plane sapphire substrate as indicated in Fig. 4.12a. The sapphire thermal dilatation coefficients are strongly anisotropic in its A plane. The deformations experienced by the GaN layer are

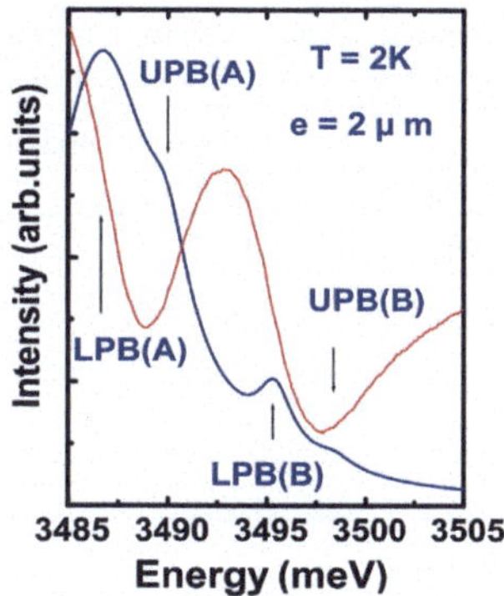

Fig. 4.11 Reflectance properties of GaN at 2K (*top*) and the corresponding photoluminescence showing polariton accumulation in the low and high energy dispersion branches (*bottom*)

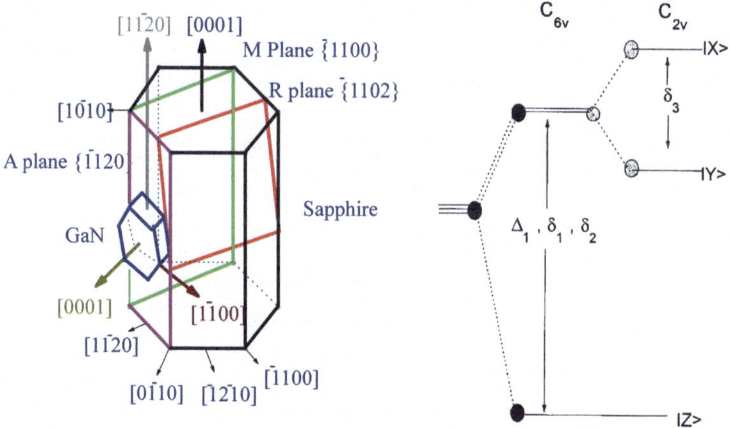

Fig. 4.12 *Left* relative orientation of the (0001)-grown GaN layer and A plane sapphire. *Right* evolution of the p-type valence band states under the influence of an orthorhombic deformation

different in the [1–100] and [11–20] directions. The thick sapphire substrate stresses the thin GaN layer. In that situation, the components e_{xx} and e_{yy} of the strain tensor are different. This we can write $e_{xx} - e_{yy} \neq 0$. As indicated left part in Table 4.3 where are plotted the relationships between the symmetrized components of the strain tensor and the irreducible representation of C_{6v}, the wurtzite symmetry is lost, as $e_{xx} - e_{yy}$ transforms like Γ_6 (The mechanical perturbation experience by the strained layer does not transform like Γ_1) and the degeneracy of p_x and p_y states is left as sketched in Fig. 4.12b.

Using elasticity theory developed in Chap. 1 and geometrical consideration, one concludes that the wurtzite symmetry is lost and that the symmetry of the strained layer is now C_{2v}.

Table 4.3 (right part) indicates the relationship between the strain tensor and the irreducible representations of C_{2v}.

Table 4.3 Left symmetrized components of the strain tensor for wurtzitic semiconductors. *Right* symmetrized components of the strain tensor for orthorhombic crystals

Wurtzite C_{6v}			Orthorhombic C_{2v}		
Symmetry	Polar vector	Strain tensor	Symmetry	Polar vector	Strain tensor
Γ_1	z	$e_{zz}, e_{xx}+e_{yy}$	Γ_1	z	e_{zz}, e_{xx}, e_{yy}
Γ_2			Γ_2	x	e_{xz}
Γ_3			Γ_3		e_{xy}
Γ_4			Γ_4	y	e_{yz}
Γ_5	(x,y)	(e_{xz}, e_{yz})			
Γ_6		$(e_{xy}, e_{xx}-e_{yy})$			

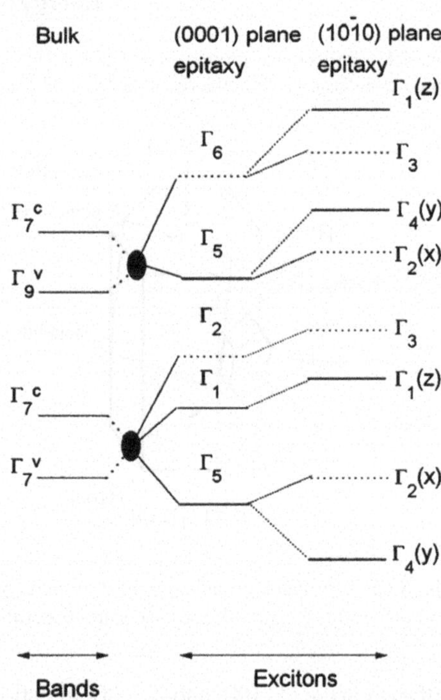

Fig. 4.13 Evolution of the excitonic dissociation when reducing the symmetry from C_{6v} to C_{2v}

According to the compatibility relations between C_{6v} and C_{2v}, the two-fold degeneracy of Γ_5 excitons is lift into $\Gamma_2 \bigoplus \Gamma_4$ as indicated in Fig. 4.13. We remark that that the two-fold degeneracy of the Γ_6 excitons is lift.

The eigen vectors for the exciton states of different symmetries are given in Table 4.4.

We introduce quantities δ_1, δ_2, and δ_3 which respectively transform like the symmetries Γ_1, Γ_1 and Γ_6 of C_{6V}. Quantity δ_1 represents the energy shift due to the hydrostatic part of the strain, δ_2 represents the variation of the crystal field splitting parameter Δ_1 under strain, and δ_3 represents the strength of the orthorhombic distortion.

Table 4.4 Eigenvectors for excitons in othorhombic symmetry. p_+ represents $-(p_x + ip_y)/\sqrt{2}$ and p_- represents $(p_x - ip_y)/\sqrt{2}$

State	Basis functions
Γ_1	$(p_+ \uparrow \beta + p_- \downarrow \alpha)/\sqrt{2}$
Γ_1'	$(p_- \uparrow \beta + p_+ \downarrow \alpha)/\sqrt{2}$
Γ_1''	$p_z(\downarrow \beta + \uparrow \alpha)/\sqrt{2}$
Γ_3	$(p_+ \uparrow \beta - p_- \downarrow \alpha)/\sqrt{2}$
Γ_3'	$(p_- \uparrow \beta - p_+ \downarrow \alpha)/\sqrt{2}$
Γ_3''	$p_z(\downarrow \beta - \uparrow \alpha)/\sqrt{2}$
Γ_2	$(p_+ \uparrow \alpha - p_- \downarrow \beta)/\sqrt{2}$
Γ_2'	$(p_+ \downarrow \beta - p_+ \uparrow \alpha)/\sqrt{2}$
Γ_2''	$p_z(\downarrow \alpha - \uparrow \beta)/\sqrt{2}$
Γ_4	$(p_+ \uparrow \alpha + p_- \downarrow \beta)/\sqrt{2}$
Γ_4'	$(p_+ \downarrow \beta + p_+ \uparrow \alpha)/\sqrt{2}$
Γ_4''	$p_z(\downarrow \alpha + \uparrow \beta)/\sqrt{2}$

$$\delta_1 = m_3(e_{xx} + e_{yy}) + l_2 e_{zz}$$
$$\delta_2 = (l_1 + m_1)(e_{xx} + e_{yy}) + m_2 e_{zz}$$
$$\delta_3 = (l_1 - m_1)[e_{xx} - e_{yy} + 2ie_{xy}]$$

This representation is general and by replacing the components of the strain field e_{ij} by a quadratic form in k: $k_i k_j$, one can treat the quantum well problem in case of growth on M-plane or A plane. Then symetrized quantities $e_{xx} - e_{yy}$ (or $k_x^2 - k_y^2$) have to be replaced by $e_{xx} - e_{yy} + 2ie_{xy}$ (or $k_x^2 - k_y^2 + 2ik_x k_y$).

The matrix representation of the Γ_1 eigentates of the exciton writes:

$$\begin{array}{ccc} |\Gamma_1> (A,z) & |\Gamma_1'> (B,z) & |\Gamma_1''> (C,z) \\ \Delta_1 + \Delta_2 + \delta_1 + \delta_2 + \gamma/2 & \delta_3 & 0 \\ \delta_3 & \Delta_1 - \Delta_2 + \delta_1 + \delta_2 + \gamma/2 & \sqrt{2}\Delta_3 \\ 0 & \sqrt{2}\Delta_3 & \delta_1 - 3\gamma/2 \end{array}$$

The exciton wave function writes: $\Psi(\Gamma_1) = a_1|\Gamma_1> + b_1|\Gamma_1'> + c_1|\Gamma_1''>$.

The oscillator strength (in $E//z$ polarization) is the square of the sum of the coefficients of the non spin-flip basis vector.

$$|\Omega(\Gamma_1'')|^2 = c_1^2$$

The matrix representation of the Γ_2 eigentates of the exciton writes:

$$\begin{array}{ccc} |\Gamma_1> (A,x) & |\Gamma_1'> (B,x) & |\Gamma_1''> (C,x) \\ \Delta_1 + \Delta_2 + \delta_1 + \delta_2 + \gamma/2 & \delta_3 - \gamma & 0 \\ \delta_3 - \gamma & \Delta_1 - \Delta_2 + \delta_1 + \delta_2 - \gamma/2 & \sqrt{2}\Delta_3 \\ 0 & \sqrt{2}\Delta_3 & \delta_1 - 3\gamma/2 \end{array}$$

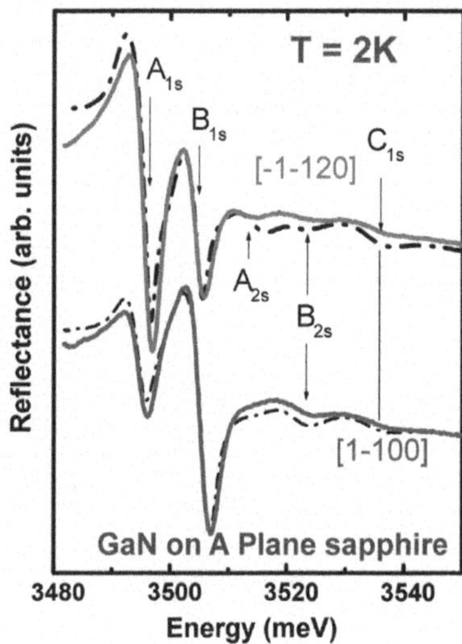

Fig. 4.14 Experimental reflectivity spectra probing Γ_4 and Γ_2 excitons in case of sample described in Fig. 4.12a (*full lines*). The fit to the date is indicated as dashed lines

The exciton wave function writes: $\Psi(\Gamma_2) = a_2|\Gamma_2 > +b_2|\Gamma_2' > +c_2|\Gamma_2'' >$.

The oscillator strength (in $E//x$ polarization) is the square of the sum of the coefficients of the non spin-flip basis vector.

$$|\Omega(\Gamma_2)|^2 = (a_2 + b_2)^2/2$$

The matrix representation of the Γ_4 eigentates of the exciton writes:

$$\begin{array}{ccc} |\Gamma_4 > (A, y) & |\Gamma_4' > (B, y) & |\Gamma_4'' > (C, y) \\ \Delta_1 + \Delta_2 + \delta_1 + \delta_2 + \gamma/2 & -\delta_3 - \gamma & 0 \\ -\delta_3 - \gamma & \Delta_1 - \Delta_2 + \delta_1 + \delta_2 - \gamma/2 & \sqrt{2}\Delta_3 \\ 0 & \sqrt{2}\Delta_3 & \delta_1 - 3\gamma/2 \end{array}$$

The exciton wave function writes: $\Psi(\Gamma_4) = a_4|\Gamma_4 > +b_4|\Gamma_4' > +c_4|\Gamma_4'' >$.

The oscillator strength (in $E//y$ polarization) is the square of the sum of the coefficients of the non spin-flip basis vector.

$$|\Omega(\Gamma_4)|^2 = (a_4 + b_4)^2/2$$

In Fig. 4.14b are reported low temperature reflectivity spectra probing Γ_4 and Γ_2 excitons in case of sample described in Fig. 4.12a (full lines). The fit to the date is indicated as dashed lines. **The line shape fitting of experimental results led to a value of 0.7 meV for the short-range spin-exchange interaction in GaN. Note the evolution of the line shape with polarization which permits to determine the excitonic binding energy (29 meV) in GaN.** also note the strong blue-shift of the experimental features with respect to the bulk (unstrained) GaN values. Also note

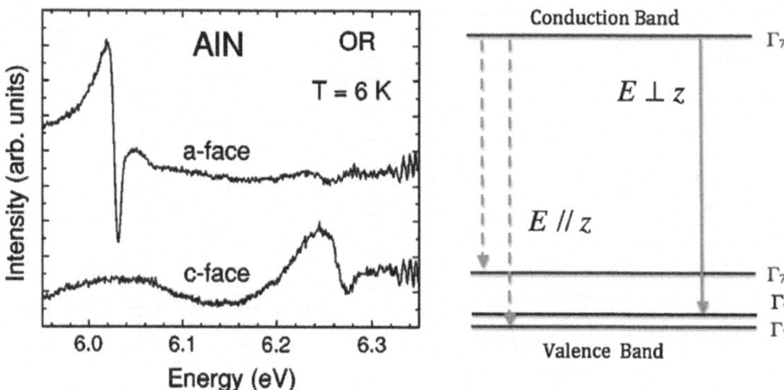

Fig. 4.15 *Left* experimental reflectivity of Aluminum nitride. After [11]. *Right* valence band ordering in AlN

the weakness of line C for these two polarization conditions ($\vec{E}\perp\vec{c}$). The anisotropic strain field is not so high and the departure from evolution of oscillator strength predicted for (0001) growth is more or less kept (if considering the couple of lines A and B as a whole split far away from C).

4.3 Aluminum Nitride

4.3.1 Optical Properties of Bulk Aluminum Nitride

In Fig. 4.15a is represented the low temperature reflectance of bulk aluminum nitride. Clearly the valence band ordering is different from the preceding case with a strong low energy (6.029 eV) transition easily detected for $\vec{E}//\vec{c}$ polarization and absolutely silent in $\vec{E}\perp\vec{c}$ polarization. At higher energy It is possible to detect a strong feature with hardly resolved doublet structure. This band structure scheme is typical of what is predicted in case of negative crystal field splitting parameter and positive spin-orbit coupling parameters and sketched in Fig. 4.15b.

The values for unstrained AlN are $\Delta_1 = -220\,\text{meV}$, $\Delta_1 = 6\,\text{meV}$, $\Delta_3 = 6.5\,\text{meV}$.

Optical spectroscopy measurements have been performed in particular by means of low temperature photoluminescence (they will be shown later) which led to estimate the exciton binding energy at 49 meV. Regarding the short range exchange interaction, the situation is very murky at the time of writing: a group published a negative value of −4 meV [10] while another one [11] published a positive value of about 6.8 meV. The discrepancy resides in different interpretation of fluorescence lines. This is in the way of being solved.

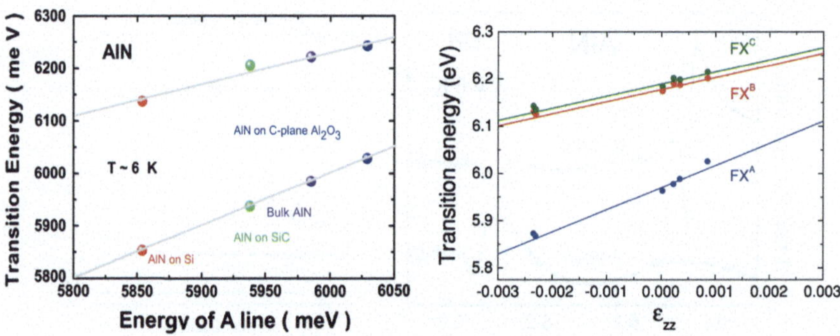

Fig. 4.16 *Left* plot of the transition energies of AlN epilayers grown on a large variety of foreign substrates versus energy of A-line. *Right* plot of the transition energies in AlN epilayers versus deformation in the c direction. After [12]

4.3.2 Strain-Fields in Aluminum Nitride Epilayers

In Fig. 4.16 is the analog of Fig. 4.8. Clearly the growth on foreign substrates can drastically shift the energy of the fundamental band gap but the large value of the crystal field splitting parameter compared with the values of the spin-orbit interaction parameter prevent to report large strain-induced variations of the oscillator strengths. The crystal field splitting parameter varies from a growth substrate to another one but clearly the valence band ordering can not be reversed. The topmost valence band is always a Γ_7 level and it is dominantly built from p_z state.

4.3.3 Excitons and Polaritons in AlN

In Fig. 4.17 has been plotted a reflectivity spectrum computed for AlN from the measurement of the real (ϵ_1) and imaginary (ϵ_2 parts of the dielectric constant of AlN. From these measurements a value of about 6 meV is found for the value of the logitudinal-transverse splitting for the 6.031 eV reflectivity structure in AlN at 4 K in π ($\vec{E}//\vec{c}$) polarization. This is a record value but it is not surprising: the Rydberg energy increases with the value of the band gap, so does the oscillator strength of the 1s state and subsequently the longitudinal transverse splitting.

4.4 Zinc Oxide

4.4.1 Optical Properties of Bulk Zinc Oxide

In Fig. 4.18 have been plotted some absorption spectra taken in both polarization conditions for thin ZnO layers at liquid helium temperature. Note the high value

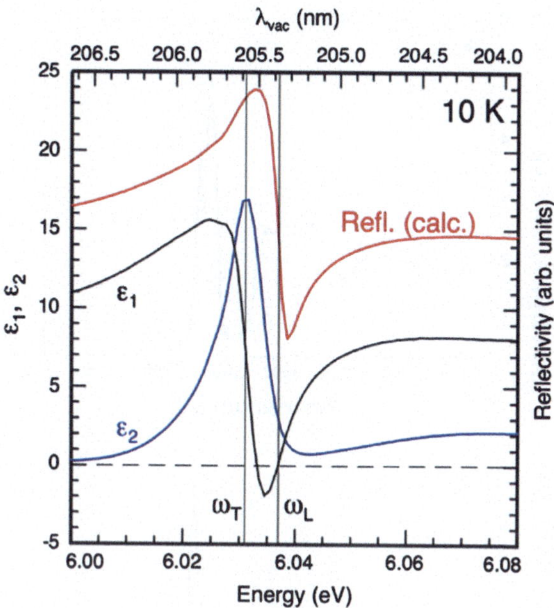

Fig. 4.17 Experimental values of the real (*middle*) and imaginary (*bottom*) parts of the low temperature dielectric constant of AlN as well as the corresponding calculated reflectivity (*top spectrum*). Courtesy Matthias Feneberg University of Magdeburg

of the absorption at the energies of the excitonic peaks and the high value of the absorption at the energies above them (200,000 cm^{-1}). From the splittings between 1s, 2s and 3s excitons, one deduces an exciton binding energy of 60 meV. We remark transitions labelled L_1 on the absorption spectra. They correspond to a three-particle interaction process: in that case a photon of energy equal the excitonic energy plus the energy of an LO phonon (75 meV in ZnO) is absorbed, with simultaneously creating an exciton and an LO phonon. This is an evidence of an efficient exciton-phonon interaction. Moreover the observations of weaker structures labelled L_2 at higher energies are the evidence of a very efficient exciton-phonon interaction: an exciton being in that case simultaneously created with two LO phonons.

4.4.2 Optical Properties of Zinc Oxide Heteroepitaxies

In Fig. 4.19a have been reported low temperature reflectivity experiments which indicate the selection rules and demonstrate that energies of both γ_5 and γ_1 excitons are discriminately measured. In Fig. 4.19b have been reported some reflectivity experiment in normal incidence conditions on a series of (0001) oriented ZnO samples so that are probed γ_5 excitons. The bottom sample is bulk ZnO, the middle sample was

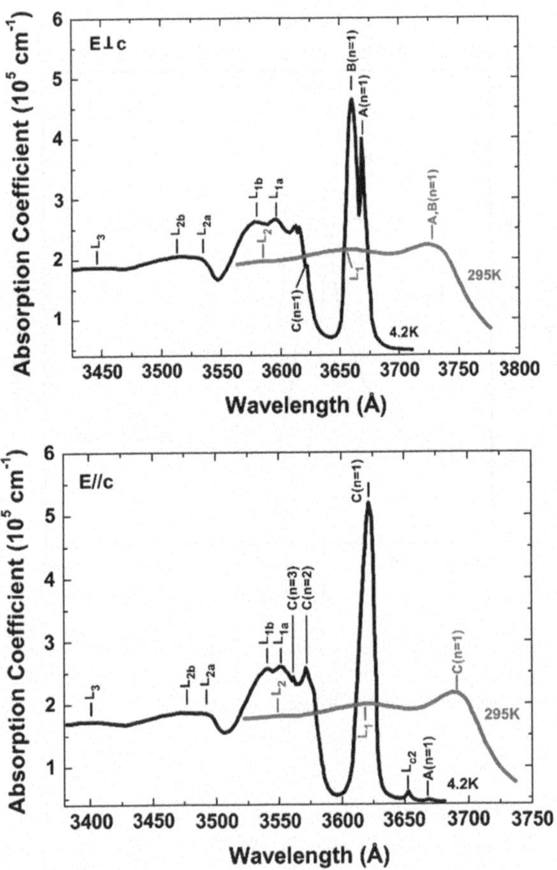

Fig. 4.18 Absorption coefficients for thin bulk ZnO samples. Reproduced from [13]

grown on saphirre and the top sample was grown on GaN on saphirre. Slight energy shift reveal the influence of stress on the reflectance energies.

The observation of 2s excitonic states at higher energy than the strong 1s transitions suggest an exciton binding energy of 59 meV. The fit to the data (energies and oscillator strength) has been performed in the context of an excitonic model. We remind from Sect. 4.1.2 that the exciton wave function writes: $\Psi(\Gamma_5) = |\Gamma_5 > +b|\Gamma_5' > +c|\Gamma_5'' >$.

The oscillator strength (in σ polarization) is the square of the sum of the coefficients of the non spin-flip basis vector.

$$|\Omega(\Gamma_5)|^2 = (a+b)^2/2$$

Then the full set of data could be fitted using a short-range exchange interaction parameter of 4.73 meV.

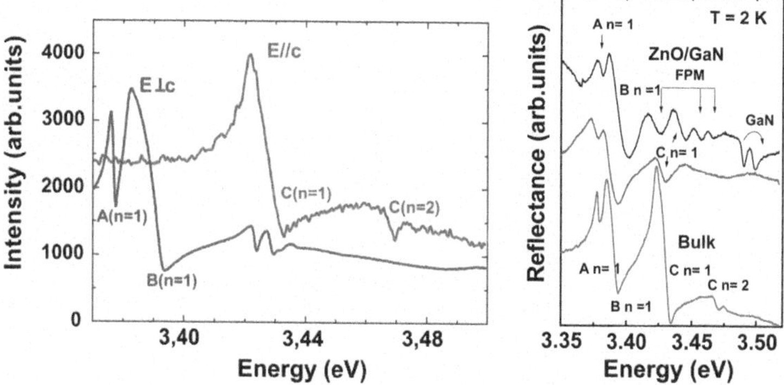

Fig. 4.19 *Left* low temperature reflectivity experiments which indicate the t energies of both γ_5 and γ_1 excitons after [14]. *Right* some reflectivity experiment in normal incidence conditions on a series of (0001) oriented ZnO bulk and heteroepitaxies to reveal the shift of the reflectance energies with stress. After [15]

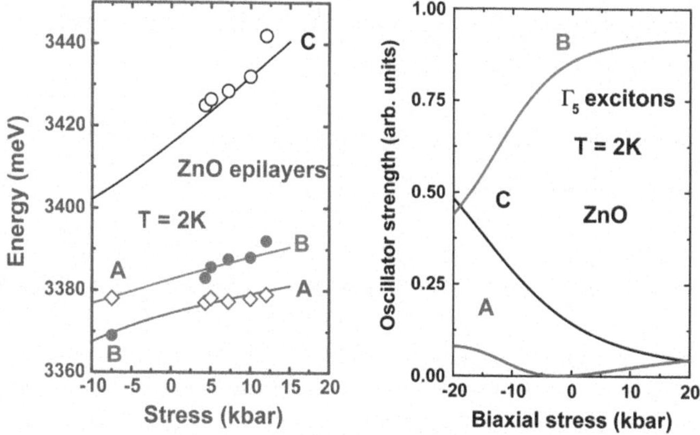

Fig. 4.20 *Left* stress-induced evolutions of the γ_5 excitons in (0001)-grown ZnO epilayers after [15]. *Right* stress-induced evolution of the oscillator strength for γ_5 excitons in (0001)-grown ZnO epilayers. After [16]

We remark that the strain has the possibility to change the ground state valence band from Γ_9 to Γ_7 (Fig. 4.20). if not, please suggest an alternate the citation.

The parameters the author prescripts for ZnO are $\Delta_1 = 30.5$ meV; $\Delta_2 = 4$ meV; $\Delta_3 = 11.5$ meV. The 3d states of Zn interact with the p states of oxygen and they may change the sign of the spin-orbit interaction.

4.4.3 Polaritons in ZnO

From the line shape fittings of the reflectivity of zinc oxide, one can obtain record values for the longitudinal-transverse splittings about 1 and 5 meV for A and B excitons (Γ_5) in [17]. More recently a value of 11 meV was proposed for the longitudinal transverse splitting of B exciton (Γ_5) in [18]. Finally, values of 1.5 and 5 meV for A and B excitons (Γ_5) in [19].

4.5 Indium Nitride

The value of the band gap of indium nitride is not accurately known, due to the impossibility to grow undoped indium nitride so far. However, some polarized ellipsometry measurements were recently published in [20]. They predict a small (about 10 meV) value for the crystal fiend splitting parameter.

4.6 Excitonic Binding Energies in Wurtzitic Materials: The Influence of Anisotropies

4.6.1 General Description in the Framework of the Effective Mass Approximation

The excitonic binding energies and oscillator strengths in materials of bandgap energy E_g which are anisotropic along one crystal axis (z axis), are here computed in the context of the effective mass approximation. This problem for which both kinetic energy and potential energy terms in the Schrödinger equation are anisotropic (each with with a given anisotropy) is extremely complicated compared with the situation for spherical 3D or planar 2D situations. We have chosen to solve this problem numerically in the context of the finite element methods. To do so, as usual, we first define the reduced masses parallel and perpendicular to the direction of the z axis $\mu_{//}$ and $\mu_{\perp}$, and note $\epsilon_{//}$ and $\epsilon_{\perp}$ the corresponding dielectric constants. The effective mass equation for excitons writes as a function of the operator H:

$$H\Psi^{\beta}(\vec{r}) = E^{\beta}\Psi^{\beta}(\vec{r})$$

where

$$\mathrm{H} = -\hbar^2/2\mu_{//}\partial^2/\partial z^2 - \hbar^2/2\mu_{\perp}(\partial^2/\partial x^2 + \partial^2/\partial y^2) - e^2/\left[4\pi\epsilon_0\sqrt{\epsilon_{//}\epsilon_{\perp}(x^2+y^2)+\epsilon_{\perp}^2 z^2}\right]$$

The method consists in first re-arranging this equation through through the coordinates x, y, z of the international basis used to describe the physical properties of the wurtzite materials as follows:

Table 4.5 Eigenvalues of the long range Coulomb energy computed in reduced units, for different anisotropies γ, for the four lowest energy states of $\sum_g^+$ symmetry

γ	$\sum_g^+(1)$ 1s	$\sum_g^+(2)$ 2s	$\sum_g^+(3)$ $3d_0$	$\sum_g^+(4)$ 3s
2	0.783	0.162	0.073	0.073
1.5	0.870	0.195	0.088	0.087
1	1	0.250	0.111	0.111
0.8	1.077	0.269	0.126	0.119
0.75	1.099	0.276	0.130	0.121
0.5	1.247	0.318	0.160	0.139
0.4	1.326	0.345	0.178	0.150
0.3	1.443	0.386	0.203	0.166
0.1	1.851	0.595	0.337	0.253

We define x′ = x; y′ = y and z′ = z $\sqrt{\mu_{//}/\mu_{\perp}}$.

Further defining the reduced Rydberg Energy $R^* = m_0c^2\alpha^2/2\mu_{\perp}/\mu_{//}$ (here α represents the fine structure constant), the in-plane radius $\rho' = \sqrt{x'^2 + y'^2}$ and the cylindrical Laplacian $\Delta(\rho', z')$, we are left to solving a simple second order differsntial equation of the kind:

$$\left[\Delta(\rho', z') + 1/\sqrt{\rho'^2 + \gamma z'^2}\right]\xi^{\beta}(\rho', z') = E^{\beta}\xi^{\beta}(\rho', z')$$

In this equation the energies are measured with respect to E_g, and they read E^{β} in units of R^* and the eigenstates are the quantities $\xi^{\beta}(\rho', z')$.

Both eigenstates $E^{\beta}$and eigenstates $\xi^{\beta}(\rho', z')$ can be classified according to the irreducible representations of group $D_{\infty h}$. The general eigenstates of this cylindrical Schrödinger equation are labelled $\Xi_i^{\pm}(n)$. In these notations Ξ stands through $\Sigma, \Pi, \Delta, \Phi, \ldots$ if the absolute value of the z′ component of the angular momentum is 0, 1, 2, 3,... respectively. The sign + (respectively −) means even (respectively odd) state under reflection in a plane through the z′ axis. The subscript i which runs between g and u (for gerade and ungerade), means even and odd states under inversion through the origin for g and u respectively. Last, states having the same symmetry are labelled in terms of the index n when increasing energy.

As an example the ns states of spherical symmetry are always labelled as $\sum_g^+$. In Table 4.5 are given some results obtained for the four lowest energy states of $\sum_g^+$ symmetry. These states are the four low energy ones with non vanishing oscillator strength. In our case these states are those for which the absolute value of the angular momentum in the z direction equals 0 and which are even states under reflection in a plane through the z axis. These states are the ns states in case of spherical symmetry, the axial symmetry enlarges them to levels like $3d_0, 4d_0, \ldots$

Resolution of equation gives us the eigensates $\xi^{\beta}(\rho', z')$. Those for which $\beta = \sum_g^+(i)$ are coupled with the electromagnetic field, or said with another phrasing correspond to excitonic states with non vanishing oscillator strengths. By strict application of quantum mechanics from the normalization condition, one can write:

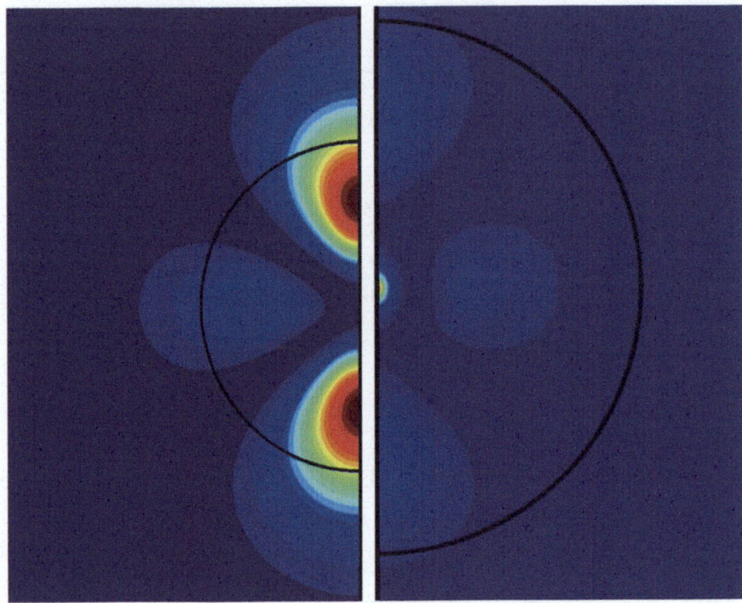

Fig. 4.21 *Left* plot of $|\xi^{\Sigma_g^{+}(3)}(\rho', z')|^2(3d_0)$ for $\gamma = 1$. The Bohr radius is plotted in *black*. *Right* plot of $|\xi^{\Sigma_g^{+}(3)}(\rho', z')|^2(3d_0)$ for $\gamma = 0.25$. The Bohr radius is plotted in *black* Courtesy Didier Felbacq

$$\int 2\pi \xi^{\Sigma g^{+(i)}}(\rho', z')\xi^{*\Sigma_g^{+}(j)}(\rho', z')\rho' d\rho' dz' = \delta_{ij}$$

which gives for the oscillator strength of state $\sum_g^{+}(i)$ computed in the (ρ', z') space the value: $|\xi^{\Sigma_g^{+}(i)}(0, 0)|^2$.

The exact oscillator strength of the exciton exciton state living in the real (ρ, z) world has to be computed as:

$$|\Psi^{\Sigma_g^{+}(i)}(0, 0)|^2 = \frac{\mu_{//}}{\mu_{\perp}}|\xi^{\Sigma_g^{+}(i)}(0, 0)|^2 = \frac{1}{\gamma}\frac{\epsilon_{\perp}}{\mu_{//}}|\xi^{\Sigma_g^{+}(i)}(0, 0)|^2.$$

4.6.2 Wave Functions

In Fig. 4.21 have been plotted $|\xi^{\Sigma_g^{+}(3)}(\rho', z')|^2)$, the density of probability for the $3d_0$ state $(\sum_g^{+}(3))$ for γ= 1 (left) and for γ= 0.25 (right). In the left-hamd side we recognize the well known textbook density of probability for the $3d_0$ state in spherical symmetry. The z′ axis is vertical. Explicitly shown as a red area is the getting of oscillator strength for the $3d_0$ state with anisotropy.

Table 4.6 Comparison between the computed values of the 1s binding energy, values of 1s–2s splitting and the experimental ones (printed between parenthesis)

γ	$m_{h//}$	$m_{h\perp}$	$m_{e//}$	$m_{e\perp}$	$\epsilon_{\perp}$	$\epsilon_{//}$	γ	$\frac{1s-2s}{1s}$	1s (meV)	1s–2s (meV)
GaN	2	0.57	0.23	0.23	9.5	10.4	0.72	0.75	25.0	18.8
GaN									**(25–26)**	**(19)**
ZnO	2.73	2.66	0.246	0.329	7.8	8.75	1.16	0.76	55.6	42.4
ZnO									**(59.6)**	**(44.8)**
CdS	5	0.7	0.2	0.2	8.45	9.12	0.75	0.75	30.0	22.7
CdS									**(29.5)**	**(22)**
CdSe	1	0.45	0.13	0.13	8.75	9.25	83	0.754	18	13.6
CdSe									**(17.5)**	**(13)**
ZnS	1.4	0.49	0.28	0.28	8.32	8.32	0.76	0.75	38.2	28.9
ZnS									**(40)**	**(30)**
InN	1.98	0.44	0.06	0.06	13.1	14.4	0.82	0.754	4.1	3.1
AlN	0.2	4.2	0.23	0.24	7.76	9.32	1.72	0.78	35.3	27.6
AlN	0.25	3.68	0.33	0.25	7.76	9.32	1.37	0.77	39.5	30.4
AlN	0.26	4.05	0.35	0.35	7.76	9.32	1.80	0.78	49.3	38.7
AlN	0.26	3.47	0.32	0.34	7.76	9.32	1.80	0.78	47.5	37.2
AlN									**(49–50)**	**(36–38)**

4.6.3 Numerical Values for Wurtzitic Semiconductors

In Table 4.6 are summarized the comparison between the computed values of the 1s binding energy, values of 1s–2s splitting and the experimental ones. Obviously the variation of the binding energy expected from the value of γ is composted by the value of the reduced Rydberg energy. However when a set of parameters is proposed from the theoretical side, in the literature, the check of the binding energy given by this set of parameters and the comparison with experiment can lead to severe conclusions.

4.7 Influence of Temperature

4.7.1 Bulk Materials

When increasing the lattice temperature, the number of phonons increases as predicted by Bose-Einstein statistics:

$$n(\hbar\omega) = \frac{1}{\left[e^{\frac{\hbar\omega}{kT}} - 1\right]}$$

where $\hbar\omega$ represents the phonon energy, k is the Boltzmann constant and T represents the temperature. This produces an enhancement of the oscillations of atoms from their equilibrium positions. The features that are measured if performing a temperature-dependent x-ray scattering experiment broaden and the diffraction angles indicate that the lattice parameters increase. The behavior is is general non monotonous, but the statement holds for high enough temperatures after a few Kelvins. The overlaps between wave functions of close neighbor atoms decrease. So does the bandgap. Increasing the number of phonons increases the number of collisions with electrons (electron-phonon interaction) and all optical features will broaden as shown in Fig. 4.17 for instance where are reported the 77 and 2 K absorption features of ZnO thin films. This can be accounted for phenomenologically by considering that the damping parameter γ (non radiative homogeneous broadening) varies with temperature proportionally to the phonon population. The wavelength of the light emitted at the band gap energy in vacuum is less than one half micrometer (and is even, smaller in the crystal, since divided by the refractive index).This value is very small with respect to the inverse of the wave vector at the edge of the Brillouin zone (about one angstrom). The phonons which wave vector is similar to the wave vector of the light are phonon modes at Brillouin zone center. The phonon modes that dominantly collide with the electrons are the LO phonons and acoustic phonons.

$$\gamma(T) = \gamma_0 + \gamma_{acoustic}\frac{1}{\left[e^{\frac{\hbar\omega_{acoustic}}{kT}} - 1\right]} + \gamma_{LO}\frac{1}{\left[e^{\frac{\hbar\omega_{LO}}{kT}} - 1\right]}$$

The energy of acoustic phonons is very small near Brilloiun zone center, which leads to:

$$\gamma(T) = \gamma_0 + \frac{k\gamma_{acoustic}}{\hbar\omega_{acoustic}}T + \gamma_{LO}\frac{1}{\left[e^{\frac{\hbar\omega_{LO}}{kT}} - 1\right]}$$

that we reorganize as

$$\gamma(T) = \gamma_0 + \gamma_1 T + \frac{\gamma_{LO}}{\left[e^{\frac{\hbar\omega_{LO}}{kT}} - 1\right]}$$

where γ_0 is the low temperature homogeneous broadening, γ_1 represents the interaction with acoustic phonons which is important at low temperature, and γ_{LO} accounts for the electron–LO phonon interaction.

The third term $\frac{\gamma_{LO}}{\left[e^{\frac{\hbar\omega_{LO}}{kT}} - 1\right]}$ rapidly dominates when the temperature increases.

Values for GaN are typically $\gamma_0 = 2.4$ meV $\gamma_1 = 16 \times 10^{-6}$ eV K^{-1} and $\gamma_{LO} = 390$ meV according to Femtosecond four-wave-mixing studies of nearly homogeneously broadened excitons in GaN, [1] who took a phonon energy $\hbar\omega_{LO} = 91.7$ meV.

For ZnO, using an average phonon energy $< \hbar\omega_{LO} >= 33 \pm 7.4$ meV, it was proposed that $\gamma_0 = 0.65 \pm 0.6$ meV; $\gamma_1 = 16 \pm 13 \times 10^{-6}$ eV K^{-1}; and $\gamma_{LO} = 47 \pm 12$ meV [21]. The broadening of the excitonic feature is less than 1 meV at 2K and about 40 meV at room temperature.

4.7.2 Reduction of the Band Gap of Bulk Materials When Increasing Temperature

The reduction of the band gap with increasing temperature has been modeled empirically using a two-parameter meaningless law called the Varshni's law [22].

$$E_g(T) = E_g(0) - \alpha T^2/(\beta + T)$$

α and β are the Varshni's coefficients. They have no physical meanings but this law has been word widely used, and it is still used.

An alternative law is based on Bose-Einstein statistics. It writes the evolution of the band gap with T according to:

$$E_g(T) = E_g(0) - \frac{2\alpha_B}{[e^{\frac{\theta_B}{T}} - 1]}$$

α_B is a constant and and $k\theta_B$ represents the average phonon energy in the crystal. Both quantities are fitting parameters but they have (at least θ_B) a physical meaning.

In Table 4.7 are given the typical values of these parameters for GaN and AlN.

The Bose-Einstein description gives better results than Varshni's modeling for wide band gap semiconductors.

At cryogenic temperatures a more sophisticated description is required, called the Pässler modeling [23]. In Fig. 4.22a, b are plotted the variations of the energy gap of bulk GaN and AlN. One remarks that increasing the temperature from liquid helium to room temperature produces a red-shift of several tens of meV.

4.7.3 Epilayers

In case of epilayers or in case of multilayered structures, the temperature-induced dilatation of the epilayer or the multilayer is a complicate function of the whole contribution of the substrate-epilayer (or multilayer) ensemble and the red-shift with temperature can be very different from the bulk one. This is illustrated in Fig. 4.23 for several GaN bulk layers and epilayers. The red shift also varies with the orientation of the surface substrate when the substrate has strongly anisotropic dilatation coefficients.

Table 4.7 Average values of the constants for Varshni 's model α and β) and for the Bose-Einstein like description (α_B is a constant and and θ_B)

Parameters	α (meV/K)	β (K)	α_B (meV/K)	θ_B (K)
GaN	10	1,100–1,400	100	350–400
AlN	0.97	1000	130	450

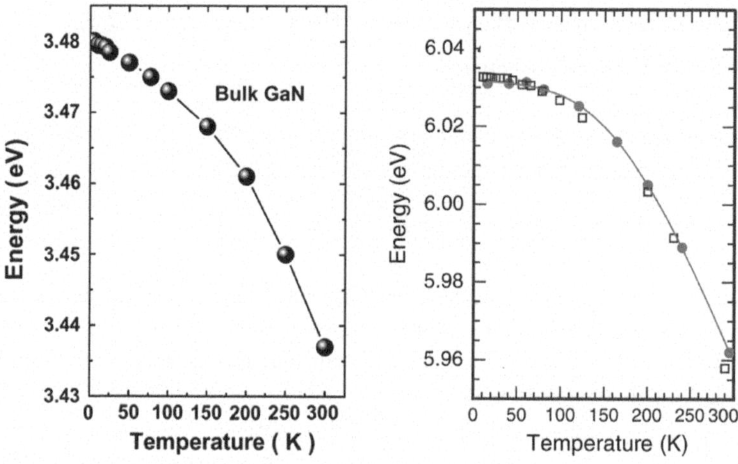

Fig. 4.22 *Left* temperature dependence of the energy gap of GaN. After [24]. *Right* temperature dependence of the energy gap of AlN. After [10]

4.8 Photoluminescence of Wurtzitic Semiconductors

4.8.1 Classical Description of the Photoluminescence Process for Free Excitons and Free Carriers

In such experiment electron-hole pairs are created by absorption of photons with an energy higher than the value of the band gap. The photo created carriers then relax from their excited state energy by emission of acoustic phonons until they reach the bottom of the conduction band (for electron) and the top of the valence band (for holes). Then they recombine radiatively via photon emission or non radiatively (sample heating,...).

Momentum conservation prescripts $\vec{k}_e - \vec{k}_h = \vec{k}_{photon} \approx \vec{0}$ for the radiative recombination process. The long range Coulomb interaction between electron and hole leads to formation of an exciton which travels through the crystal with wave vector $\vec{K} = \vec{k}_e + \vec{k}_h$. From the reflectivity section, and thanks to interaction of the exciton with the radiation field only exciton polariton momenta $\vec{K}$ of value comparable to the value $\vec{k}_{photon}$ of the wave vector of the photon outside the crystal can couple efficiently to the electromagnetic field and give light outside the crystal interaction. This radiative recombination of the exciton by crossing the semi conductor to air interface

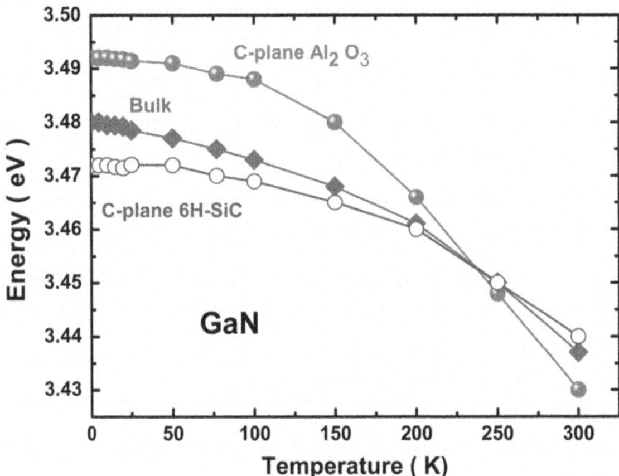

Fig. 4.23 Evolution of the band gap of bulk GaN, of GaN on c-plane sapphire and of GaN on c-plane 6H SiC

requires severe criteria at the scale of the orientations of the electron and hole wave vectors and at the scale of their lengths. They have to be both comparable in size to the momentum of the photon. Thus radiative recombination of an electron hole pair is highly unprovable and this related fluorescence is in general weak. Un-doped semiconductors do not emit much light. The low temperature photoluminescence spectrum of GaN has been previously shown in Fig. 4.11. Signatures of both polariton dispersion relations are detected as well as A-related and B related doublets. Due to polariton accumulation in the low energy part of the low polariton branch (polariton bootle neck region), the energy splitting between the photoluminescence band maxima is is general larger than the value of the longitudinal transverse splitting. The high energy feature correspond to the longitudinal exciton polariton mode, the low energy photoluminescence maximum occurs at an energy lower than the energy of the low branch of the transverse exciton polariton. It is very important to compared a reflectivity experiment to a photoluminescence one in order to discriminate signature of free exciton photoluminescence from recombination of bound exciton as we shall see it later. both experiments are complementary of each other. When exciton polariton propagation occurs in a direction away from the [0001] direction, there are complementary effects linked to the birefringence of wurtzite.

When a semi conductor is n-doped and degenerated, at low temperature, the electron relaxes its kinetic energy till reaching the Fermi energy where it remains blocked in relation with Pauli's exclusion principle; every conduction state at energy below E_F being already occupied (by two electrons with antiparallel spins) relaxing to it is a strict symmetry-forbidden process. The wave vector is then k_F. "The concept of exciton detailed before fails and has to be replaced by the concept of Mahan-exciton resonance [25]. In such description the expansion wave function of the exciton in momentum space cannot go through electron states of wave vectors smaller than k_F.

The kinetic energy term of the Schrödinger equation for the electron-hole interacting pair is higher than for excitons and the Coulomb binding energy is reduced. Doping screens the exciton binding energy. The first consequence we can anticipate from doping is a blue-shift of the "excitonic" photoluminescence energy due to screening of the long range Coulomb interaction between electron and hole.

The electron may recombine with a hole of wave vector $\vec{k_F}$ which energy is $E_V + E_F \frac{m_e}{m_h}$ in terms of the electron and hole effective masses m_e and m_h (they are taken isotropic here). The recombination energy is then theoretically $E_c + E_g - E_{exc} + E_F(1 + \frac{m_e}{m_h})$. The emission energy occurs at an energy higher by an amount $E_F(1 + \frac{m_e}{m_h})$ than the average free exception luminescence $E_c + E_g - E_{exc}$. Such blue-shift is called the Moss-Burstein shift. It theoretically permits to determine the n-doping of the semiconductor.

In the real world, the situation is more complex: the lattice temperature has a finite value. Thus the Fermi Diract statistics rules the electron population in the neighborhood of the Fermi energy. The probability for the electron to have a wave vector higher than $k_F(T = 0)$ increases with temperature. Then the sharp theoretical high energy photoluminescence line sitting at the energy said above changes to more or less broad band with a high energy wing which extends at higher energies than the 0 K photoluminescence energy. The high energy shape can be quite well described using a Maxwellian distribution of the intensities. Finite temperature also holds concerning holes. Radiative recombination can also occur between an electron at energy lower than E_F and one of the photo created holes, the one of adhoc momentum. Photoluminescence can thus occurs from the bottom of the conduction band to the top of the valence band. This contributes to determining the lower possible recombination energy. These considerations indicate that full width at half maximum (FWHM) of the of the photoluminescence band, including its variation with temperature is a good indicator of the doping of the semiconductor.

This is illustrated in Fig. 4.24 where is plotted the photoluminescence line shape of indium nitride as a function of doping and temperature.

Note the formation of a high energy wing when the temperature increases and the broadening of the photoluminescence line with doping at constant temperature. In the real world increasing the doping increases the number of electrons, which leads to an enhancement of the dielectric constant (see the classical Drude Lorentz model of the dielectric constant in Chap. 3). From this enhancement it results a reduction of the band gap energy (the so-called band gap renormalization). Several models of the screening of the dielectric constant with doping are available which are not of the scope of this paper but we refer to the recent papers of Arnaudov et al. for a more complete and detailed description of this physics [27–29].

In summary: doping leads to broad photoluminescence band with asymmetrical shapes, the asymmetry increasing with the temperature. The photoluminescence line shape and blue-shift permits to estimate the doping of the semiconductor (in the degenerate case). when the doping is too high, these simple arguments fail and the redshift of the photoluminescence due to band gap renormalization effects prevents

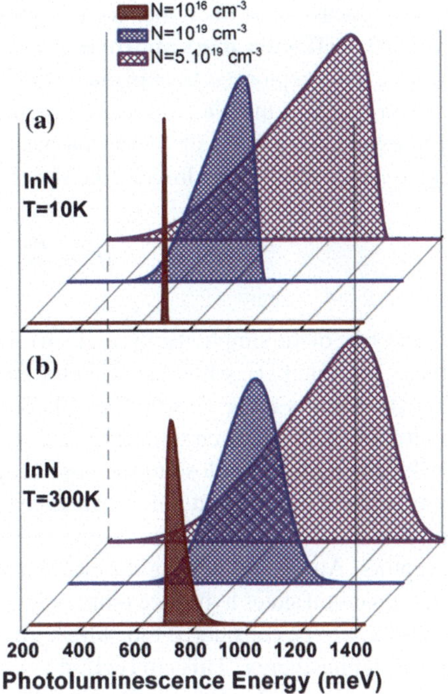

Fig. 4.24 Evolution of the Photoluminescence of InN versus doping and temperature. After [26]

to treat it with a simple model. What was said above in case of n-doping also holds for p-doping.

4.8.2 Photoluminescence of Bound Excitons and Other Extrinsic Recombination Processes

When free excitons (or free carriers) move through the crystal with wave vector $\vec{K}$ (or $\vec{k_e}$, $\vec{k_h}$) their probability to be scattered by the breaking of the translational crystal potential provoked by a foreign atome or a topological defects. They can bind in the neighborhood of such departure from periodicity, to form a stable state. The localization energy will vary with the strength of such localization potential (by strength are meant here both depth and spatial extension in the crystal). Depending with the chemical nature of the impurity and the site of the crystal are defined donor impurity (one excess electron with respect to the substituted atom in the shell that contributes to the chemical bond), acceptor (one missing electron with respect to the substituted atom in the shell that contributes to the chemical bond), isoelectronic impurities (same column of Mendeleev table, but different atomic size, different energy level from the electronic levels of the core shells of the substituted atom).

The eigenstate of the donor (or acceptor) is computed in text book in the frame work of the **effective mass approximation**. The quantum state of the excess electron (or hole for acceptors) is treat in the frame work of the hydrogen model. The electron with an effective mass m_* orbits around the positive center of the ionized donor (Z protons and Z − 1 inner shell electrons) in a medium with dielectric constant ϵ_*. The eigen-values follow a hydrogen-like series with an effective Rydberg:

$$R^* = \frac{m_0 c^2 \alpha^2}{2} \frac{m_*}{\epsilon_*^2}$$

For the sake of the simplicity, the dielectric constant is taken isotropic here. The donor energy are noted D_0, while the acceptor energies are noted A_0. There are impurities that fit with this naive description: silicon and oxygen while substituting nitrogen in nitrides, gallium when replacing zinc in ZnO behave like a shallow donors (they produce bound states close to the conduction band minimum).The dispersion relations being in general flatter in the valence band than in the conduction band, binding energies of effective mass donors are smaller than binding energies of effective mass acceptors. Atoms are non rigid objects that are described by quantum mechanics. Their quantum states hybridize more or less efficiently with the Bloch states and core levels have to be considered too, which give a measurable correction to the effective mass binding energy. **This correction is called the chemical shift.**

Free excitons (X) form many particle complexes with impurities, which when impurities are not ionized are labelled D_0X and et A_0X. They recombine according to the sketch of Fig. 4.25a: One of the two electrons (in a particle-permutation anti symmetric state) of the bound exciton complex recombine with the hole and produces a photon measured with an energy ED_0X. Fig. 4.25b one of the two holes (in a particle-permutation anti symmetric state) of the bound exciton complex recombine with the electron and produces a photon measured with an energy ED_0X.

The donor bound exciton localization energy EDB_0X is defined as the difference between the energy of the free exciton luminescence E_{exc} and the energy ED_0X: $E_{exc} - ED_0X$. This definition of the localization energy also holds for the acceptor bound exciton and for any other bound exciton.

In Fig. 4.26 is reproduced a near band edge photoluminescence spectrum of aluminum nitride. One distinguishes the free exciton photoluminescence line X_A (and its 2s state at higher energy) and three sharp lines are located at 6.0127, 6.0190, and 6.0280 eV. The corresponding localization energies, that is, the energy separations from the free exciton, are 28.7, 22.4, and 13.4 meV. The donor bound exciton line with localization energy of 22.4 meV, most prominent sample having moderately high silicon concentration is assigned to the shallow donor silicon. The line with localization energy of 28.7 meV is attributed to shallow donor oxygen. The energetic order of silicon and oxygen donors then would be the same as in GaN. The third line with localization energy of 13.4 meV is attributed to an unidentified donor. The localization energies are influenced by the local environment (strain-

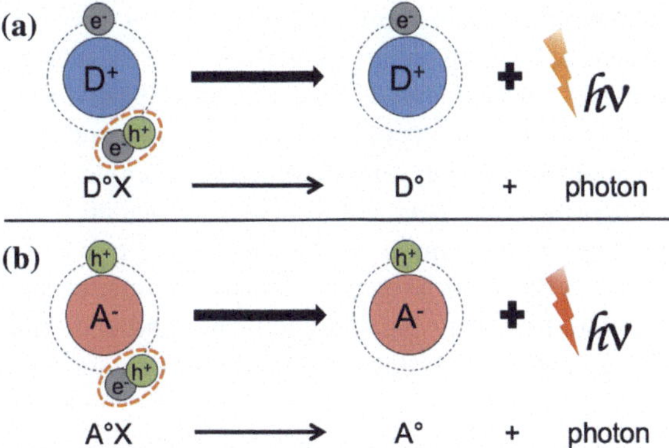

Fig. 4.25 Recombination scheme of a donor bound exciton **a** of an acceptor bound exciton. After Aurelie Pierret-Phd Thesis—University Paris VI-2013

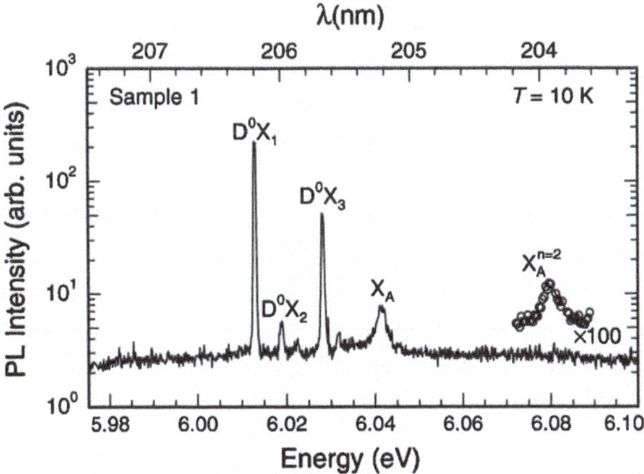

Fig. 4.26 Photoluminescene of Aluminum nitride showing different bound exciton lines. After [30]

fields, ...) and external perturbations (electric fields, magnetic fields, high densityes of photo-created electron-hope pairs).

Effective mass impurities are a very specific kind of impurities for which the binding energy is well accounted for by a description in terms of effective mass and effective dielectric constant. Their advantage is that, being quite shallow they easily ionize furnishing a complementary free electron and hole as soon as the temperature increases. The situation is in general more complex than: what is doing carbon (a grow IV element like silicon) for instance in gallium nitride: is it a donor-like impurity? (then it should replace gallium) or is it an acceptor (if replacing nitrogen?). Arsenic in gallium nitride does not substitute nitrogen to behave like an isovalent impurity would do; it substitute gallium and then behaves like as double donor.

Atomic levels of aluminum, indium hybridize with the Bloch waves of the host crystal to form an alloy instead of forming impurity levels. Atomic vacancies also form levels deeper in the band gap. There is thus a large variety of behaviors connected with the incorporation of impurities or crystal imperfections (vacancies, dangling bonds produced by dislocations) in crystals. the situation even complexities in line with the propensity some impurities may have not to distribute randomly in the crystal but to coalesce so that they can form multi-site fee exciton (or free-carrier) localization centers. Photoluminescence (and its cousin cathodo-luminescence which utilizes the energy of an electron beam to break a chemical bond and thus to produce an electron hole pair) are extremely efficient methods for diagnosing the doping of a semiconductor.

In nitrides the most common impurity being oxygen, these materials are generally non-intentionally grown n-type (with a residual doping of about 10^{17} atoms cm^{-3}. Indium nitrite is degenerated).

Free to bound transitions can be observed which correspond for intense to the recombination of a free electron with a hole at a neutral acceptor: $A^0 + e \rightarrow A_- + \hbar\omega(eA_0)$ or a free hole can recombine radiatively with the hole of a neutral donor: $D^0 + h \rightarrow D_+ + \hbar\omega(hD_0)$. Such transitions are labelled $e - A_0$ or $h - D_0$ and correspond to photon energies: $\hbar\omega(eA_0) = E_g - E_A^b$ and $\hbar\omega(hD_0) = E_g - E_D^b$ where E_A^b (resp. E^bD)) are the binding energy of the hole (resp. electron) to the acceptor (resp.donor).

The donor binding energy E_D^b being in general smaller than the acceptor binding energyE_A^b the electron to accepor recombination is easier to observe than the hole to donor recombination. Donor ionization being easier than acceptor ionization, temperature produces an increasing of the electron concentration, which leads to an enhancement of the electron to acceptor transition with respect to the deeper recombination band that may be observed and correspond to donor-acceptor pair recombinations.

Donor-acceptor pair transitions correspond to the radiative recombination of the electron of a neutral donor with the hole of a neutral acceptor: $A_0 + D_0 \rightarrow A_- + D_+ + \hbar\omega_{DAP}$. The donors and acceptors are spatially separated by a quantity $\vec{R}_{DA}$, and the recombination occurs as a consequence of the overlap of electron and hole wave functions.

$$\hbar\omega_{DAP} = E_g - E_D^b - E_A^b + \frac{e^2}{4\pi\epsilon_0\epsilon_*}\frac{1}{\|\vec{R}_{DA}\|}$$

For the sake of the simplicity, the dielectric constant is taken isotropic here, again. The possible values for $\vec{R}_{DA}$ lead to transitions that may have very different values and general one observes a broad asymmetrical band or a series of well defined sharp lines that correspond to the different possibilities for $\vec{R}_{DA}$. The asymmetry of the shape of the broad band is correlated to the reduction of the recombination probability of electron and holes for distant donor and acceptors. In Fig. 4.27 are sketched the principal recombination processes that can be observed in wurtzitic semiconductors

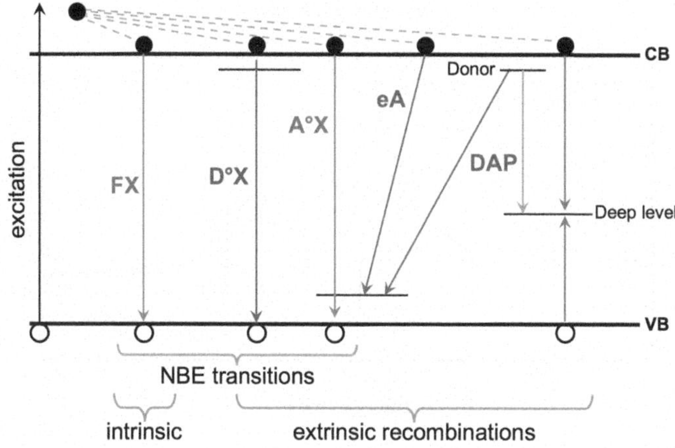

Fig. 4.27 Different recombination processes in a semiconductor: free exciton, donor-bound exciton, acceptor-bound exciton, electron to acceptor, donor-acceptor pairs, free electron to deep level, and free hole to deep level

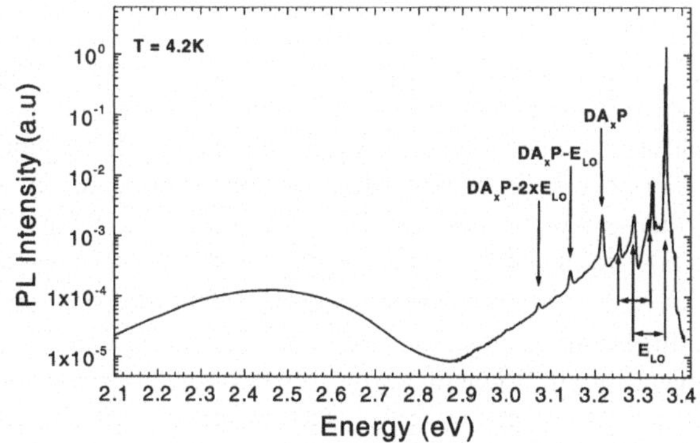

Fig. 4.28 Photoluminescence spectrum of bulk ZnO showing excitonic, donor acceptor pair (DAP) and deep level emission. The corresponding phonon replica with longitudinal optical phonons (LO) are indicated (HeCd excitation). After [31]

In Fig. 4.28 has been reported a large energy scale spectrum for ZnO. One distinguishes at high energy sharp needles that correspond to the Near Band Edge transitions sketched in Fig. 4.27. Then DAP and a broad band at 2.3–2.4 eV. Some transitions are reported as phonon replicas or NBE recombinations or DAPs. Such phonon replicas are observed when radiative recombination has a substantial probability to occur either with producing a photon an energy E, or photons of energy $E - n\hbar\omega_{LO}$ (then the electron-hole recombination produces a photon of energy E-$n\hbar\omega_{LO}$ plus n LO phonons of energy $\hbar\omega_{LO}$). In that case the relative intensities of these phonon replicas can be fitted by a Poisson Law:

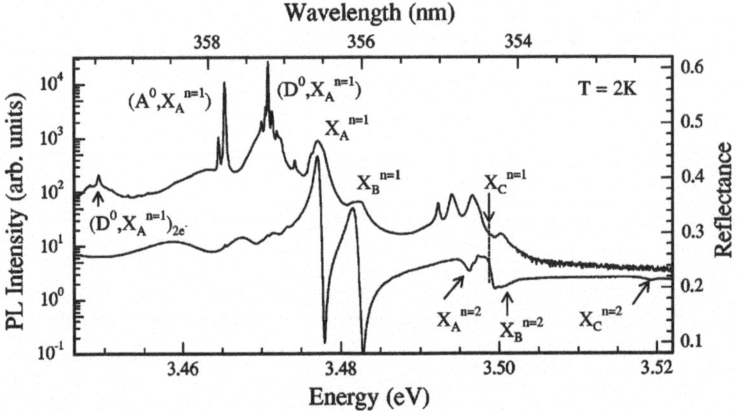

Fig. 4.29 Photoluminescence and reflectance spectroscopy of excitonic transitions in high-quality homoepitaxial GaN film

$$I(E - n\hbar\omega_{LO}) = I(E)S^n/n!$$

S is called the Huang-Rhys factor. The probability to observe phonon replicas increases with the amount of electrons in particular for DAPs.

The broad band at low energy involves deep levels and are often found in wide band gap semiconductors. specific impurities (deep donors and deep acceptors), vacancies can contribute to these recombinations. The prototype of such band is the famous yellow band in GaN. Eradication of these broad band is a challenge and a goal for crystal growers since they have deleterious influences on materials and devices.

Finally, crystalline defects may efficiently trap carriers. The capture cross section of such defects may be high. When increasing temperature, ones increases the mobility of free carriers and this favors their tramping to such defects, strongly quenching the photoluminescence intensity: the non radiative recombination may dominate the radiative one at ambient temperature. Increasing the density of excitation, it may be possible to saturate these levels and the radiative recombination may prevail again with respect to the non radiative one.

4.8.3 High Resolution Spectroscopy in Wurtzite Semiconductors: The GaN Case

In Fig. 4.29 is reported a high resolution, low temperature photoluminescence spectrum of GaN with its corresponding reflectivity analog [32].

There are a lot of informations to extract from the comparison of both spectra: the energy of A, B, C excitonic transitions, the exciton binding energy, the localization energies of donor and acceptor bound excitons and a complementary weak line at

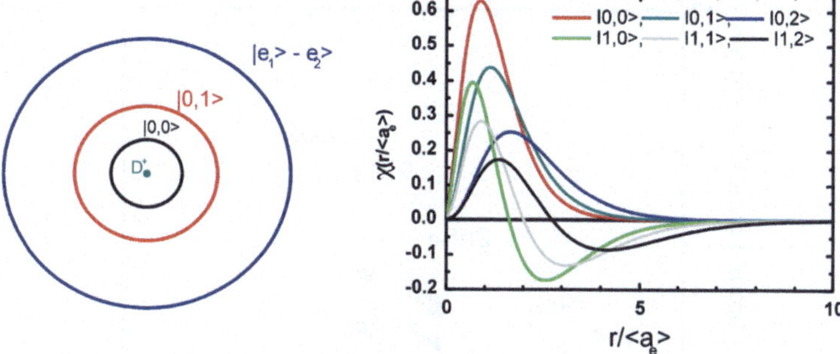

Fig. 4.30 *Left* relative representation of the position of electrons (in the *outer circle*) and of the hole for two low-energy levels of the rotator state with respect to the donor nucleus. *Right* wave functions of the hole in the context of the self-consistent non-rigid rotator model including mass anisotropies. After [33]

21.6 meV below the energy of thee donor bound exciton line which is measured in this sample at 3.4709 eV. This line is interpret as a two electron transition. In the classical recombination scheme, a donor bound exciton is detected through the recombination of one of the two electrons of the multiple particle complex with the hole. The initial state consists in a neutral donor plus an electron-hole pair, or a neutral donor plus an exciton, or a fixed positive nucleus, the two mobile electrons and a mobile hole. All these guys are experiencing Coulomb interaction, and the balance between attractions, repulsions, and also spin-exchange related contributions gives a stable state. The final state (neutral donor without bound exciton) consists in a positive fixed nucleus and a mobile electron. The energy levels of this electron and its allowed wave functions, if the neutral donor is envisioned like a hydrogenic atom, are the analogs of the hydrogen levels. At low temperature, and in the most probable situation, the donor bound exciton state is built from antisymmetric combination of two 1s electron states in electrostatic interaction with a hole. One cannot exclude that instead of this the electron occupies another state of the hydrogen series, the 2s state for example. This situation corresponds to an excited state of the donor. When a photon is detected at 3.4709 meV this corresponds to a recombination process where the final state is the donor ground state with the donor electron occupying a 1s state. When a recombination occurs 3.4493 eV, the remaining electron is promoted to a 2s state and the donor is left in an excited state. The energy of the emitted photon is 3.4709 eV diminished from the energy amount required to promote the donor electron from a 1s state to 2s. The corresponding energy (21.6 meV) time 4/3 gives the donor binding energy (here 28.8 meV). The 3.4493 eV transition is called **two-electron transitions.**

We also remark that the donor bound exciton transitions displays weak complementary lines. These lines are for some of them interpreted in terms of rotational states of the hole. Arguments based on physics of hydrogen atoms and expressions of Bohr radius indicate that the Bohr radius of the electrons (a_e) is larger than the Bohr radius of the hole (a_h) (the hole effective mass is higher) See Fig. 4.30a. The

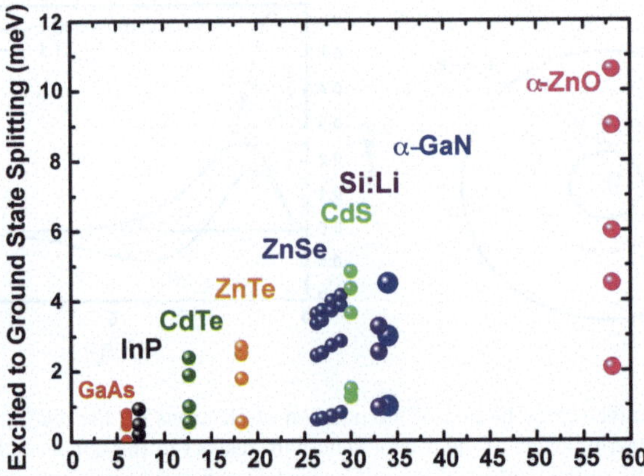

Fig. 4.31 Energy splittings of the donor bound exciton line versus donor binding energy computed in the context of the non rigid self-consistent model versus donor binding energy. After [33]

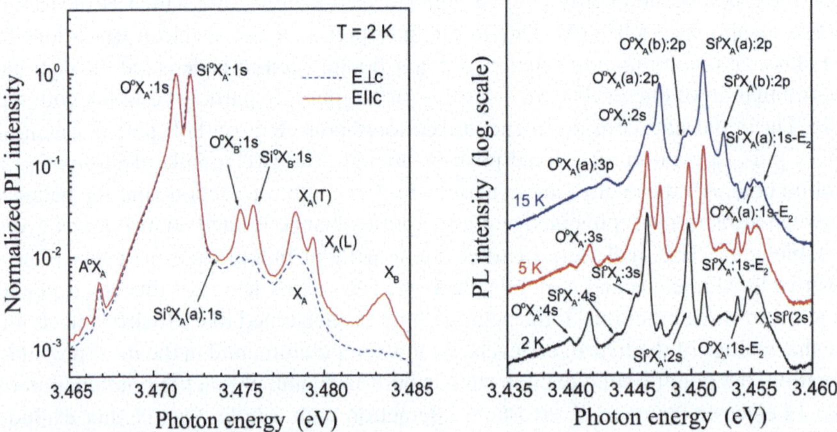

Fig. 4.32 *Left* polarized photoluminescence of GaN in the donor-bound exciton region. The Pyonting vector of the collected photon is orthogonal to [0001] direction so that σ and π polarizations of the emitted photons can be detected. *Right* photoluminescence of GaN in the two-electron transitions energy region. After [34]

hole behaves like a three-dimensional non-rigid rotator with ground state and excited states. Some typical wave function of the hole are plotted in Fig. 4.30b and the energy splittings due to this specific mechanism are plotted versus donor binding energy in Fig. 4.31. We refer the reader to [33] for a more detailed treatment of this excitonic fine structure problem.Some typical wave function of the hole are plotted in Fig. 4.30b and the energy splittings due to this specific mechanism are plotted versus donor binding energy in Fig. 4.31.

Figure 4.31 indicates a general trend versus binding energy. As discussed by Gil et al, there are selection rules between the lines of the fine structure splitting of the donor bound exciton and the two electron line (which also exhibit fine structure splitting). This was finally treated by Monemar et al. in [34]. Figure 4.32 details the intricate shape of the photoluminescence spectra in the donor bound exciton energy region (left) and in two-electron transition energy region (right) in case of GaN. Note two-electron replicas from oxygen-related and silicon-related donor bound excitons.

To conclude, we will give some words regarding acceptor bound excitons. In that case the multiple particle complex consists in a negative nucleus, two holes and one electron.The initial state has to be computed in the framework of angular momentum algebra taking into account of the sermonic nature of all those particles. the ground state is a negative fixed nucleus and a hole. The full treat meant of that problem has been performed by Gil et al. in [35].

Although it is now established that p dopant is magnesium for GaN, the spectroscopy of Mg-doped GaN is still in way of being consolidated. Regarding zinc oxide, nothing is clearly established at the time of writing.

4.9 Semiconductor Alloys

The simplest way for alloying a binary compound semiconductor is to replace one element by a statistical distribution of this element with another element of Mendeleev's table. Then one novel material is obtained which lattice parameter can be tuned as well as its band structure and the value of the bandgap. Such alloys have been determinant for realizing semiconductors heterostructures. Examples of such alloys are ZnS_xSe_{1-x}, CdS_xSe_{1-x}, $Zn_xCd_{1-x}S$, $Zn_xCd_{1-x}O$, $Al_xGa_{1-x}N$, $In_xGa_{1-x}N$, $Al_xIn_{1-x}N$, where x runs between 0 to 1 in order to cover the whole composition range. These are called ternary alloys.

Alternatively, to tune simultaneously the band gap and the value of the lattice parameter, or the refractive index or band offsets in heterostructures, quaternary alloys (four different atoms) and even quinary alloys (five different elements) may be grown. In that case, the alloy composition is extremely difficult to control but the growth of such materials being dictated by devices applications researchers involve in this tricky topic.

Thermodynamics does not always predict the stability of such alloys to be possible through the whole composition range, but epitaxial growth technics when the growth is ruled by kinetics of chemical species permit to grow materials beyond the strict limits of thermodynamical predictions.

There are some cases which are redhibitory situations among which are for instance $Zn_{1-x}Mg_xO$ alloys: ZnO is a wurtzitic semiconductors while MgO prefers to grow as a rocksalt type crystal. $Zn_{1-x}Mg_xO$ can be obtained as wurtzitic crystal up to about $x = 0.25$.

Although the lattice parameters of alloys of the $A_{1-x}B_xC$ can be measured using simple X-ray diffraction systems for "gentle" solid solutions as $(1-x)$ time the lattice parameter of AC + x time the lattice parameter of BC, optical properties immediately deviates from this virtual crystal approximation leading to simple behaviors called

Vegard's laws. High performance X-ray examinations like Extended X-Ray Absorption Fine Structure (EXAFS) indicate that the real situation is not so simple and that in reality, alloys are obtained with local fluctuations of compositions which lead to physical properties that can be, for instance in the didactical case of ternary alloys significantly different from an average of the properties of the binary compounds sitting at both ends of the whole composition range. In the case of Al_xGa_{1-x} alloys, the GaN distance decreases by an amount of 0.25

The band gap of random alloys is measured by photoluminescence of reflectivity and one obtains a quadratic dependence with x:

$$E_g(A_{1-x}B_xC) = (1-x)E_g(\mathrm{AC}) + \mathrm{xE_g(BC)} - \mathrm{bx(1-x)}$$

The quantity b is called the bowing parameter. Bowing parameter b is a positive quantity which equals typically 1 eV, but it can be composition-dependent that is to say it can vary with x in very specific situations. The value of b is ruled by many physical quantities including the chemical alloy disorder and the experimental values available in the literature are in general proposed with a large scattering.

In a perfect alloy $A_{1-x}B_xC$, considering the sublattice dedicated to chemical species A and B, each node of this sub-lattice is statistically occupied by atom A (with a probability $1-x$) or atom B (probability x). This gives a chemical disorder, which is probed by the exciton and the photoluminescence broadens. The theoretical treatment of the inhomogeneous band gap broadening E_0 has been treated by [36]. Using a binomial statistical description of the random atomic distribution,they obtain:

$$E_0 = \frac{1}{178}\frac{\alpha^4 M^3 x^2(1-x)^2}{\hbar^6 N^2}$$

where $\alpha = dE_g/dx$, M is the exciton mass and N is the volume density of atomic sites. The value of E_0 varies like the third power of the exciton mass. As the Bohr radius, and exciton binding energy too are correlated with the exciton mass, the equation indicates that small size (high binding energy) excitons are very sensitive to fluctuations of alloy composition. Quantity α requires some attention. In the absence of bowing, it would be a constant quantity, and the maximum of alloy broadening would occur for x = 0.5. The existence of a bowing shifts the maximum of the broadening towards compositions near the high band gap semiconductor. This is illustrated in Fig. 4.33 for $Al_xGa_{1-x}N$ alloys.

There is a possibility that chemical disorder locally fluctuates stronger than predicted by the binomial statistics. Carriers may be strongly localized in specific regions of the crystal as indicated in Fig. 4.34a.Then instead of one broad line it is possible to observed several lines as an evidence of bimodal or spinodal decomposition of the alloy at the scale of the size of the laser beam used to probe the optical properties of material. The evolution of the photoluminescence is ruled by a very complex scenario when the temperature increases. The exciton population in the disordered alloy can be viewed as the sea surrounding an archipelago of slightly decoupled islands and atolls at low tide. The atolls and the islands are not so high and the sea can totally

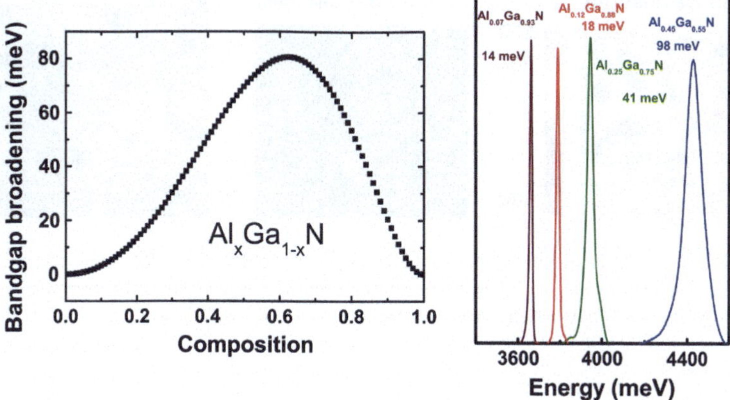

Fig. 4.33 *Left* inhomogeneous broadening of photoluminescence in AlGaN alloy due to chemical disorder. *Right* photoluminescence of AlGaN for various aluminum compositions

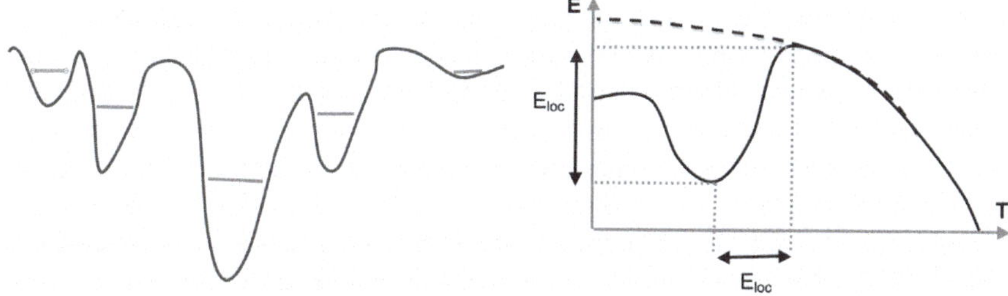

Fig. 4.34 *Left* spatial fluctuations of the band gap in disordered alloys. the fluctuation of compositions are supposed to be strong enough to trap excitons. *Right* S-shaped photoluminescence energy in disordered materials. Note the way we define the localization energy E_{loc}

recovers them at high tide. Of course the number of such archipelagos should be high enough to observe a pronounced effect on the photoluminescence. When the temperature increases, that is to say when the tide is getting higher and higher, sea surrounds more and more efficiently the islands. At high tide, there is not signature of the complex structure of the ground behind the sea level. The situation is more or less the same in alloys. At low temperature, excitons are localized in low band gap energy regions. When increasing the lattice temperature, excitons get more and more mobile. They may escape from their trapping regions and get localized in the deep potential regions of the crystal, and the average photoluminescence energy decreases. When further increasing the temperature, they kinetic energy increases again until they become fully delocalized and the photoluminescence energy increases again. At that temperature, they have reached a mobility edge energy and the photoluminescence energy follows the free exciton energy. The excitons are now fully delocalized, weakly affected by details of potential fluctuations. The sea has is high tide. The evo-

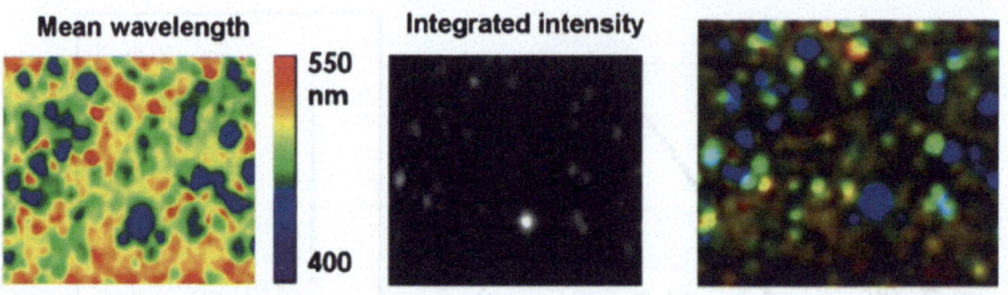

Fig. 4.35 *Left* the inhomogenity of the emitted energy obtained by cathodoluminescencemapping of an InGaN layer. *Center* The intensity of emitted light. *Right* the product of the wavelength by the intensity of emission. Courtesy Professor Kevin O' Donnell, Strathclyde University, Glasgow

lution of the photoluminescence energy with temperature is said S-shaped and it is illustrated in Fig. 4.34b.

In Fig. 4.35 (left) is represented the inhomogeneity of the emitted energy obtained by cathodoluminescence mapping of an InGaN layer. At the center of the triptych is represented the intensity of emitted light. In the right hand part is represented the product of the wavelength by the intensity. The data has been provided to the author by Professor Kevin O' Donnell, Strathclyde University, Glasgow. This experiment probes the sketch of inhomogeneous distribution of atoms in the alloys, which gives in color the evidence of an inhomogeneous distribution of the band gap as the one sketched in Fig. 4.34a.

A consequence of the spectral distribution of the band gap is that one observes an energy shift between reflectivity (or absorption) and photoluminescence experiments. Excitons thermalize in the low energy region of the energy distribution, where the density of states is not so high in general while reflectivity (or absorption) is marked at energies for which the density of state becomes high (at the delocalized free exciton energy). Low temperature photoluminescence is red-shifted in energy with respect to the reflectivity. This is illustrated in Fig. 4.36 in the specific case of $Ga_{0.93}Al_{0.07}N$ at 2K. This energy difference is called Stokes-shift and it is very comparable with the localization energy. When increasing the temperature, thermal population effects favor the high energy localized excitons and this Stokes shifts vanishes.

4.10 Photonics in High Quality Thin Films

Reflectance spectra of high-quality samples exhibit well-marked structures at the energies of the exciton-polaritons. Additional optical features can be observed on samples having discontinuous allowed values for the wave number of the center of mass $\vec{K}$. This requires samples having a finite, adapted thickness and for which one encounters quantizing boundary conditions [4–10]. The energies of these additional optical features agree with the results of a simple modelling of quantized wave-

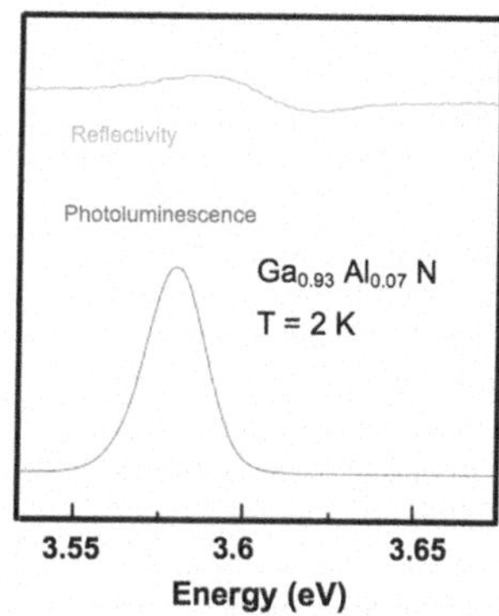

Fig. 4.36 Reflectivity (*top*) and photoluminescence (*bottom*) spectra of $Ga_{0.93}Al_{0.07}N$ at 2K

vectors. The characteristic lengths that come into play, and that are required to observe such discrete confined states are: the film thickness L_z, the wavelength of light λ, and the Bohr radius of the exciton a_B. Generally speaking, any form of quantization can be observed only if the coherence length of the confined particle is much larger than the film thickness, so that standing waves can establish.

One can distinguish three regimes of confinement:

- For $L_z \gg \lambda \gg a_B$, the distribution of the excitonic states remains quasi continuous while "optical" confinement occurs in the photon-like branch for $\omega < \omega_L + \sqrt{\omega_L \omega_{LT}}$. The quantized energies depend on an integer number N, following:

$$E(\vec{K}, N) = \frac{\hbar c}{\epsilon_B}\left[K^2 + \frac{N^2\pi^2}{L_z^2}\right]$$

  where ϵ_B is the background dielectric constant, $\vec{K}$ is the in-plane dispersion wave vector.
- Further reducing L_z ($L_z \approx \lambda \gg a_B$), the quantization can be observed in the excitonic-like branch, in the so-called "exciton center-of-mass quantization" regime. A reasonable approximation of the quantized energies is given by:

$$E(\vec{K}, N) = \frac{\hbar^2}{2M}\left[K^2 + \frac{N^2\pi^2}{L_z^2}\right]$$

where M is the on-axis translation mass of the exciton.

In the range of energies such that $\omega_L < \omega < \omega_L + \sqrt{\omega_L \omega_{LT}}$, the dispersion of the upper polariton branch is photon-like, but its wave function is mostly exciton-like. Nevertheless, Fabry-Pérot modes of this upper polariton branch can appear, if the material quality allows for it.

- A third situation can be defined, when $\lambda \gg L_z \approx a_B$. This is, at its lower limit, the quantum well situation. In any case, the observation of such effects requires large enough coherence lengths, in other words high quality samples.

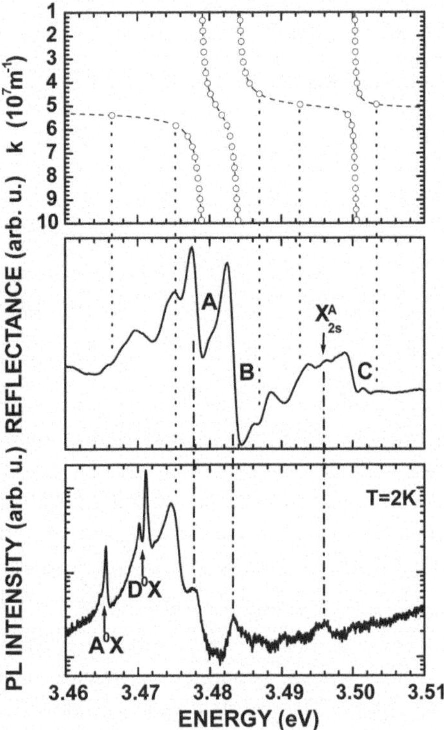

Fig. 4.37 *Upper part* the dispersion relation of exciton-polaritons in unstrained GaN (*dashed lines*) and the allowed modes for a 700 nm *thick layer* (*open circles*). *Middle part* the reflectance spectrum of the GaN homoepitaxial layer, taken at T = 2K. Clear features are seen in addition to the usual A, B and C 1s exciton resonances. *Dotted lines* show the connection of the confined polariton modes with the reflectance features and with the additional PL lines shown in the lower part of the figure (After [37] and references therein)

To model these confined modes, one has to compute the modifications of the complex dielectric function near the excitonic resonances. One can use a series of damped Drude-Lorentz oscillators. In the case of wurtzite GaN, we need to model at least three resonances corresponding to A, B and C 1s excitons. this three-resonance description, but the can be extended to inclusion of 2s excited states. In GaN, the binding energy of the A exciton is around 25 meV, which gives a 2s feature 19 meV above the main 1s A transition. The dispersion relations for the transverse modes are obtained by solving: $\omega^2\epsilon(K,\omega) = c^2K^2$. The longitudinal modes rather satisfy: $\epsilon(K,\omega) = 0$. In this work, we have not included the resonance introduced by the 2s state of the A exciton, although we observe the corresponding feature on the reflectance and PL spectra (T = 2 K) shown in the middle and lower parts of Fig. 4.37, respectively.

On the reflectance spectrum, shown in the middle of Fig. 4.37, we observe, in addition to the usual A-1s, B-1s, A-2s and C-1s excitonic resonances, several features which correspond to confined exciton-photon hybrid modes in the epilayer. This interpretation nicely agrees with the dispersion relations plotted in the upper part of the figure. In this plot, open circles correspond to the quantized values of the wave-numbers, i.e. the confined modes allowed by the present layer thickness.We observe confined modes from the photon-like branch between B and C resonances, which are not seen in reflectance spectra for thicker high-quality GaN layers. It may be argued that those features are merely Fabry-Pérot oscillations, since they correspond to the photon-like part of the dispersion. Nevertheless, "photon-like" does not mean "purely photon": the non-constant energy difference between the "fringes" testifies to the strong nonlinearity of the dispersion relation, itself related to the partial excitonic character. More significant is the observation of characteristic reflectance features slightly below the usual energy of resonance A. These features are confined modes in a regime where the wavefunction of the polariton is well shared between photon-and exciton-like characters. This explains one of the most striking observations in this reflectance spectrum: the apparent oscillator strength of exciton B is much stronger than that of exciton A. This dramatic reduction of the apparent oscillator strength of the A exciton is neither predicted by band-to-band nor by excitonic calculations, even if residual strain effects are properly included. The apparent small strength of resonance A is fully explained in the exciton-polariton picture. For such thin layers, the apparent strength of a resonance (the amplitude of the reflectance feature) can be seen as resulting from the accumulation, near a given energy, of numerous allowed confined modes that originate from the weakly dispersive A and B exciton-like polariton branches. The oscillator strengths of these accumulating modes decrease when the wavevector goes away from the photon wavevector. The result is the apparent small oscillator strength of the feature at 3.478 eV, and the appearance of an additional reflectance feature near 3.475 eV. The excitonic character of the standing waves is confirmed by the additional radiative features observed in the PL spectrum in the lower part of Fig. 4.37: in addition to the A^0X and D^0X excitonic complexes, we observe the clear splitting of the A exciton line. The usual photoluminescence peak A at 3.478 eV is widely dominated by the contribution near 3.475 eV, denoted A′ hereafter. This photoluminescence line corresponds to the above-mentioned additional reflectance feature. It also corresponds reasonably to one allowed polariton mode, as shown by a dotted line in Fig. 4.37. The hypothesis of D^0Xcomplexes near 3.475 eV is not compatible with the present lineshape: one would rather expect a doublet of narrow lines, comparable to the D^0X and D^0X lines.The A′ photoluminescence line and the apparent weakness of the A resonance do not arise from the GaN substrate due to its n-doping and/or due to a strain state different from that of the epilayer. In such cases, two different B resonances should appear on the reflectance spectrum, instead of this strong feature. The A′ line can result from a specific thermalization pattern, due to the existence of forbidden K vectors. In the case of quasi-continuous dispersion relations, like for thick layers, the polaritons access radiative states, with a wave vector close to the photon wave vector K_{photon}, by emitting acoustic phonons. This gives rise to the usual A, B and C PL lines. This thermalization is an efficient and fast process, because of the quasi-continuous dispersion of polaritons. In other words, away from K_{photon}, the thermalization is much faster (likely) than the radiative recombination. It is only when the polariton wave vector becomes close to K_{photon}, and when the group velocity becomes large, that the radiative recombination becomes much faster than the thermalization. In the case of thin layers, with discrete K vectors (i.e. quantized along the growth direction), the thermalization rate is reduced because not all phonons can fulfill the K-conservation rule and because there are only a few K-states. Nevertheless, in the weakly dispersive exciton-like branches of the polariton dispersion, the phonon emission remains quite efficient, because the confined modes are all close to each other in energy. In any case, in the exciton-like branches, the phonon emission remains faster than the radiative recombination. But the situation changes drastically, compared to the case of thick layers, near K_{photon} because there exist gaps of polariton states. Some of these gaps lie at energies where, normally (for thick layers), the radiative recombination would take place, being faster than thermalization. For a very thick layer, the polariton can reach these energies by fast phonon emission. For the present thin layer, at low temperature, the exciton-polariton can only thermalize onto the next, lower-lying, allowed (radiative) state. In conclusion photonics effects may be observed, giving optical features which interpretation are often untrivial due to the dispersion of the dielectric constant at energies near the excitonic resonances.

References

1. A.J. Fischer, W. Shan, G.H. Park, J.J. Song, D.S. Kim, D.S. Yee, R. Horning, B. Goldenberg, Phys. Rev. B **56**, 1077 (1997)
2. *Exciton Spectrum of Cadmium Sulfide*, David G. Thomas and John J. Hopfield, Physical. Review **116**, 573 (1959)
3. *Aspects of polaritons*, John J. Hopfield Proc. Int. Conf. Physics of semiconductors, Kyoto, 1966, J. Phys. Soc. Japan, 21, suppl.77-88, (1966), *Theory of the Contribution of Excitons to the Complex Dielectric Constant of Crystals*, J.J.Hopfield, Physical. Review **112**, 1555 (1958)
4. *Fano-Hopfield model and photonic band gaps for an arbitrary atomic lattice*, Mauro Antezza, Yvan Castin. Phys. Rev. A 80, 013816, (2009)
5. *Spatial dispersion effects in a mixed mode of exciton polariton in CdS*, M.V.Lebedev, V.G.Lysenko and V.B. Timofeev, Sov. Phys. JETP 59, 1277, (1984)
6. *Absorption, reflectance, and luminescence of GaN epitaxial layers*, R.Dingle, D.D.Sell, S.E.Stokowski and M. Ilegems (Physical Review B 4, 1211, (1971)
7. *Effects of biaxial strain on exciton resonance energies of hexagonal GaN heteroepitaxial layers*, S. Chichibu, A. Shikanai, T. Azuhata, and T. Sota, A. Kuramata and K. Horino, S. Nakamura, 3766 Appl. Phys. Lett. 68 (26), 3766, (1996)
8. *Reflectance and emission spectra of excitonic polaritons in, GaN*, K. Torii, T. Deguchi, T. Sota, K. Suzuki, S. Chichibu, and S. Nakamura. Phys. Rev. B **60**, 4723 (1999)
9. *Determination of the spin-exchange interaction constant in wurtzite GaN*, M.Julier, J. Campo, B. Gil, J.P.Lascaray, and S. Nakamura, Physical Review B 57, R6791, (1998)
10. Negative spin-exchange splitting in the exciton fine structure of AlN, Martin Feneberg, Maria Fatima Romero, Benjamin Neuschl, Klaus Thonke, Marcus Röppischer, Christoph Cobet, Norbert Esser, Matthias Bickermann, and Rüdiger Goldhahn, Appl. Phys. Lett. 102, 052112 (2013)
11. *Huge electron-hole exchange interaction in aluminum nitride*, Ryota Ishii, Mitsuru Funato, and Yoichi Kawakami, Phys. Rev. B 87, 161204 (2013)—*Excitonic structure of bulk AlN from optical reflectivity and cathodoluminescence measurements*, E. Silveira, J. A. Freitas, Jr., O. J. Glembocki, G. A. Slack, and L. J. Schowalter Phys. Rev. B **71**, 041201 (2005)
12. *Influence of exciton-phonon coupling and strain on the anisotropic optical response of wurtzite AlN around the band edge*, Georg Rossbach, Martin Feneberg, Marcus Roppischer, Christoph Werner, Norbert Esser, Christoph Cobet, Tobias Meisch, Klaus Thonke, Armin Dadgar, Jrgen Blasing, Alois Krost, and Rudiger Goldhahn, Phys. Rev. B 83, 195202 (2011)
13. *Transmission spectra of ZnO single crystals*, WY. Liang and AD. Yoffe. Physical Review Letters **20**, 59 (1968)

14. *Valence-band ordering in ZnO*, D. C. Reynolds, D. C. Look, and B. Jogai, C.W. Litton, G. Cantwell, W. C. Harsch. Phys. Rev. B **60**, 2340 (1999)
15. *Strain-fields effects and reversal of the nature of the fundamental valence band of ZnO epilayers*, Bernard Gil, Alain Lusson, Vincent Sallet, Said-Assoumani Said-Hassani, Robert Triboulet, and Pierre Bigenwald Japanese Journal of Applied Physics Letters, 40, L1089, (2001)
16. *The oscillator strengths of, A, B, and C excitons in ZnO films, Bernard Gil, Physical Review B Rapid, Communications***64**, 201 310 (2001)
17. *ZnO as a material mostly adapted for the realization of room-temperature polariton lasers*, M.Zamfirescu, A.Kavokin, B.Gil, G.Malpuech, M.Kaliteevski, Physical Review B **65**, 161 205–1 (2002)
18. *Polarized photoreflectance spectra of excitonic polaritons in a ZnO single crystal*, S. F. Chichibu, T. Sota, G. Cantwell, D. B. Eason, and C. W. Litton, J. Appl. Phys. **93**, 756, (2003)
19. *Accurate determination of homogeneous and inhomogeneous excitonic broadening in ZnO by linear and nonlinear spectroscopies*, E. Mallet, P. Disseix, D. Lagarde, M. Mihailovic, F. Reveret, T. V. Shubina, and J. Leymarie, Phys. Rev. B **87**, 161202(R) (2013)
20. *Optical anisotropy of A- and M-plane InN grown on free-standing GaN substrates*, P. Schley, J. Rathel, E. Sakalauskas, G. Gobsch, M. Wieneke, J. Blasing, A. Krost, G. Koblmuller, J. S. Speck, and R. Goldhahn, Phys. Status Solidi A **207**, 1062 (2010)
21. *Temperature dependent band gap and homogeneous line broadening of the exciton emission in ZnO*, R. Hauschild, H. Priller, M. Decker, J. Bruckner, H. Kalt, C. Klingshirn, phys. stat. sol. (c) **3**, 976–979 (2006)
22. *Temperature dependence of the energy gap in semiconductors*, Y. Varshni. Physica **34**, 149 (1967)
23. *Pässler modeling [Parameter Sets Due to Fittings of the Temperature Dependencies of Fundamental Bandgaps in Semiconductors*, R. Paessler, Physica Status Solidi (b) **216**, 975 (1999)
24. *Temperature dependence of the energy gap in GaN bulk single crystals and epitaxial layer*, H. Teisseyre, P. Perlin, T. Suski, I. Grzegory, S. Porowski, J. Jun, A. Pietraszko, and T. D. Moustakas, J. Appl. Phys. **76**, 2429 (1994)
25. *Excitons in Degenerate Semiconductors*, G. D. Mahan. Phys. Rev. **153**, 882 (1967)
26. *The determination of the bulk residual doping in indium nitride films using photoluminescence*, M.Moret, S. Ruffenach, O Briot, and B Gil, Applied Physics Letters, **95**, 031910, (2009)
27. *Modeling of the free-electron recombination band in emission spectra of highly conducting n-GaN*, B. Arnaudov, T. Paskova, E. M. Goldys, S. Evtimova, and B. Monemar, Phys. Rev. B **64**, 045213, (2001)
28. *Energy position of near-band-edge emission spectra of InN epitaxial layers with different doping levels, B.* Arnaudov, T. Paskova, P. P. Paskov, B. Magnusson, E. Valcheva, B. Monemar, H. Lu, W. J. Schaff, H. Amano, and I. Akasaki, Phys. Rev. B **69**, 115216, (2004)
29. *Effect of silicon and oxygen doping on donor bound excitons in bulk GaN*, G. Pozina, S. Khromov, C. Hemmingsson, L. Hultman, and B. Monemar, Physical Review B **84**, 165213 (2011)
30. *Sharp bound and free exciton lines from homoepitaxial, AlN, Martin Feneberg, Benjamin Neuschl, Klaus Thonke, Ramon Collazo, Anthony Rice, Zlatko Sitar, Rafael Dalmau, Jinqiao Xie, Seiji Mita, and Rudiger Goldhahn.* Phys. Status Solidi A **208**, 1520 (2011)
31. *Bound exciton and donoracceptor pair recombinations in ZnO, B. K. Meyer, H. Alves, D. M. Hofmann*, W. Kriegseis, D. Forster, F. Bertram, J. Christen, A. Hoffmann, M. Straburg, M. Dworzak, U. Haboeck, and A. V. Rodina, phys. stat. sol. (b) **241**, 231 (2004)
32. *Photoluminescence and reflectance spectroscopy of excitonic transitions in high-quality homoepitaxial GaN films*, K. Kornitzer, T. Ebner, K. Thonke, and R. Sauer, C. Kirchner, V. Schwegler, and M. Kamp, M. Leszczynski, I. Grzegory, and S. Porowski, Phys. Rev. B **60**, 1471, (1999)
33. *Internal structure of the neutral donor bound exciton complex in cubic zinc blende and wurtzite semiconductors*, B.Gil, P.Bigenwald, M.Leroux, P.Paskov and B.Monemar, Physical Review B **75**, 085204, (2007)

34. *Recombination of free and bound excitons in GaN*, B. Monemar, P. P. Paskov, J. P. Bergman, A. A. Toropov, T. V. Shubina, T. Malinauskas, and A. Usui, Physica Status Solidi (b) **245**, 1723, (2008)
35. *Internal structure of acceptor-bound excitons in wide-band-gap wurtzite semiconductors*, B.Gil, P.Bigenwald, P.Paskov and Bo Monemar, Physical Review B **84**, 085211, (2010)
36. *Band edge smearing in solid solutions*, S.D Baranowski and A.L.Efros, Soviet Physics. Semiconductors **12**, 1328 (1978)
37. *Confined exciton-polariton modes in a thin, homo-epitaxial, GaN film grown by molecular beam epitaxy*, Aurelien Morel, Thierry Taliercio, Pierre Lefebvre, Mathieu Gallart, Bernard Gil, Nicolas Grandjean, Jean Massies, Izabella Grzegory, and Sylvester Porowski, Materials Science and Engineering: B Volume **82**, 173, (2001)

Chapter 5
Optical Properties of Quantum Wells and Superlattices

In this chapter are reviewed the basic concepts required to compute the band structure of quantum wells. Then we discuss the variational treatments of the exciton binding energies. Quantitative applications are given in the case of GaN–AlGaN, GaInN-GaN and ZnO–ZnMgO quantum wells grown along arbitrary orientations so that Quantum Confined Stark Effect exists or not. Finally we discuss the properties of quantum dots.

5.1 Basic Theoretical Concepts Borrowed from Quantum Mechanics Text Books

5.1.1 Square Quantum Wells and One-Band Envelope Functions

A comprehensive description of the concept of semiconductor quantum wells and superlattices have been offered to us by Gérald Bastard in his seminal text book, written for zinc blende type semiconductors. The whole monograph' conclusions are applicable to wurtzitic semiconductors *mutatis mutandis*.

Quantum wells are artificial crystals. They are realized by sandwiching along a given direction ($\vec{w}$), a thin slice/layer of a given semi-conductor between thick layers of either another semiconductor or two different semi-conductors. This produces a macroscopic breaking of the crystal translation symmetry along the growth direction of the heterostructure. The band-gap mismatch $\Delta E_g(\vec{w})$ between the different semiconductors distributes in a specific manner the conduction—and the valence-band states. This writes:

$$\Delta E_g(\vec{w}) = \Delta E_c(\vec{w}) + \Delta E_v(\vec{w})$$

The quantities $\Delta E_c(\vec{w})$ and $\Delta E_v(\vec{w})$ are labelled conduction and valence-band offsets respectively. They are ad hoc indicators for representing the quantum well poten-

B. Gil, *Physics of Wurtzite Nitrides and Oxides*, Springer Series in Materials Science 197, DOI: 10.1007/978-3-319-06805-3_5,

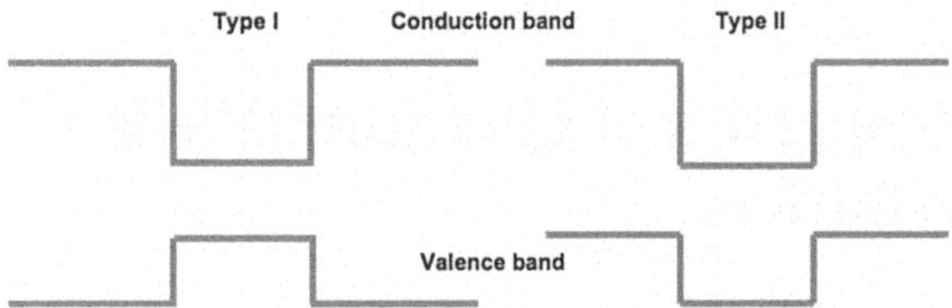

Fig. 5.1 The sketch of the conduction and valence band lineups in the case of type-I (*left*) and type-II (*right*) configurations

tial in the growth direction and are of paramount importance for calculating the band structure of the quantum well. These quantities depend on the **chemical contrast** between the quantum well and the barrier layers.

In this section, we will consider two different spatial configurations for the band lineups in the growth direction of the multilayer structure. We also take identical materials at both sides of the thin layer.

The type-I configuration occurs when the bottom of the conduction band lineup and the top of the valence band lineup are both peaking in the same material. The type-II configuration is obtained when the bottom of the conduction band lineup and the top of the valence band lineup peak in different materials. These two possibilities are sketched in Fig. 5.1.

To the best of the author's knowledge, wurtzitic semi-conductors are always met when the materials used to grow the hetero-structure possess a common anion. This is the case for pure nitride-based or pure oxide-based heterostructures. When the hetero-structure materials do not share a common anion, like GaN–ZnO for instance, the picture may be different. We will restrict ourselves to type-I heterostructures in this monograph. Then, the thin band-gap material constitutes the quantum well layer whereas the wider band-gap semiconductors are the barrier layers.

The electronic levels of the heterostructures are very different from the bulk energy levels. There are two asymptotic cases:

- In the limit of an infinite well width (no barrier layers), they are the smaller band gap semiconductor states.
- In the opposite case of a vanishing well width one (no well layer), they are the wider band gap semiconductor states.

Between these two asymptotic cases, there quantum states show a continuous dependence on the thicknesses of the different layers of the hetero-structure. To calculate them, we will adopt the envelope-function approach, in the framework of the effective-mass approximation.

The two major assumptions governing this model are the following ones:

- Each wave-function of the heterostructure projects itself in the basis of the periodic Bloch waves of each of the semi-conductors forming the hetero-structure.
- The Bloch waves are supposed to be identical in all semi-conductors along the growth direction.

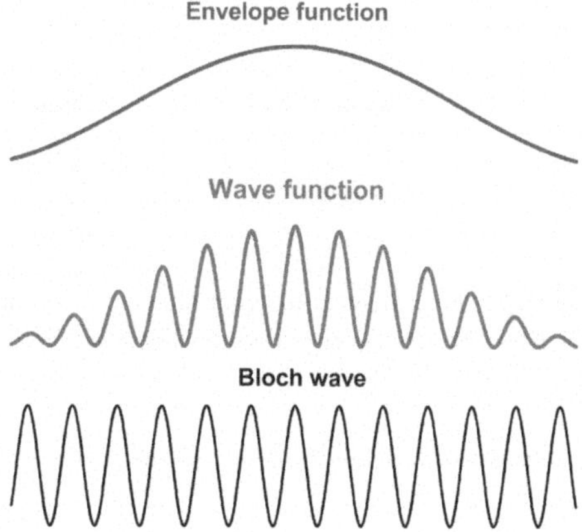

Fig. 5.2 From *top* to *bottom*: the Bloch wave, wave-function and envelope function

In this context, the wave-functions are the product of two different contributions: a rapidly-varying Bloch-type part possessing the lattice periodicity, and another one describing the three-dimensional quantum structure with the abstraction of its periodicity. The latter which characteristics size is the real size of the heterostructure is called the envelope function.

In Fig. 5.2 have been sketched the wave-function of a trappped particle (middle in the figure) as the convolution of the Bloch wave (bottom) and the envelope function (top), illustrating their relative spatial variations.

For a particle trapped in a quantum well, at the Brillouin zone center i.e. for wave-vectors vanishing away from the growth direction, and for a high-symmetry growth direction or for the s-type conduction band, its wave-function is the product of a contribution which accounts for the quantification along the growth direction, times a in-plane or plane-wave contribution. These two contributions are not coupled. For low symmetry growth directions and for the valence band states, Bloch states of the individual semi-conductors displaying different symmetries in the bulk crystal may be coupled by the macroscopic quantum well potential. Then, a multi-band version of the elementary one-band envelope function approach should be developed, but we disregard this situation in this section.

The important properties of the envelope function χ_i, subscript i representing either electrons or holes, to outline here are the following ones:

- *The envelope function is the solution of the effective mass equation*:

$$\left[-\frac{\hbar^2}{2}\frac{\partial}{\partial w}\left(\frac{1}{m_i(w)}\frac{\partial}{\partial w}\right)+V_i(w)\right]\chi_i(w)=E_i\chi_i(w)$$

where $V_i(w)$ is the potential experienced by particle i at position w, and $m_i(w)$ is the effective mass of particle i at w. This simple writing permits to take into account the variety of different materials along the growth direction.

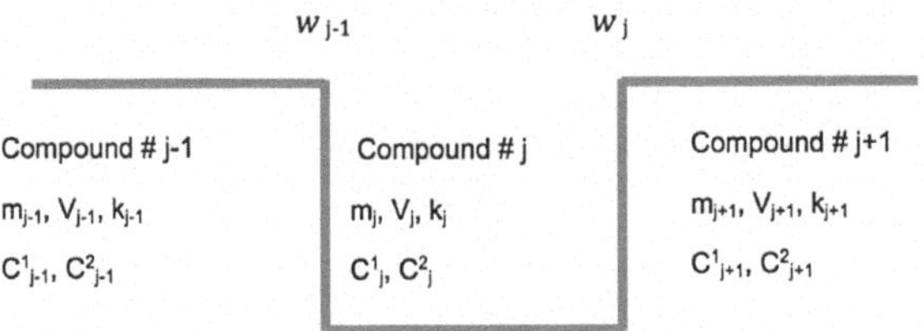

Fig. 5.3 The sketch of the band line-ups and indications of the relevant parameters utilized to calculate eigen-states and envelope functions

- *The envelope function is normalized to unity*:

$$\int_{-\infty}^{\infty} |\chi_i(w)|^2 dw = 1$$

- *The envelope function is continuous through every interface* w_α:

$$\chi_i(w_\alpha - o(w_\alpha)) = \chi_i(w_\alpha + o(w_\alpha))$$

This identity may be written as

$$\chi_i(w_\alpha)^- = \chi_i(w_\alpha)^+$$

- *The current of probability is continuous through every interface* w_α.

$$\left[\frac{1}{m_i(w_\alpha)}\frac{\partial}{\partial w}\chi_i(w_\alpha)\right]_{w_\alpha^-} = \left[\frac{1}{m_i(w_\alpha)}\frac{\partial}{\partial w}\chi_i(w_\alpha)\right]_{w_\alpha^+}$$

This expression differs from what is proposed in common books of quantum mechanics, due to the spatial dependence of the effective mass $m_i(w)$.

A convenient method to obtain the eigen-values and eigen-vectors solutions of the Schrödinger equation is to work in the context of the transfer-matrix formalism. We consider a model system of n layers of finite thickness embedded between two semi-infinite layers, at the left and right hand sides of the ensemble of n layers. This model contains $n+2$ layers and $n+1$ interfaces and a typical basic building block is sketched in Fig. 5.2.

The active layer α is limited by interfaces w_{j-1} and w_j, where the confining potential is noted $V_{i,\alpha}$ (Fig. 5.3).

In this layer α, the envelope function for particle i writes in the most general way:

$$\chi_{j,i,\alpha}(w) = C^1_{j,i,\alpha}e^{k_{j,i,\alpha}w} + C^2_{j,i,\alpha}e^{-k_{j,i,\alpha}w}$$

with

$$k_{j,i,\alpha} = \sqrt{2m_{i,\alpha}(E_{j,i} - V_{i,\alpha})}/\hbar$$

and $E_{j,i}$ is the energy of particle i, j being a quantum number. The envelope function has a plane-wave behavior ($k_{j,i} \in \mathbb{R}$) in well layers, and a vanishing wave behavior in barrier layers ($k_{j,i} \in Im(\mathbb{C})$).

Let us forget for a moment the quantum confined number j to handle light equations.

From continuity relations of the envelope function and density of probability through each interface, one can express the relations between the amplitudes $\left[C^1_{i,\alpha}, C^2_{i,\alpha}\right]$ and $\left[C^1_{i,\alpha+1}, C^2_{i,\alpha+1}\right]$ in the framework of a matrix notation. This is the basis of the transfer matrix method.

$$\begin{bmatrix} C^1_{i,\alpha} \\ C^2_{i,\alpha} \end{bmatrix} = \begin{bmatrix} M^\alpha_{i,11} & M^\alpha_{i,12} \\ M^\alpha_{i,21} & M^\alpha_{i,22} \end{bmatrix} \cdot \begin{bmatrix} C^1_{i,\alpha+1} \\ C^2_{i,\alpha+1} \end{bmatrix} = M^\alpha_i \cdot \begin{bmatrix} C^1_{i,\alpha+1} \\ C^2_{i,\alpha+1} \end{bmatrix}$$

Then, the transfer matrix for the whole structure is defined by:

$$M_i = \prod_{\alpha=1}^{\alpha=n+1} M^\alpha_i$$

Let us consider the relation between the amplitudes of the left- and right-hand-side semi-infinite layers. Since the envelope function is a vanishing wave in the barrier layers:

$$\begin{bmatrix} C^1_{i,1} \\ C^2_{i,1} \end{bmatrix} = \begin{bmatrix} C^1_{i,1} \\ 0 \end{bmatrix}$$

and

$$\begin{bmatrix} C^1_{i,\alpha+2} \\ C^2_{i,\alpha+2} \end{bmatrix} = \begin{bmatrix} 0 \\ C^2_{i,\alpha+2} \end{bmatrix}$$

$$\begin{bmatrix} C^1_{i,1} \\ 0 \end{bmatrix} = \begin{bmatrix} M_{i,11} & M_{i,12} \\ M_{i,21} & M_{i,22} \end{bmatrix} \cdot \begin{bmatrix} 0 \\ C^2_{i,\alpha+2} \end{bmatrix}$$

To determine the eigen-states of the structure, one simply has to browse the energies E and to keep those that leading to $M_{i,22} = 0$.

For a single quantum well of thickness L with identical barrier layers, one has to solve the transcendental equation:

$$\cot(kL) - 0.5(\xi - 1/\xi) = 0$$

where $\xi = (k \cdot m_B)/(\rho \cdot m_W)$, $k = \sqrt{2m_W(V_{Well} - E)}/\hbar$ and $\rho = \sqrt{-2m_B E}/\hbar$.

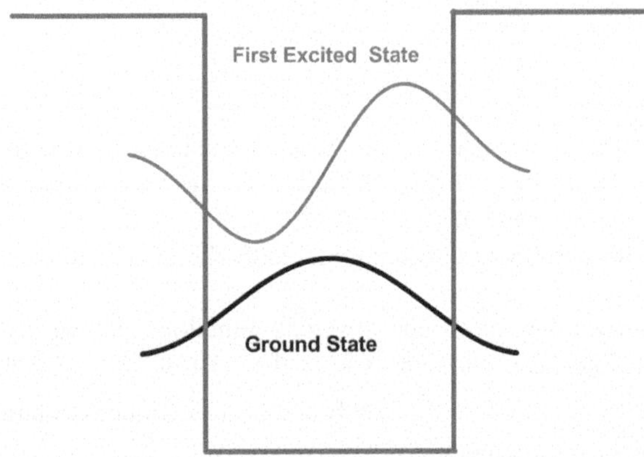

Fig. 5.4 Plot of the ground and first excited states' envelope functions in a square well of finite depth. Note the different functions' parities with respect to a reflection plane located at the center of the well layer and parallel to it

In our notation, the energies are negative quantities as well as the potential energy V_{Well} in the well layer, while the potential $V_{Barrier}$ is vanishing in the barrier layers. Effective masses m_W and m_B respectively correspond to well layer and barrier layers. Several values of the energy E may fulfill the condition $M_{22} = 0$, if there are several confined levels: this gives birth to the confined quantum number j introduced above. There is at least one confined level in a uni-dimensional quantum well problem. A similar transcendental equation permits to calculate the energies of un-confined states but we will disregard them (Fig. 5.4).

In a quantum well, transitions are observed at energies $E_g(Barrier) + E_i(electron) + E_j(hole)$, i and j being quantum numbers for confined electron and hole levels, that are higher than $E_g(Well)$. Quantum mechanics prescripts, as shown in Fig. 5.5, that, in a symmetric/square quantum well, states with odd quantum numbers, starting from the first confined, have even parity with respect to the center of the well, while states with even quantum numbers have odd parity with respect to this point.

The oscillator strength for a band to band transition between confined conduction and confined valence states is proportionnal to the square of matrix element:

$$\int_{-\infty}^{\infty}\int_{-\infty}^{\infty} \chi_{i,e}^{*}(w_e)u_c^{*}(w_e)H_{int}u_v(w_h)\chi_{j,h}(w_h)dw_e dw_h$$

where H_{int} is the light-matter interaction Hamiltonian. Taking into account the rapid variation of the Bloch waves $u_c(w_e)$ and $u_v(w_h)$ with respect to the envelope functions $\chi_{i,e}(w_e)$ and $\chi_{j,h}(w_h)$, the oscillator strength may be taken as proportional to the squared overlap integral:

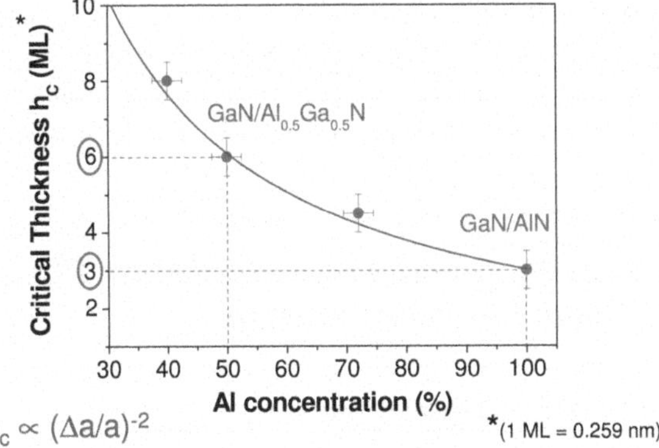

Fig. 5.5 Critical thickness for coherent growth of various AlGaN on GaN. Courtesy Julien Brault, Thomas Huault and Benjamin Damilano

$$\int_{-\infty}^{\infty} \chi_{i,e}^{*}(x)\chi_{j,h}(x)dx$$

For finite-depth potentials, this integral vanishes when electron and hole envelope functions have different parities or when i and j quantum numbers have different parities. One can demonstrate that the overlap integral for an infinitely deep well differs from zero only when i and j quantum numbers are identical.

We have to remark here that the type-II quantum well configuration leads, in general, to electron and hole envelope functions peaking in different layers. Therefore, the oscillator strength for an optical process is in general much weaker for type-II configurations than for type-I band line-ups.

Multiple quantum wells are stackings of several identical or different single wells separated by barriers. Among the diversity of possible stacking configurations, superlattices are studied separately.

Superlattices are specific multi-quantum well structures with thin enough barrier layers so that wave-functions are distributed over multiple wells.

For an infinite superlattice, of period L, with well and barrier thickness L_W and $L_B = L - L_W$ respectively, the application of Born-Von Karman cyclic conditions leads to:

$$\begin{bmatrix} C_{i,1}^{1} \\ C_{i,1}^{2} \end{bmatrix} = \begin{bmatrix} C_{i,\alpha+2}^{1} \\ C_{i,\alpha+2}^{2} \end{bmatrix} \cdot e^{i\phi}$$

Thus, the matrix (M) is the identity matrix and the transcendantal equation to solve is:

$$\cos(qL) = \cos(kL_W)\cosh(\rho L_B) + 0.5(\xi - 1/\xi)\sin(kL_W)\sinh(\rho L_B)$$

here, q is the superlattice Bloch wave-vector ($q \in \{-\pi/L, \pi/L\}$). Scanning q through its definition range, one obtains energy bands. Since there may exist several confined states in a quantum well, we can also evidence several energy bands for a superlattice. Super-lattices can be formed with type-I or type-II band lineups.

5.1.2 Strained Layers, Quantum-Confined Stark Effect and One-Band Envelope Functions

5.1.2.1 Strained Layers and Energy Shifts

The semiconductors are in general lattice-mismatched and when all monolayers of the heterostructures are assembled coherently, each of them experiences a strain field in its growth plane. This strain-field is called the built-in strain field. The strain induced modification of the band gap of each material can be determined as discussed in Chap. 2 in the framework of the deformation potential theory, once the components of the strain field are known relatively to the international orthogonal basis. These deformations can be evaluated by assuming lattice matching to a growth substrate, or by global minimization of the total elastic energy stored in the whole heterostructure (we know after Chap. 1 that the density of elastic energy stored in the epilayers is $1/2C_{ij}\epsilon_i\epsilon_j$). Another possibility is to measure them directly by X-ray diffraction. When the amount of elastic energy strored in a strain layer, exceeds a critical value, part of this strained elastic energy is relaxed via formation of dislocations. These dislocations have deleterious influences on optical properties and carriers mobilities. Such threshold is in general reached after growth of a strained layer with thickness beyond a critical value. Such value is called the critical thickness for coherent growth. Although some models exist to calculate it, this quantity is difficult to describe theoretically for many reasons, also including its dependence with the growth orientation of the strained-layer heterostructure. The best way to determine it is to grow layers of different thicknesses and to measure the density of dislocation using and adapted experimental method like electronic microscopy for instance. Figure 5.5 illustrates the variation of the critical thickness for coherent growth of various AlGaN alloys on GaN

The built-in strain field the conduction band is at first order shifted by a given amount and the conduction band offset $\Delta E_c(\vec{w})$ has to be renormalized taking into account of the influence strain-induced shifts of the conduction bands of the materials that constitute the heterostructure.

Regarding the valence band states the situation is much more tricky since we are facing a six-fold p-type multiplet and the strain tensor may couple the different valence band states of the unstrained wurtzitic materials used to form the heterostructure. To describe the valence band physics requires a more complex theory than in the case of the conduction band.

What we will keep in mind at this stage is that the distribution of band offsets into valence band offset and conduction band offset varies with the strained states of the epilayers. For most of the researchers who interpret experimental data or conceive devices, it is introduced in computational codes as a fitting parameter.

5.1.2.2 Quantum Confined Stark Effect

Wurtzitic materials exhibit a spontaneous polarisation (see Chap. 1, Sect. 1.9) which depends on the chemical nature of the material (see Table 1.4).This spontaneous polarization $\vec{P}_{sp}$ is oriented along the $\vec{c}$ direction of the crystal. The spontaneous polarization in the direction $\vec{W}$orthogonal to an $(hk \cdot \ell)$ reticular plane is:

$$P_{spW} = \sqrt{3}\ell a B_{ac} P_{sp}$$

where a and c are the wurtzite lattice parameters and

$$B_{ac} = \frac{1}{\sqrt{3\ell^2 a^2 + 4c^2(h^2 + hk + k^2)}}$$

This will impact drastically the optical properties of quantum wells: in heterostructures based on wurtitic semiconductors, there may exist electric fields in the different layers, which will produce modifications of the band line-ups. The wave functions (and the envelope functions in particular) will be distorted, the macroscopic parity of the envelope function will be destroyed, the overlap integrals will be reduced or enhanced, and the energies will be shifted.

The ground state band to band transition between first confined electron and hole will be in particular significantly red-shifted. To progressively describe this phenomenon, we begin by first considering a thin slice of a wurtzitic semiconductor W in vacuum. According to Maxwell equations, in line with continuity of the electric displacement vector $\vec{D}$ at the wurtzite-vacuum interfaces, the existence of the spontaneous polarisation manifests itself through formation of two charged planes with densities of charges of opposite signs at both sides of the wutzite.

Let $\vec{P}_W$ this polarization field and $\vec{n}$ the unity vector normal to the semiconductor surface. Then the density of charges accumulated at a given air-semiconductor interface is $\sigma_W = \vec{P}_W \cdot \vec{n}$.

This is illustrated in Fig. 5.6 when the spontaneous polarization is aligned with the crystal direction. The surface density of charges σ equals $\pm\|\vec{P}\|$ at the edges of the crystal, the positive charge being at the surface where $\vec{P}$ exits of the crystal. There are no charges at the surfaces orthogonal to $\vec{P}$. For another surface, σ would be the projection of $\vec{P}$ through this surface.

If we now embed this slice of semiconductor into two thick slabs of another wurtzitic semiconductor for which the polarization field is $\vec{P}_B$, the interfacial density of charge will write:

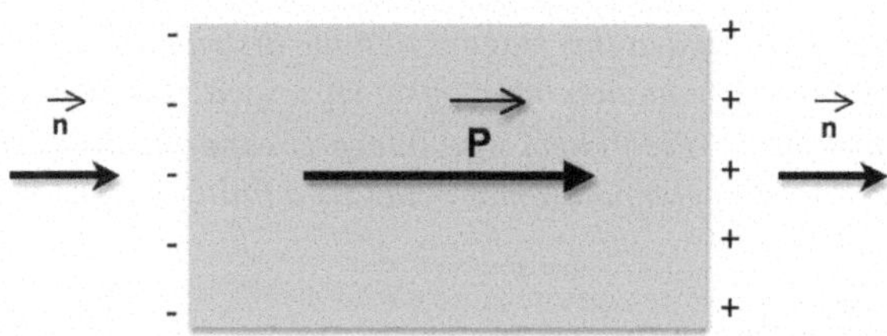

Fig. 5.6 Conventional orientation of the spontaneous polarization and the signs of the charges at the interfaces of the crystal

$$\sigma_{W-B} = [\vec{P}_W - \vec{P}_B] \cdot \vec{n}$$

The conservation of the electric displacement vector $\vec{D}$ through the interface leads to write:

$$[\vec{D}_W - \vec{D}_B] \cdot \vec{n} = [\epsilon_W \vec{F}_W - \epsilon_B \vec{F}_B] \cdot \vec{n} = [\vec{P}_B - \vec{P}_W] \cdot \vec{n}$$

where the quantities ϵ_W and ϵ_B are the dielectric constants in materials W and B (they include ϵ_0).

To express the discontinuity of the electric field F at the interface between W and B materials we applicate Born Von Karman cyclic conditions and the continuity of the electric field through a closed boucle:

$$\vec{F}_W \cdot \vec{n} L_W + \vec{F}_B \cdot \vec{n} L_B = 0$$

Then the electric field in materials W and B write:

$$\vec{F}_W \cdot \vec{n} = \frac{L_B}{\epsilon_W L_B + \epsilon_B L_W} [\vec{P}_B - \vec{P}_W] \cdot \vec{n}$$

$$F_B \cdot \vec{n} = -\frac{L_W}{\epsilon_W L_B + \epsilon_B L_W} [\vec{P}_B - \vec{P}_W] \cdot \vec{n} = -\frac{L_W}{L_B} \vec{F}_W \cdot \vec{n}$$

These equations qualitatively indicate that the electric field in the well layer and barrier layers vary with the different thicknesses of all materials. These expressions quantitatively indicate that the electric field vanishes when the normal to the surface is orthogonal to the spontaneous polarization, that is to say for $(hk \cdot 0)$ *reticular planes.*

In the case of the **polar quantum well**, the envelope functions to solve are the eigenfunctions of the Shrödinger equation below:

$$\left[-\frac{\hbar^2}{2} \frac{\partial}{\partial w} \frac{1}{m_i(w)} \frac{\partial}{\partial w} + V_i(w) + q_i F_i(w) \right] \chi_i(w) = E_i \chi_i(w)$$

where $F_i(w)$ is the electric field at w and q_i represents the charge of particle i.

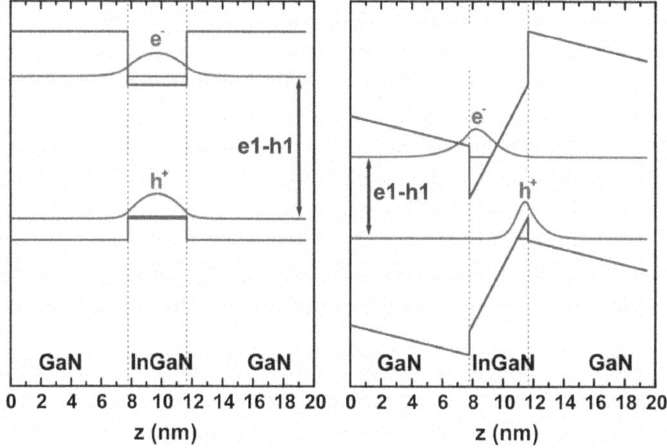

Fig. 5.7 Typical band band line ups for two model situations of 4 nm thin wurtzitic quantum wells grown on a $(hk \cdot 0)$ oriented plane (*left-hand side*) and on a (0001) polar orientation (*right hand side*). The wave functions have been also plotted as well as the positions of the first confined electron and confined hole levels. Transition energies are labeled e_1h_1

To compute the energy levels in the case of the quantum well grown along polar orientation, the layers can be each divided into one hundred identical pieces in which the potential is constant but is shifted in energy by one hundredth of the total potential drop through the layer. By using the transfer matrix model the problem can be accurately solved without having to handle Airy functions.

In Fig. 5.7 have been plotted the typical band band line ups for two model situations of 4 nm thin wurtzitic quantum wells grown on a $(hk \cdot 0)$ oriented plane (left-hand side) and on a (0001) polar orientation (right hand side). The wave functions have been also plotted as well as the positions of the first confined electron and confined hole levels. Transition energies are labeled e_1h_1. For the growth on polar reticular plane (or said alternatively along the polar direction), an electric field of 3 MV cm^{-1} was introduced in the quantum well region. In the barrier layers, the electric field is taken the opposite sign of the electric field in the well layer and ten times smaller in magnitude. The important point to outline concerning **non polar quantum wells** is the high degree of overlap between the electron and hole envelope functions which are located in the same region of the heterostructure, in the well layer. Both of them peak at the center of the layer. We expect **good oscillator strengths** for band to band recombination process which occur at energies higher than the energy of the band gap of the semiconductor used as quantum well layer (this band gap energy may eventually modified buy built-in strain).

$$E(e_1h_1) = E_g(Well) + e_1 + h_1$$

In the case of the **polar quantum well**, the electric field pulls apart from each other the electron and hole. Their envelope functions do not overlap so efficiently than in case of the square well situation. One can easily imagine that this overlap rapidly decreases when increasing well width. This is a first important impact of built-in electric field in wurtzitic heterostructures: the overlap integral between electron and hole envelope functions (this is the oscillator strength for band to band recombination) decreases with well width. This will impact the radiative recombination rate of electron and hole and in contrast with growth along non polar orientations, **the optical properties of thick quantum wells will rapidly degrade when increasing the well width. This is a first manifestation of the Quantum Confined Stark Effect**.

Last, the electron confinement energy is measured relatively to the left-hand bottom corner of the conduction band potential and the hole confinement energy is measured relatively to the right-hand top corner of the valence band band. The recombination energy is the sum of the band gap of the material used as quantum well layer (eventually modified buy built-in strain) energy, of the confinement energies, and it is in addition reduced from the potential drop (about $eL_W F_W$):

$$E(e_1h_1) = E_g(Well) + e_1 + h_1 - eL_W F_W$$

The redshift increases almost linearly with well. This is a second manifestation of the Quantum Confined Stark Effect. When the potential drop through the well $eL_W F_W$ is high enough, the optical transition between first electron and hole confined band occurs at energy below the band gap energy of the material used as quantum well layer.

Built-in strain fields may produce a piezoelectric polarization in the strained layers as shown in Chap. 1. This may strongly impact the band line-ups which may drastically depart from the sketch of Fig. 5.1 via the polarization mismatch through each of the interfaces of the heterostructure. All equations to calculate the total polarization in a strained wurtzite layer with given orientation have been given in Chap. 1, Sects. 1.9 (1.8) and 1.11 (1.10). *The total polarization in a strained layer is the sum of the spontaneous polarization with the piezoelectric one. It should replace the spontaneous polarization terms in the equations above.*

Equation 1.10, Chap. 1 indicates that the component of the piezo electric field vanishes in the growth direction for $(hk \cdot 0)$ reticular planes.

$$P_{pzW} = B_{ac}\left[Ac(2h+k)2e_{15}e_{xz} + \sqrt{3}Ack2e_{15}e_{yz} + \ell a[e_{13}(e_{xx}+e_{yy}) + e_{33}e_{zz}]\right]$$

where:

$$A = \frac{1}{\sqrt{h^2 + hk + k^2}}$$

As:

$$P_{spW} = \sqrt{3}\ell B_{ac} a P_{sp}$$

Then the total polarization in the direction $\vec{W}$ orthogonal to plane $(hk \cdot \ell)$ is:

$$P_{tot\,W} = B_{ac}\left[Ac(2h+k)2e_{15}e_{xz} + \sqrt{3}Ack2e_{15}e_{yz} + \ell a[e_{13}(e_{xx}+e_{yy}) + e_{33}e_{zz} + P_{sp}]\right]$$

There is no quantum Confined Stark Effect for wurtzitic heterostructures grown on $(hk \cdot 0)$ reticular planes. These orientations are called non polar orientations.

5.1.3 Exciton Binding Energy in Quantum Wells

Optical transitions in case of undoped semiconductors reveal signatures of excitons which result of the long-range Coulomb interaction between the photo-created electron and hole pair. Such excitons can be reasonably described using an effective hydrogen-like model and a series of lines that are split as expected in case of an hydrogen series can be detected below the band to band value of the band gap. Further including electron and hole spin one introduces spin-exchange related interaction to account for fine structure effects. In case of quantum wells one expects this excitonic interaction to be modified. One expects an enhancement of the excitonic interaction when the spatial separation of the electron hole pair is reduced, a reduction is the other case. There are various ways to compute the exciton binding energies in quantum wells, which all require heavy computing. Here, as in Bastard' book, we will treat it using a variational approach. This method does not give access to excited states and its accuracy depends of the choice of the trial function used to describe the excitonic interaction. However it is very well appropriate for a tutorial and it gives good results. This motivates our choice here. Let us suppose that the envelope functions χ_e and χ_h of the confined electrons and holes are known, and let us treat the coulomb interaction as a perturbation of the band to band interaction. Let us in addition assume an isotropic dielectric constant ϵ_r to deal with light equations, and let us also suppose a mass isotropy in the growth plane of the heterostructure so that we can use a cylindrical approximation and get rid of integrations over the azimuthal angle ϕ. Let us note the growth direction as $\vec{z}$ and let us note ρ the radial coordinate. The band to band Hamiltonian have been previously solved:

$$H_e\chi_e(z_e) = \left[-\frac{\hbar^2}{2}\frac{\partial}{\partial z_e}\frac{1}{m_e(z_e)}\frac{\partial}{\partial z_e} + V_e(z_e) - eF_e(z_e)\right]\chi_e(z_e) = E_e\chi_e(z_e)$$

$$H_h\chi_h(z_h) = \left[-\frac{\hbar^2}{2}\frac{\partial}{\partial z_h}\frac{1}{m_h(z_h)}\frac{\partial}{\partial z_h} + V_h(z_h) - eF_h(z_h)\right]\chi_h(z_h) = E_h\chi_h(z_h)$$

The Hamiltonian of the system is $H = H_e + H_h + H_{exc}$ with:

$$H_{exc} = -\frac{\hbar^2}{2\mu}\left(\frac{\partial^2}{\partial\rho^2} + \frac{1}{\rho\partial\rho}\right) - \frac{e^2}{4\pi\epsilon_0\epsilon_r}\frac{1}{\sqrt{\rho^2 + (z_e - z_h)^2}} + \frac{1}{\rho^2}\frac{\partial^2}{\partial\phi^2}$$

The excitonic wave function is written:

$$\Psi(z_e, z_h, \rho, \phi) = \chi_e(z_e\chi_h(z_h)\psi_{\rho,z,\phi}(\rho, z, \phi)$$

where $z = z_e - z_h$ and $\psi_{\rho,z,\phi}(\rho, z, \phi)$ is a trial function that we take of the $e^{-f(\rho,z)}$ kind with:

$$f(\rho, z) = \frac{\sqrt{\rho^2 + \alpha^2 z^2}}{\lambda}$$

Quantities α and λ will be variational parameters; λ is the in-plane Bohr radius, and α is the dimensionality of the excitonic trial function. Let

$$\Psi_{\rho,z,\phi}(\rho, z, \phi) = (\int\int\int \chi_e^2(z_e)\chi_h^2(z_h))e^{-2f(\rho,z)}dz_e dz_h \rho d\rho)^{-1/2}\chi_e(z_e)\chi_h(z_h)e^{-f(\rho,z)}$$

One can easily demonstrate that $\langle\Psi|\Psi\rangle = 1$

Then, the exitonic binding energy is obtained by a minimization procedure through α and λ:

$$E_B = min_{\alpha,\lambda}\langle\Psi|H_{exc}|\Psi\rangle$$

$$E_B = min_{\alpha,\lambda}(A/B)$$

with

$$\begin{aligned}
A = &\left(\frac{\hbar^2}{4}\right)\int \chi_e^2(z_e)dz_e \int \chi_h^2(z_h)dz_h \left[\alpha^2\left(\frac{1}{m_e}+\frac{1}{m_h}\right)+\frac{1}{2\mu}-\frac{2\alpha|z|}{\lambda}\right]e^{-2\alpha|z|/\lambda}\\
&-\frac{e^2}{4\pi\epsilon_0\epsilon_r}\int \chi_e^2(z_e)dz_e \int \chi_h^2(z_h)dz_h \int_0^\infty \frac{e^{-2f(\rho,z)}}{\sqrt{\rho^2+z^2}}\rho d\rho\\
&-\frac{\hbar^2\alpha^2}{2\lambda^4}\int \chi_e^2(z_e)dz_e \int \chi_h^2(z_h)dz_h \int_0^\infty z^2\frac{f(\rho,z)+1}{f^3(\rho,z)}\left[\alpha^2\left(\frac{1}{m_e}+\frac{1}{m_h}\right)-\frac{1}{\mu}\right]^{-2f(\rho,z)}\rho d\rho\\
&-\alpha^2\left(\frac{\hbar^2}{2m_e}\right)\int \chi_e(z_e)\chi_e'(z_e)dz_e \int \chi_h^2(z_h)dz_h z e^{-2\alpha|z|/\lambda}\\
&-\alpha^2\left(\frac{\hbar^2}{2m_h}\right)\int \chi_h(z_h)\chi_h'(z_h)dz_h \int \chi_e^2(z_e)dz_e z e^{-2\alpha|z|/\lambda}
\end{aligned}$$

and

$$\begin{aligned}
B &= \int\int\int \chi_e^2(z_e)\chi_h^2(z_h)e^{-2f(\rho,z)}dz_e dz_h \rho d\rho\\
&= \left(\frac{\lambda^2}{4}\right)\int \chi_e^2(z_e)dz_e \int \chi_h^2(z_h)dz_h \left(1+\frac{2\alpha|z|}{\lambda}\right)e^{-2\alpha|z|/\lambda}
\end{aligned}$$

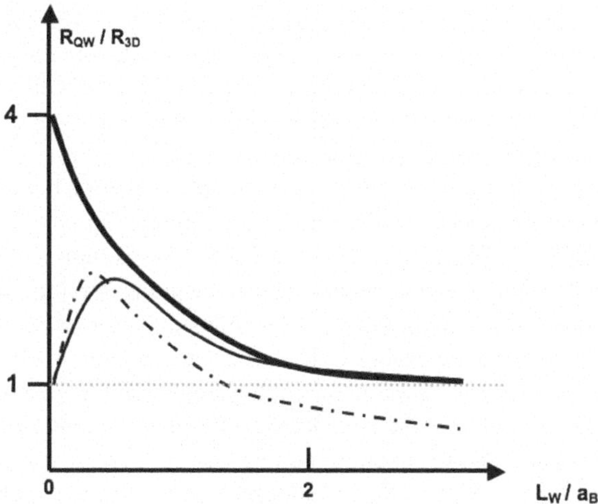

Fig. 5.8 Typical and schematic variations of the exciton binding energy against well-width expressed in term of the 3D excitonic Bohr radius a_B. *Bold line* represents a 2D case, *full line* is used to draw the behavior for a finite depth semiconductor square quantum well, and *dash-dotted line* represents the variation of the exciton binding energy for a semiconductor quantum well with internal electric field. The asymptotic 3D value is plotted used a *light dot line*

The minimization procedure and these equations are considerably simplified if using a two-dimensional trial function ($\alpha = 0$) or to a less extend a three dimensional one ($\alpha = 1$).

The oscillator strength for the excitonic transition is obtained for $z_e = z_h$ and $\rho = 0$ proportional to:

$$\frac{4}{\lambda^2}\frac{|\int \chi_e(x)\chi_h(x)dx|^2}{\int \chi_e^2(z_e)dz_e \int \chi_h^2(z_h)dz_h(\frac{2\alpha|z|}{\lambda}+1)e^{-\frac{2\alpha|z|}{\lambda}}}$$

We will compute these quantities later in the experimental sections. Note however that in the case of a two dimensional trial function ($\alpha = 0$), the oscillator strength is just:

$$\frac{4}{\lambda^2}\left|\int \chi_e(x)\chi_h(x)dx\right|^2$$

In Fig. 5.8 are plotted some typical and schematic variation of the exciton binding energy against well-width expressed in term of the 3D excitonic Bohr radius a_B. Bold line represents a 2D case, full line is used to draw the behavior for a finite depth semiconductor square quantum well, and dash-dotted line represents the variation of the exciton binding energy for a semiconductor quantum well with internal electric field. The asymptotic 3D value is plotted used a light dot line. For infinitely deep conduction band and valence band wells, the exciton binding energy continuously increases up to four time the 3D value, which is the exact 2D value. For a finite size quantum well with square potential profiles, when reducing the well width, the expected behavior is a continuous increasing when reducing the well width, reaching a maximum value (below 4) and then decreasing down to the value of the exciton binding energy in the 3D semiconductor used as material for the barrier layers.This description is qualitatively obtained by the variational calculation in the context of the one band approximation; there is a discrepancy in particular for wide wells. However the physics is here. More interesting is the variation of the exciton binding energy for quantum wells with triangular potential shapes. In the thin well limit, the asymptotic value is the exciton binding energy in the 3D semiconductor used as material for the barrier layers. Then when increasing well width a first regime is observed for thin wells when both carriers are experiencing high enough confinement and when their envelope functions are not too much impacted by the electric field. The band to band overlap integral is good and the exciton binding energy increases. When increasing well width the influence of the electric field increases and the envelope functions start to peak more and more clearly near opposite interfaces. The overlap integral decreases (we will see later that it can scale through more than 7 decades); so does the excitonic binding energy which reaches values below the 3D bulk value in the semiconductor used to form the quantum well layer. However in contrast with the overlap integral of envelope function which is ruled by the well width and value of the electric field, the Coulomb interaction is a longe range interaction which includes on axis and in-plane contributions. It cannot decreases as dramatically as the overlap integral does.

5.1.4 Effects of High Photo Injection Densities in Quantum Wells

Under photo illumination there a lot of manybody effects that can be produced: exciton–exciton collisions, excitonic bleaching, band gap renormalization, hole burning,... We will interest here to the most dramatic one: the blue shift induced by photo-injection in case of polar quantum wells. When such quantum well is shined by a laser or an electron beam of energy high enough to be absorbed, electron and holes are created which, before they recombine may accumulate at different interfaces as indicated in Fig. 5.6. From Poisson's equation it can be easily demonstrated that these accumulations of electron and holes at different interfaces constitute a dipole which produces an electric field that tends to compensate the existing one: Then the Shrödinger equation writes:

$$\left[-\frac{\hbar^2}{2}\frac{\partial}{\partial w}\frac{1}{m_i(w)}\frac{\partial}{\partial w}+V_i(w)+q_iF_i(w)+q_i\Phi_i(w)\right]\chi_i(w)=E_i\chi_i(w)$$

where $F_i(w)$ is the electric field at w and q_i represents charge of particle i and $\Phi_i(w)$ is the electric field photo created by the accumulation of charges at the interfaces. At $T = 0$, assuming for the carriers a Fermi-Dirac distribution of the Heavyside kind.

The photo induced electric field is defined as:

$$\Phi(w+dw)-\Phi(w)=e\sigma\int_{w}^{w+dw}\sum_{i,m}\int_{0}^{k_\perp(\sigma)}\alpha_{i,m}(k_\perp,\sigma)\chi^2_{h,i,m}(k_\perp,u)dk_\perp\frac{du}{\epsilon(u)}$$
$$-e\sigma\int_{w}^{w+dw}\sum_{j,n}\int_{0}^{k_\perp(\sigma)}\beta_{j,n}(k_\perp,\sigma)\chi^2_{h,j,n}(k_\perp,u)dk_\perp\frac{du}{\epsilon(u)}$$

In this busy equation, indices e and h refer to electron and holes respectively. For a given surface injection density σ, the integration over $k_\perp$ occurs for the i hole levels (respectively j electron levels) that are occupied. These levels are indicated via coefficients α and β. The indices m and n account for the lift of Kramers degeneracy away from $k_\perp = 0$ which results from the lack of macroscopic parity symmetry at the scale of the shape of the quantum well.

Electrical neutrality occurs through conditions:

$$\int_{-\infty}^{\infty}\sum_{i,m}\int_{0}^{k_\perp(\sigma)}\alpha_{i,m}(k_\perp,\sigma)\chi^2_{h,i,m}(k_\perp,u)dk_\perp du=1$$

$$\int_{-\infty}^{\infty}\sum_{j,n}\int_{0}^{k_\perp(\sigma)}\beta_{j,n}(k_\perp,\sigma)\chi^2_{h,j,n}(k_\perp,u)dk_\perp du=1$$

At this stage it is necessary to self-consistently resolve both Poisson's equation and Schödinger equation. This is tricky.

We remark that the Fermi wave vector k_f which constitutes an upper values for the integration over $k_\perp$ is correlated to σ by well known equation: $k_f = \sqrt{2\pi\sigma}$ and that the Fermi energy E_f is in the simplest of its possible form, that is to say neglecting anisotropies and non-parabolicity: $E_f = \frac{\pi\hbar^2}{m_\perp}\sigma$. This permits to grossly and qualitatively evaluate how many confined bands are occupied by photo created carriers.

To summarize this subsection here we emphasize that high photo-injection densities produce an electric field that counter balances the existing one. From this screening results a blue-shift of the ground state transition and a total redistribution of the electronic structure of the quantum well. In Fig. 5.9a we illustrate this blue-shift in case of a 6.5 nm thick $Zn_{0.78}Mg_{0.22}O$–ZnO–$Zn_{0.78}Mg_{0.22}O$ single quantum well grown along the polar direction. The photo-induced blue shift is 96 meV in this sample. Exact determination of the transition energy requires to use particularly low photo-injection densities.

In Fig. 5.9b are plotted some typical envelope functions for GaN–AlGaN single quantum well under various photo injection conditions.

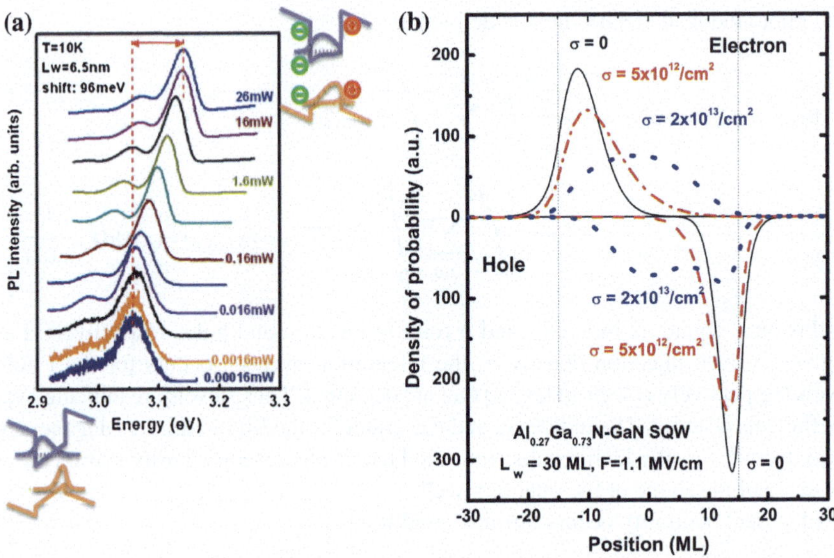

Fig. 5.9 **a** *Blue-shift* of the photoluminescence for a 6.5 nm thick $Zn_{0.78}Mg_{0.22}O$–ZnO–$Zn_{0.78}Mg_{0.22}O$ single quantum well grown along the polar direction. The schematic potential profiles as well as schematic envelope functions for low injection and high injection conditions are respectively given at *left-hand bottom corner* and *right-hand top corner* of the figure. **b** evolution of electron and hole envelope function in GaN–AlGaN quantum well under various photo-excitation conditions

In case of dopings the equations above also hold, one just has to include the electron contribution for n-doping, the hole one in case of p-doping.

5.2 Optical Properties of Polar Quantum Wells

Growth of semiconductors is an experimental science (an indisputable statement), which requires a lot of different and complementary experimental processes, which involve many actors. These processes include growth of substrates, their epi-ready surface preparation using chemical routes, manipulation of chemical growth precursors, control of the growth process using ad-hoc instruments, material structural, electrical, optical characterizations. At the end one associates some performances to specific semiconductor architectures. The aim of this book being to form master students, instead of browsing the huge literature of the field, to review what has been done, I will restrict this chapter to illustrating samples grown as much as possible in my specific partner laboratories. A strong motivation for that choice is that I have easy access to complementary growth information.

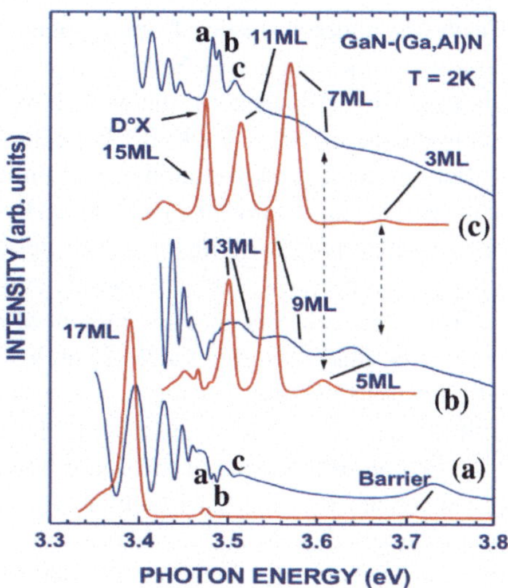

Fig. 5.10 Reflectance (*dotted line*) and PL (*solid line*) spectra taken from Samples 1 (*a*), 2 (*b*) and 3 (*c*). *Thick lines* allow to connect reflectance features to the corresponding PL peaks. *Dashed arrows* show, on two examples, the matching of PL energies, for a given nominal well width, with the reflectance features for widths larger by 2 monolayers

5.2.1 GaN–AlGaN Polar Quantum Wells

In this section are treated the impact of quantum well thickness, barrier thickness and chemical contrast between the well layers and the barrier layers on the energy of optical transitions in the case of polar GaN–AlGaN quantum wells with low aluminum compositions. The oscillator strengths and variation of the exciton binding energy versus well width are also addressed.

Samples grown along polar orientations will be made of strained layers but the wurtzite symmetry will be locally kept: the only non vanishing symmetrized components of the strain field that may experience a layer are $(e_{xx} + e_{yy})$ and e_{zz}. Both transform like Γ_1 in the language of group theory. The spontaneous and piezo electric polarization are strictly collinear to the growth axis. The different values of the spontaneous polarization in all materials that can be associated and the values of the components of the piezo electric tensor are such as there is polarization mismatch at the hetero-interfaces and thus different electric fields in the different layers. One expects to observe Quantum Confined Stark Effect except if non-intentional residual dopings were unfortunately screening these electric fields.

5.2.1.1 Quantum Confined Stark Effect

In Fig. 5.10 are represented a series of low temperature reflectivity (upper curves for each couple of spectra) and cw photoluminescence spectra collected on a series of GaN–AlGaN samples with well layers (GaN) of different thicknesses and different compositions in the barrier layers. Top in the figure is a sample with four different quantum wells, with thicknesses of 3 monolayers (MLs), 7, 11 and 15 MLs. the growth was realized starting from a c-plane sapphire substrate carefully prepared so that a thick high quality GaN thick layer could be deposited for serving as growth template. The thicknesses of the wells have been deduced at the growth step by following the evolution of the RHEED (Reflection High Energy Electron Diffraction) oscillations in the growth chamber. The aluminum composition was also calibrated at the growth set-up, a value of 11 % is given by the grower. The corresponding low temperature reflectivity experiment is indicated above the photoluminescence spectrum. Reflectivity structures can be observed at higher energy than the photoluminescence spectrum, and a one by-one assignment of reflectivity and photoluminescence features can be made assuming a 20–25 meV Stokes-shift between photoluminescence and reflectivity features for the 3, 7 and 11 ML samples. In this sample the photoluminescence peak that we attribute to the 15 ML sample overlaps with the energy of the photoluminescence of the donor bound exciton (D^0X in the GaN buffer. The optical signature of the GaN template is revealed by signature of three polaritonic features labelled A, B, and C, according to the standard notation. Reflectivity oscillations occur below the band gap of GaN when the sample becomes transparent. To strengthen this initial identification, a second sample was grown in identical conditions, except that 3 wells only were grown with thicknesses of 5, 9 and 13 MLs respectively. The reflectivity and photoluminescence features intercalate in between the optical signatures given by the preceding sample. The 13 ML sample emits light at the energy of the GaN template which impacts the reflectivity line shape. A third sample was grown following the same procedure. It consists of a single quantum well 17 ML thick which emits light clearly below the energy of GaN (see the 3.48 eV line). The thick barrier layers contribute to both reflectivity and photoluminescence as the GaN template but we cannot detect the signature of the quantum well in reflectivity in that latter case. All these behaviors are indications of Quantum confined Stark Effect as expected.

The fit to the data in the context of the envelope function approximattion requires to know the the depths of the conduction and valence band potentials. A type I, 70/30 distribution was retained between the conduction band and heavy-hole band line-ups on our calculations. By doing so, we can fairly satisfactorily fit the optical properties of the samples presented here. To obtain the best agreement between theory and experiment, we need to introduce an electric field of some 450 kV/cm for the MQW samples having 11 % aluminum in the barrier layers, and to compensate it by an opposite one in the barrier layers, chosen so that there is no potential drop through the whole MQW structure. In Fig. 5.11 have been plotted two sets of data, the reflectance ones (high energy set) and the photoluminescence ones (low energy set of data) for the series of MQWs with 11 % aluminum in the barrier layers. The result of the square well calculation (the exciton binding energy is included) is shown together with the result of a triangular well calculation with an electric field of 450 kV/cm in the well layers.

The application of cyclic conditions at the size of the sample indicates that the electric field in quantum wells increases with the thickness of the barrier layers. This we will demonstrate now. To show the influence of the thickness of barrier layers samples including single and multiple QWs were grown in which the barrier width

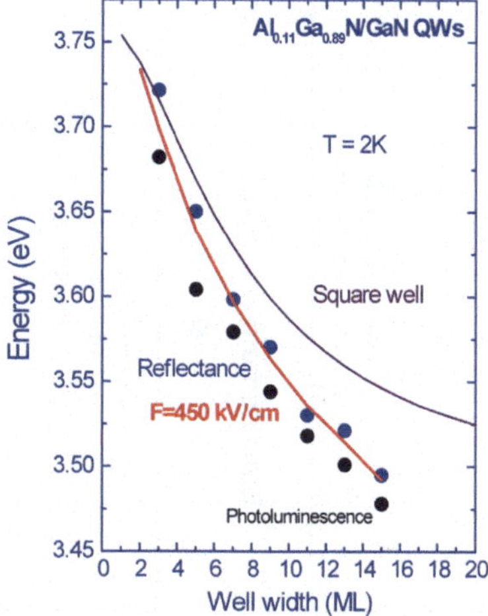

Fig. 5.11 Experimental transition energies in the MQWs with 11 % aluminum and 5 nm thick barrier layers (high energy data is reflectance, low energy ones are photoluminescence data) together with the result of a square well calculation (*thin line*), and with the inclusion of an internal electric field in the well layers (here 450 kV/cm). The agreement between theory and experiment together with the crossing of the confined excitonic transitions with the transition in the GaN template indicate the existence of quantum confined stark effect in these samples. After [1]

was varied between 5 and 60 nm. The main result is that a MQW with thin barriers (say 5 nm) is blue shifted relative to a single similar QW but, as expected, if the barriers are wide enough (say 300 nm), the multiple and single QWs have about the same energy as illustrated in Fig. 5.12a. The 4-ML-wide wells are the least sensitive to the electric field due to their narrowness and they all emit at 3.650 ± 0.007 eV. The ± 7 meV energy dispersion illustrates the low scatter of alloy compositions and well widths that is achievable in the MBE growth of nitride hetero-structures. Variations of ± 1 % of the Al com-position and ± 1 ML of the well width may correspond to QW transition energy variations of ± 11 and ± 34–39 meV, respectively. Turning now to wider wells for which energies are more dependent on the strength of the electric field, one observes in Fig. 5.12a a clear redshift of the transition energies with increasing barrier width. We have checked the possible screening of the internal electric field due to injected carriers by decreasing the laser intensity over two decades. This did not change the PL energy of narrow wells 4 and 8 ML wide, but could induce a weak We have checked the possible screening of the internal electric field due to injected carriers by decreasing the laser intensity over two decades. This did not change the PL energy of narrow wells (4 and 8 ML wide), but could induce a weak (± 10 meV) redshift for the wider ones.

Now, to evaluate the impact of the chemical contrast which contributes to both mismatch of spontaneous polarization and magnitude of the piezoelectric polarization, a series of MQWs samples, with GaN well width of 4, 8, 12 and 16 MLs were grown with varying the aluminum composition up to 27 % in the barrier layers which

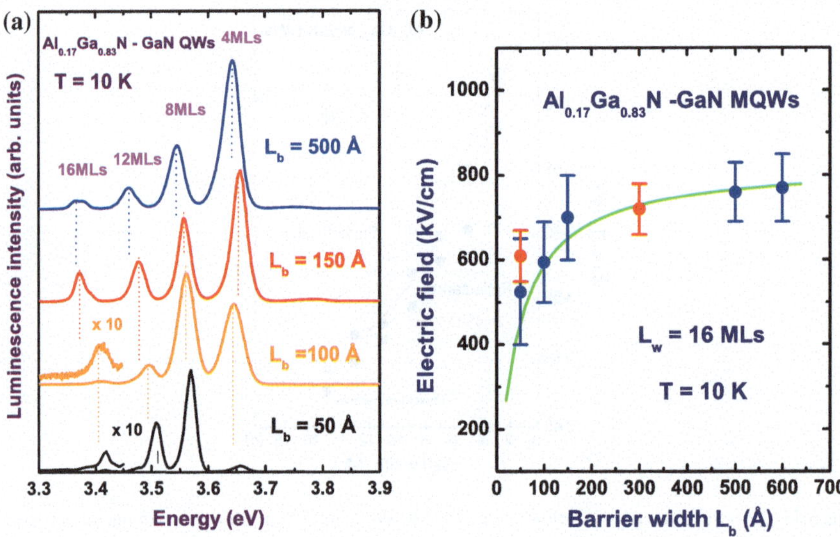

Fig. 5.12 **a** Photoluminescence spectra at 10 K of samples including QWs of various widths 4, 8, 12, and 16 MLs as a function of the $Al_{0.17}Ga_{0.83}N$ barrier width. Note the redshift of the PL energy of the wide QWs with increasing barrier width. **b** Barrier-width dependence of the electric field in the GaN QWs deduced from the PL data. After [2]

were thicknesses were kept constant (10 nm). Some typical PL spectra are given in Fig. 5.13a. A linear increasing of the electric field in the well at rate of 40–50 kV/cm per aluminum percent in the barrier layers. The variation of the electric field in the QW is plotted in Fig. 5.13b versus aluminum composition in the barrier layer.

5.2.1.2 Oscillator Strengths and Exciton Binding Energies

We have earlier indicated in Fig. 5.10 that, for a single quantum well 17 ML thick which emits light clearly below the energy of GaN (see the 3.48 eV line), thick barrier layers contribute to both reflectivity and photoluminescence as the GaN template but it is not possible to detect the signature of the quantum well in reflectivity in that latter case. This we attribute to the vanishingly small oscillator strength in line with the Quantum Confined Stark Effect and the resulting particularly small overlap of the electron and hole envelope functions for that sample. In Fig. 5.14a are reported the plot of the densities of probability (square moduli of the envelope functions) for of different widths of for various aluminum composition in the barrier layers (thus we change the magnitude of the electric field in the barrier layers). In Fig. 5.14b have been plotted the evolutions of the oscillator strength obtained using these different approaches, versus well width for $GaN/Al_{0.27}Ga_{0.73}N$ QWs. The calculation was also realized in the absence of electric field. Results for the (open symbols) or including the experimental values deduced in the limit of infinite width of the barrier

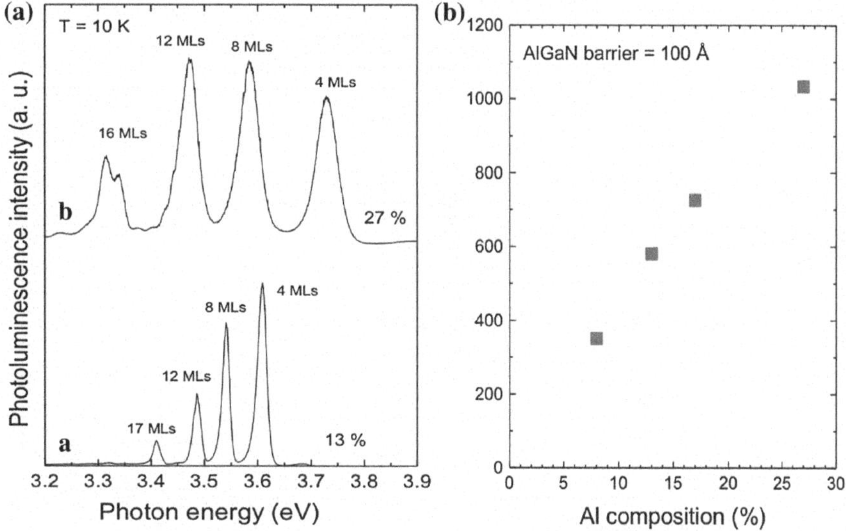

Fig. 5.13 **a** 10 K photoluminescence spectra of AlGaN/GaN QWs with an Al mole fraction in the barriers of (*a*) 0.13, and (*b*) 0.27. Note that the QW transition energies of the thinner wells are *blue-shifted* when increasing the Al composition while that of the larger wells are *red shifted*. **b** The variation of the built-in electric field as a function of the aluminum mole fraction in the AlGaN barrier layers. After [3]

layers(solid symbols). The result of a non-excitonic (band to band) model is shown by diamonds, while squares and circles correspond to the 2D and 3D excitonic models, respectively. Although the oscillator strength is not really changed (it is slightly) when changing the well width, it decreases exponentially with the well width for well widths greater than 10 monolayers (2.6 nm) whereas the binding energy remains significant. For the square well, the oscillator strength only varies by a factor of 2, in the present range of QW widths, whereas it looses five orders of magnitude, from narrow wells to 30 ML wide QWs, for F = 1.1 MV/cm. The effect is similar for excitonic calculations, but, for the most accurate of them (3D model), the oscillator strength looses six orders of magnitude, rather than five. It is thus important to include excitonic effects, to describe correctly the well-width dependence of radiative lifetimes in such QWs.

In Fig. 5.15a have been plotted the evolutions of the excitonic binding energies in cases of of 2D and 3D calculations for GaN/ $Al_xGa_{1-x}N$ QWs with different x values. When the internal electric field is ignored (the so-called "square-well model" case,), we find the trend already commented for GaAs–AlGaAs QWs. The 2D calculation underestimates the exciton binding energy E_B, down to values under the bulk effective Rydberg, for wide enough QWs. The 3D calculation is more accurate (larger E_B) and valid for the whole range of QW widths. We we include an internal electric field F = 1.1 MV/cm (for x = 0.27), we find that the 3D model in-creases E_B by up to 40 %, with respect to the 2D model, for narrow wells. Paradoxically, the improvement brought by the 3D model is reduced when increasing the QW width.

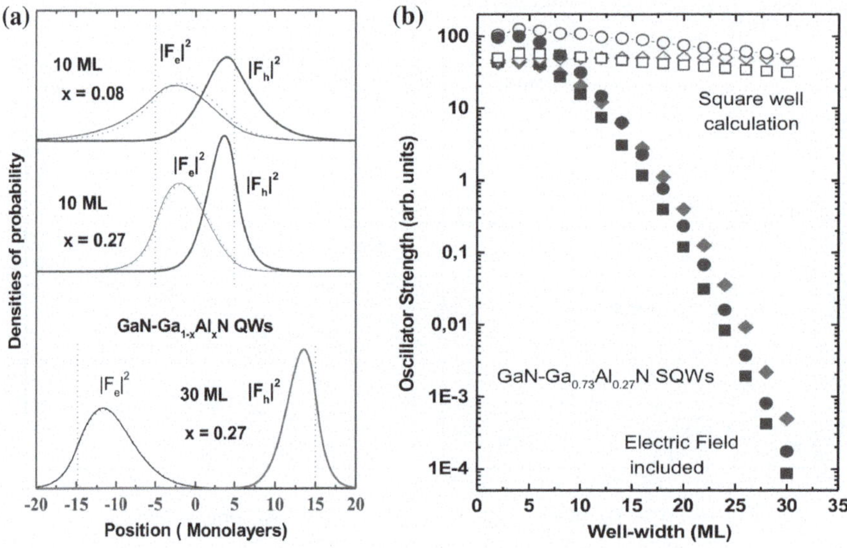

Fig. 5.14 **a** Square moduli of the envelope functions for electron and heavy hole in polarized SQW for different widths and/or Al contents. **b** Oscillator strength versus well width for GaN–$Al_{0.27}Ga_{0.73}N$ QWs. The result of a nonexcitonic (band to band) model is shown by *diamonds*. *Squares* and *circles* correspond to the 2D and 3D excitonic models, respectively. Results for the square well model are shown by *open symbols*, *solid* ones are used for the realistic potential line-up. After [41]

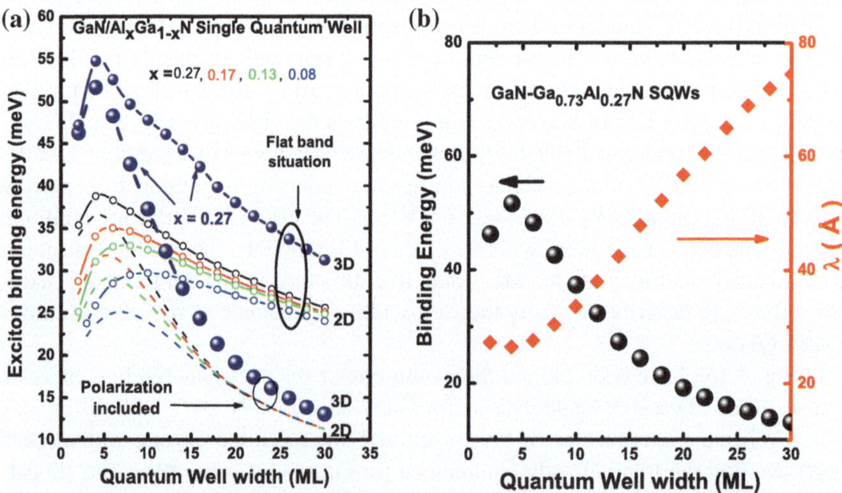

Fig. 5.15 **a** Width dependence of the exciton binding energies for GaN–AlGaN QWs with *square potential line* ups and for realistic ones, including internal electric fields. The results of the 2D model are given for all four aluminum compositions, whereas the results of the 3D model are restricted to $x = 0.27$. **b** Exciton binding energy and variational parameter λ obtained by the 3D calculation for a series of GaN–$Al_{0.27}Ga_{0.73}N$ single QWs. After [41]

This is an effect of the strong electron-hole separation induced by the field. It is this field, rather than the Coulomb attraction, which controls the on-axis extension of the exciton wave function. E_B collapses from 52 meV, for a 1 nm QW, down to 12 meV for a 8 nm QW. The 3D calculation provides a better result, i.e. larger binding energies.

Figure 5.15b shows the QW width dependence of E_B, for x $=$ 0.27, and the corresponding variation of the in-plane extension parameter λ. Due to the loosening of the Coulomb attraction, for wide QWs, the in-plane extension gains a factor of two, from narrow QWs to a 30 ML wide QW. This means that the critical density for exciton-exciton interaction is divided by four. Indeed, this density is given by $N_s = \beta/\lambda^2$, where $0.01 \leqslant \beta \leqslant 0.04$, depending on the particular conditions of the system. The Coulomb interaction is much easier to screen in wide QWs than in narrow QWs. However, it is important to remark, from the values in Fig. 5.15b, that N_s reaches a few 10^{10}–10^{11} pairs per cm^2, i.e. nearly two orders of magnitude smaller than what is necessary to screen internal electric fields by carrier accumulation. Finally the excitonic model gives lower oscillator strengths than the band to band model in straightforward relation with the λ^{-2} dependence in complement to the variation of the overlap integral.

5.2.1.3 Excited State Transitions

At this stage we can wonder about detection of excited, states, that is to say states between excited states confined electron and excited states confined holes.Room-temperature (RT) photoreflectance (PR) spectroscopy is the ad hoc technique, do detect them. Photoreflectance is a refined differential spectroscopy method which consists in simultaneously shining the sample with an energy-varying weak cw light beam and with time-modulated laser beam which energy is higher than the band gap energy. Excess carriers are thus periodically created which modulate periodically the dielectric constant, periodically changing the reflectivity. The reflected beam is analyzed by standard lock-in technique, experimental spectra are composed of derivative features which are often more pronounced than those obtained by direct reflectance. Thus, PR is really useful for detecting transitions having very weak oscillator strengths. We also stress the fact that PR is only concerned with free excitons, rather than with localized excitons or excitonic complexes, in contrast to PL. For room-temperature PR measurements in the UV range, we have used a conventional set-up with a 325nm HeCd laser used to modulate the optical response of the samples.

Top of Fig. 5.16a is shown the 2 K reflectance spectrum of a MQW with ten repeats of 1.5 nm GaN cladded between 5.0 nm $Ga_{0.89}Al_{0.11}N$ barrier layers. Beneath this active region had been deposited a 2 μm thick GaN buffer layer. In the 3.48 eV range are detected reflectance features corresponding to the A, B and C excitons in this GaN buffer. At 3.612 eV, a strong reflectance feature is detected which correspond to the MQW. At the bottom of the figure is shown the corresponding PL spectrum which exhibit a weak and narrow line at the energy of D^0X in GaN and a broader PL

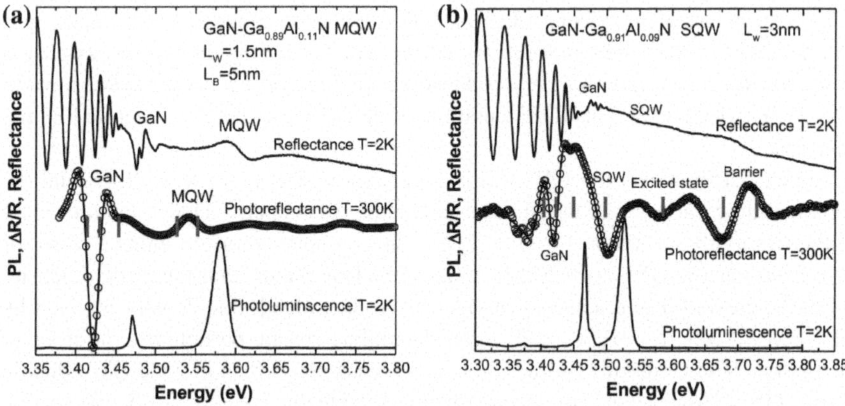

Fig. 5.16 **a** 2 K Reflectance (*top*), photoluminescence (*bottom*) and 300 K photoreflectance spectra for a GaN–$Ga_{0.89}Al_{0.11}N$ multiple quantum well with 1.5 nm well layers and 5 nm barrier layers. Photoreflectance spectrum is represented using *open circles*, fit is represented using *straight line*. **b** The analogue of Fig. 5.1 but for a single quantum well having a well width of 3 nm. The aluminum composition is 9 %. Note the detection of excited states in the photoreflectance experiment. After [45]

line that are assigned to excitons localized in enlarged-well regions of the MQW. The PR spectrum taken at room temperature is plotted in the middle of the figure. Dots correspond to the experiment while the bold line is the result of the line-shape fitting. The RT values we obtain for the energies of A, B and C excitons in GaN are 3.4147, 3.4147, and 3.4299 eV, respectively. Their 2 K counterparts are at 3.4762, 3.4834 and 3.5024 eV, respectively. It is important to note that we cannot detect the signature of transitions related to the 5 nm thick $Ga_{0.89}Al_{0.11}N$ barrier layers. There is a 60 meV red-shift for the A line between the 2K and RT spectra. Concerning the MQW structure the 2 K value is 3.612 eV. Photoreflectance line-shape fitting gives two oscillators energies at 3.526 and 3.5520 eV that we attribute to transitions between the first confined electron level and unresolved (A,B)-type and C-type confined hole states states.

In Fig. 5.16b are reported similar spectra, for a wider single GaN quantum well with 30 nm-thick $Ga_{0.91}Al_{0.09}N$barriers, grown on a 2 μm thick GaN buffer layer. In the 2 K reflectance experiment, well marked transitions are detected at the energy of the GaN buffer layer, at the energy of AlGaN and at the energy of confined transitions in the GaN well. The energies of these transitions at $T = 2$ K are: 3.480, 3.488, 3.506eV, for the GaN buffer, 3.550 eV for the QW, and 3.720 eV for the barrier layer. The lineshape fitting of the room temperature photoreflectance data gives 3.4977 eV for the QW, and two contributions at 3.6778 and 3.7285 eV for the barrier layer. The PR also allows us to observe an excited state, at 3.5854 eV. The redshift between the 2 K reflectance and RT photoreflectance energy is 54 meV for the QW. The line shape fitting of the GaN-related PR feature yields a 27 meV splitting that we attribute to the A–C splitting (the A–B splitting is not resolved). In conclusion excited state transition can be observed which involve excited electron states with a 60 meV splitting.

In conclusion some excited state transitions are observed in thick enough quantum wells.

5.2.1.4 Localization Phenomena

In all these quantum wells we have observed a more or less strong localization of the exciton, that is to say a Stokes-shift between low temperature photoluminescence and low temperature reflectivity. This we attribute to the roughness of the interfaces: controlling their flatness is really tricky, because the materials are strained and then corrugation may relax the strain, but also just because of the chemical alloy disorder. To illustrates this quantitatively we consider the excitonic function that we write using a 3D excitonic trial function in order to get analytical expressions. So, let the exciton wave function:

$$\Psi(z_e, z_h, \rho) = N\chi_e(z_e)\chi_h(z_h)e^{-\frac{\rho}{\lambda}}$$

Let us choose for origin the center of gravity of the electron and hole in the plane of the quantum well. Let: $\rho_e = \rho\frac{m_h}{m_e+m_h}$ and $\rho_h = \rho\frac{m_e}{m_e+m_h}$, then; $\rho = \rho_e + \rho_h$.

Averaging the Ψ over the hole (respectively electron) positions we get the electron and hole pseudo-envelope functions:

$$\Psi_\xi(z_\xi, \rho_\xi) = N_\xi f_\xi(z_\xi)e^{-\frac{\rho_\xi}{\lambda_\xi}}$$

The in-plane extensions of the electron and hole are: $\lambda_e = \lambda\frac{m_h}{m_e+m_h}$ and $\lambda_h = \lambda\frac{m_e}{m_e+m_h}$.

We have computed these quantities in case of a 3 nm wide GaN–$Ga_{0.83}Al_{0.17}N$ single quantum well. Then the in-plane extension of the electron and hole pseudo wave functions become comparable in size with the average distance between two aluminum atoms for aluminum compositions up to 30 %. typically one gets $\lambda_h = 0.7$ nm for a 30 ML wide quantum well and an aluminum composition of 17 % in the barrier layers. a similar calculation for GaAs-AlGaAs quantum wells gives 10–12 nms! In GaN AlGaN quantum wells (and for other wide bandgap-based quantum wells with carrier confining layers made with binary compounds) the interfacial excitonic localization results of the comparable values of the Carrier Bohr radii with the characteristics length of the alloy disorder. The reasons for this are the exceptionally high values of the effective masses, in particularly the very high value of the hole effective mass. In Fig. 5.17 we represent in plane extensions of the electron and hole densities of probability in case of a 3 nm wide GaN–$Ga_{0.83}Al_{0.17}N$ single quantum well. Aluminum atoms in AlGaN are represented using large yellow spots to illustrate the chemical alloy disorder. The in-plane extension of the hole wave function is clearly smaller than the average aluminum-aluminum distance. The chemical alloy disorder is at the origin of the interface roughness experienced by the exciton.

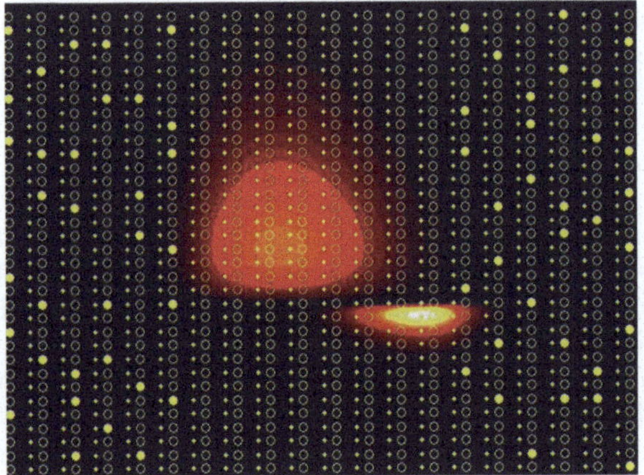

Fig. 5.17 The in plane extensions of the electron and hole densities of probability in case of a 3 nm wide GaN–$Ga_{0.83}Al_{0.17}N$ single quantum well. Aluminum atoms in AlGaN are represented using *large yellow spots* to illustrate the chemical alloy disorder. After [4]

5.2.2 ZnO–ZnMgO Polar Quantum Wells

As a second illustration we now consider ZnO–ZnMgO quantum wells. The specificity of the alloy layer is that it is not possible to grow it wurtzite up to a magnesium concentration of about 25 %. Second, the excitonic oscillator strength is ver high as well as the exciton binding energy in the bulk. Therefore such effects may be more dramatic than for nitrides. In Fig. 5.18 are represented a series of photoluminescence spectra taken for ZnO well thicknesses thickness ranging from 1.6 to 9.5 nms. The photoluminescence energy crosses the energy of the ZnO band gap for a layer thickness of about 3 nm. The strength of coupling of excitons to LO-phonons appears to be a growing function of LW, as shown by the growing relative intensities of phonon replicas. This is another effect of the electric field.

In Fig. 5.18b is represented a 2D X-ray diffraction mapping to indicate the lattice matching of the ZnO and $Zn_{0.78}Mg_{0.22}O$ layers via the similarity of the q_{100} in-plane reciprocal vectors. The on axis values of the lattice parameters of ZnO and $Zn_{0.78}Mg_{0.22}O$ as indicated by different values of the q_{001} on axis reciprocal vectors, the c parameter of ZnMgO being smaller than that of the ZnO buffer layer.

The fit of the evolution of the photoluminescence energies versus well width has been realized assuming an electric field of 900 kV/cm in the ZnO quantum well layers, in the context of an excitonic modeling. Figure 5.19a illustrates the the agreement between theory and experiment. Two main regimes can be distinguished as the well width is increased. For QWs thinner than 3 nm, the PL energy is above the bulk ZnO band gap; we call this the quantum confinement regime. For wider QWs, the PL energy decreases linearly with the well width, this is called the QCSE regime. Figure 5.19b illustrates the evolution of the exciton binding energy with well width (left-hand scale) and the evolution of the variational in-plane Bohr radius

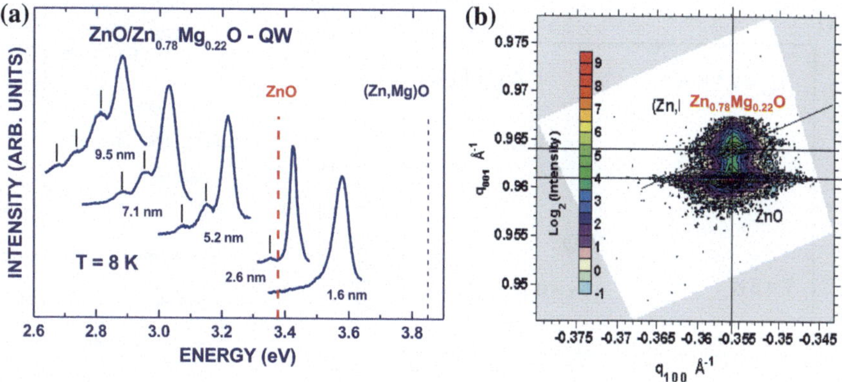

Fig. 5.18 **a** Continuous-wave PL spectra of ZnO–$Zn_{0.78}Mg_{0.22}O$ QWs of various widths, as indicated, taken at T = 10 K. The pump-power density was $100\,mW/cm^2$. The PL energies from the $Zn_{0.78}Mg_{0.22}O$ barrier layers and those of the ZnO buffer layers are shown by *dashed lines*. **b** X-ray reciprocal space map in the vicinity of the $(\bar{1}0.5)$ source reflection, in grazing incidence geometry. After [70]

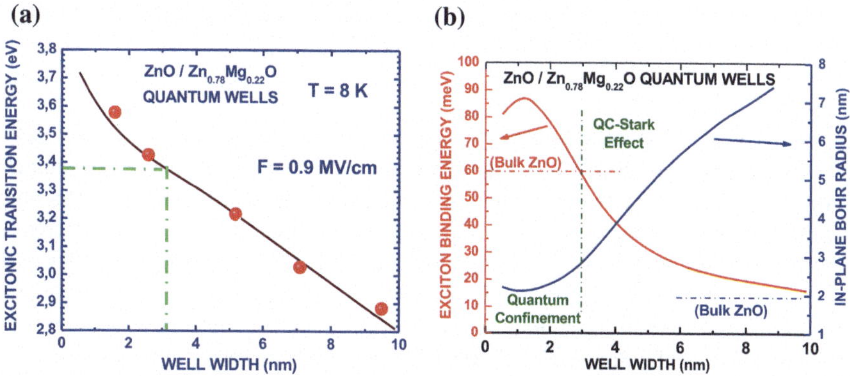

Fig. 5.19 **a** *Squares* show the PL peak energies, compared to the result of our variational calculation (*solid curve*) for an in-built electric field of 0.9 MV cm^{-1}. *Dash-dotted lines* show the excitonic gap of ZnO and the critical well width separating the two regimes of confinement in the *square* or *triangular parts* of the quantum well layers. **b** Calculated binding energy (*left scale*) and pseudo-Bohr radius (*right scale*) versus well width, obtained by assuming an electric field of 0.9 MV/cm. The two regimes described in the text are indicated. After [70]

(right-hand scale). The same two regimes can clearly be distinguished: in the quantum confinement regime, the exciton binding is enhanced by up to 40 % compared to that for bulk ZnO, but it is strongly weakened in the QCSE regime due to the electron–hole separation. Therefore only QWs thinner than about 3 nms really have exciton binding energies stronger than bulk ZnO.

Figure 5.20a displays the low temperature PL spectra of 7.1 nm wide QWs corresponding to two Mg compositions of the barriers, as shown by two different PL

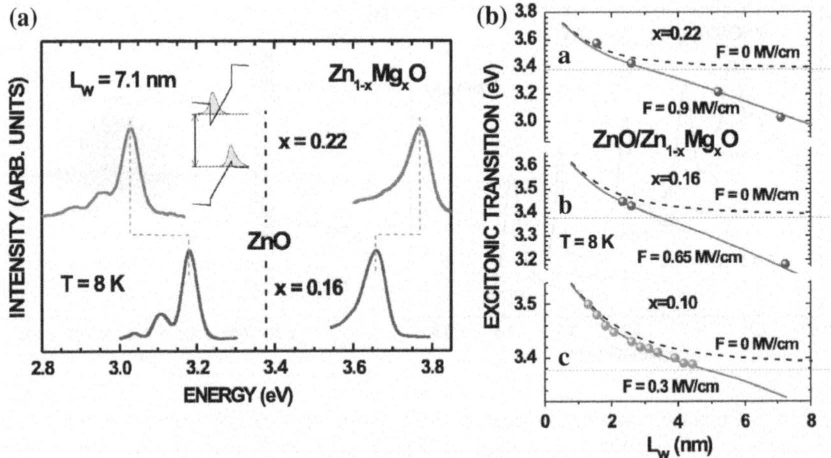

Fig. 5.20 **a** Photoluminescence spectra of 7.1 nm width ZnO-$Zn_{1-x}Mg_xO$ QWs and $Zn_{1-x}Mg_xO$ barrier layers with a Mg concentrations in the barriers of 0.16 and 0.22. **b** Comparison of the calculated exciton transition energy neglecting (*dashed line*) and including (*full line*) the electric field for different Mg contents with low temperature experimental PL energies. *Green dashed lines* indicate the energy of free-exciton in bulk ZnO. After [67]

energies. The QW spectra exhibit PL features at energies below that of the bulk ZnO-free exciton (dashed line). This is characteristic of the existence of a quantum confined Stark effect: for wide enough QWs, the electric field pushes the fundamental transition below the excitonic gap of the constituting material. Moreover, increasing the Mg concentration decreases the PL energy, which is readily assigned to an increase of the internal electric field. The recombination energy, computed taking into account the variation of binding energy of the ground-state exciton are compared to experimental data in Fig. 5.20b, the calculated excitonic transition energies including an electric field solid lines) or not including it (dashed lines).

The results shown above indicate that the built-in electric field increases linearly with x, as shown in Fig. 5.21a. Finally, the variation of the calculated PL transition energy versus the well thickness is plotted in Fig. 5.21b, for different values of x. We note a remarkable crossing of all calculated excitonic energies near 3.43 eV, for a well thickness of L_W ~2.3 nm independently of x. For the sake of the completeness we wish to indicate that similar behavior was reported in [25].

Finally the properties of GaN–AlGaN quantum wells and ZnO–ZnMgO quantum wells are rather similar. For single GaN–AlGaN quantum wells the average values of the internal electric field is of 50 kV/cm per aluminum percentage in the barrier layers and about 40 kV/cm per percentage of magnesium in the barrier layer for ZnO-ZnMgO quantum wells. In case of GaN–$Al_{0.83}In_{17}N$ (for an aluminum composition of about 17 % so that the ternary alloy is lattice matched to GaN) it has been found in [5] a huge value of 3.64 MV/cm.

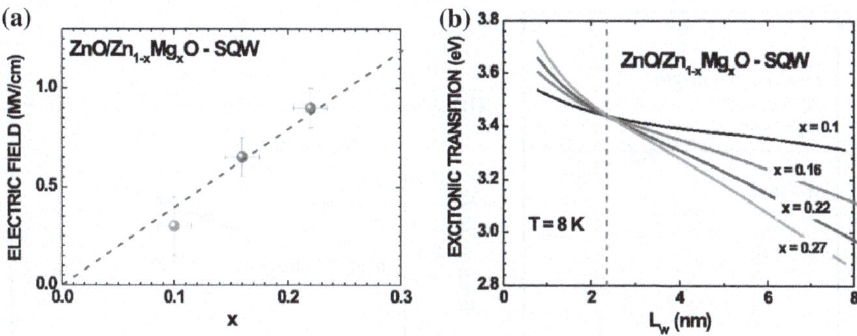

Fig. 5.21 **a** Variation of the internal electric field as a function of Mg content in the barriers. **b** Calculated excitonic transition energies in SQWs taking into account the built-in electric field for different Mg composition. After [67]

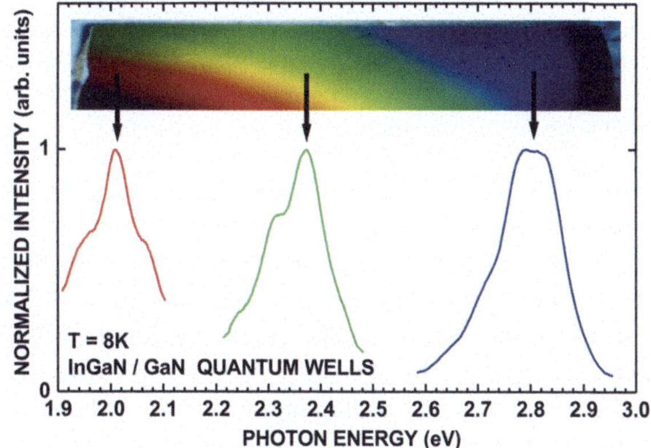

Fig. 5.22 Light emission from a GaInN quantum well with graded thickness. Courtesy Benjamin Damilano, Nicolas Grandjean, Stephane Vezian, Philippe Vennegues and Jean Massies,

5.2.3 GaInN-Based Polar Quantum Wells

In that case, the confining layer is made with a ternary compound and while increasing the indium composition one could grow quantum wells with optical properties ranging from the ultraviolet to the infrared. This has been demonstrated by Benjamin Damilano, Nicolas Grandjean, Stephane Vezian and Jean Massies who grew a $Ga_{0.83}InN_{17}$–GaN quantum well with inhomogeneous width and targeted to observed light emission from the red to the violet [6]. A photograph of the light emitted by such sample is given in Fig. 5.22. Transmission electron microscopy realized by Philippe Vennegues have revealed the thickness to continuously range between 2 nms in the ultraviolet emission region and 6 nms in the region of red emitting light. This indicates with naked eye the existence of a substantial quantum confined Stark Effect in this sample. Photoluminescence spectra recorded at different positions of the

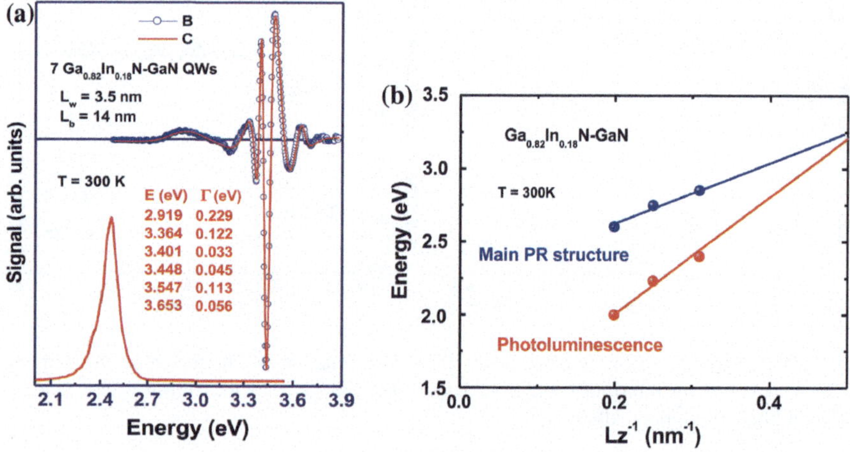

Fig. 5.23 **a** Photoluminescence spectrum (*bottom*) and photoreflectance spectrum (*top*) of a $Ga_{0.82}In_{0.18}N$–GaN multiple quantum well. Well width and barrier width are 3.5 and 14 nm, respectively. **b** Relative evolutions of the photorelectance and photoluminescence energies versus reciprocal well width. After [7]

sample peak at different energies but display similar Full Width at Half Maximum, the undulations being correlated to macroscopic light diffraction effects in the sample. Fitting the maxima of the photoluminescence peak versus well width requires to consider and electric field of about 2.5 MV/cm in this sample. This is obviously a huge value, and comparing the GaN–$Al_{0.83}In_{17}N$ data (3.64 MV/cm) with this one one can come to the conclusion that adding indium results of heterostructures with huge internal electric field. Both mismatch of spontaneous polarizations in the bulks and piezo-electric contributions have to be considered to explain this, which indicates that there is no general statement.

We now wish to calibrate how big the inhomogeneous broadening effects typical of bulk semiconductor alloys are in heterostructure when the confining layer is an alloyed material and in presence of quantum confined Stark Effect. To do so, we compare photoluminescence and reflectance features collected on a series of $Ga_{0.82}In_{0.18}N$ multiple quantum wells with varying well width (L_z) and constant (14 nm) thicknesses of the GaN barrier layers. Data corresponding to the long-wavelength regions of the light spectrum for which the investigated effects are the most dramatic are presented. For a 3.5 nm thick QW that radiates at 2.4 eV at room temperature (green light), a reflectance structure is fitted at 2.92 eV with a broadening of 229 meV as indicated in Fig. 5.23a. The reflectance structure of interest is found at 2.76 eV for a second (orange light emitting) 4 nm-thick sample and for a 5.5 nm wide multiple QW for which the weak photoluminescence peaks at 2 eV, at 10 K, one can hardly resolve some very broad (900 meV) photoreflectance feature centered at 2.8 eV. We have fitted the evolution of the room temperature PL peak maximum Ep with well width in our samples and we got Ep (eV) = 3.302

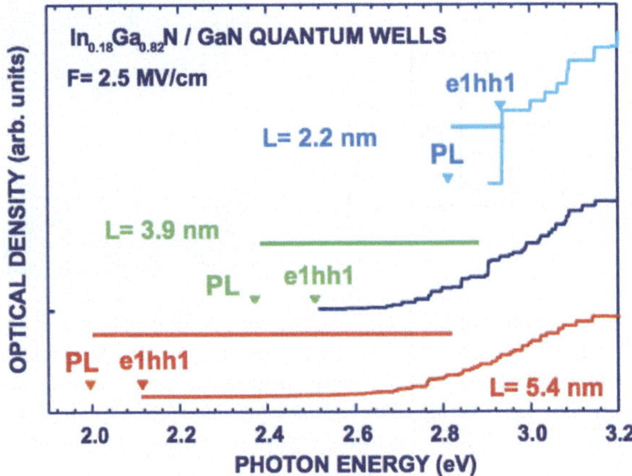

Fig. 5.24 The PL spectra taken at three different spots on the graded-width QW sample are plotted along with interband absorption spectra calculated for InGaN QWs with $x = 0.175$, for the three corresponding well widths. The part of the stokes shift resulting from the disorder-induced fluctuations was kept constant and equal to 120 meV. This corresponds to the case at the *top* of the figure (*narrow well*), where the e1-hh1 transition (*down triangles*) has a substantial oscillator strength. In the two other cases, the absorption onset is rather made of transitions between excited levels. The experimental stokes shifts are shown by horizontal segments. After [54]

$-0.2618\ L_z$ (nm). This diminution of the PL energy with the well width is the consequence of the QCSE. Similar treatment per-formed for the reflectance signal gave Eg(eV)$= 3.414 - 0.164 L_z$ (nm). In the limit case of very thin wells, the energy difference Eg–Ep tends to the localization energy. This result indicates that, for these quantum wells, the Stokes shift increases when the well width increases as reported in Fig. 5.23b.

Next, an envelope function determination of the transition energies and band-to-band oscillator strengths is then performed and we plot, as a function of their energy, the amplitudes of different band-to-band transitions computed for several quantum well designs. By doing so, we mimic the absorption coefficient as proposed in [8].

Due to the QCSE, the oscillator strength of the fundamental transition which is one of the dominant transitions for thin-well samples diminishes when the well width increases. Subsequently, due to the absence of oscillator strength for the low energy transition, one observes a strong Stokes shift between the PL energy and the PR feature for red samples (wide wells) and a weak one for violet light emitters (thin wells). The long-lived fundamental states are a bottleneck for thermalisation of the carriers and a PL signal is detected at their energy. The main band that is detected by PR always corresponds to the energy at which the onset of strong transitions occurs. When increasing the well width, this onset shifts slower than the fundamental energy transition (the PL data). Because the hole mass is extremely heavy, we have a fairly large number of hole states in the case of wide wells which gives the increase of the broadening of the photoreflectance feature with increasing well width (Fig. 5.24).

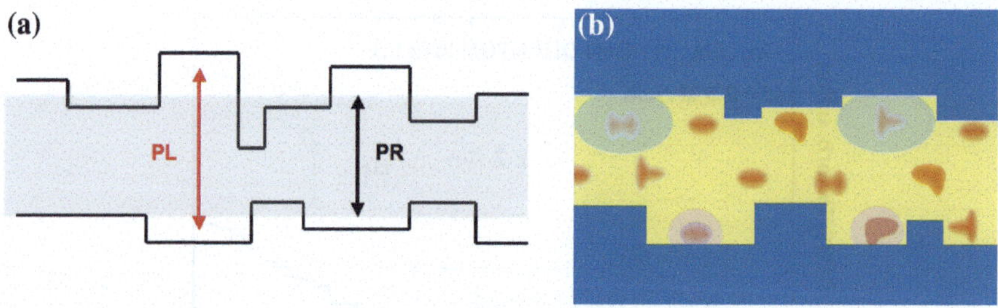

Fig. 5.25 **a** Sketch of a quantum well with binary compound confining layer and an alloy compound used as barrier layers. From the chemical disorder the apparent width of the quantum well may be subject to local variations leading to exciton confinement in the "wide" well regions. **b** Sketch of a quantum well with an alloy layer acting as confining layer and a binary compound used as barrier layers. Superimposed to interfacial roughness may occur localization in "low-bandgap" regions of the confining layer

The contribution of the QCSE superimposes on the intrinsic localization phenomena of the carriers in the InGaN alloy, and is larger by one order of magnitude. Interpretation of data for samples that emit from the blue to the red can provide only partial conclusions if both localization effects and QCSE are not taken into consideration, but it is very difficult to disentangle both effects without complementary experiment like temperature-dependent measurement of the photoluminescence intensity and time-resolved dependence of the photoluminescence intensity. When the confining layer is a binary compound, the localization of the carriers is produced by chemical alloy in the barrier layer which produces an apparent thicker layer as indicated in Fig. 5.25a and were the photoluminescence mainly comes from (this being further enhanced thanks to the help of the QCSE which localizes efficiently the carriers to the interfaces), while the photo reflectance is centered at an average energy and results from every where contributions. In the case of alloys there is also localization due to fluctuation of composition in the confining layer as indicated in Fig. 5.25b. The photoluminescence exhibits different behaviors for these different quantum heterostructures as we shall show it in the next section.

5.3 Temperature Dependent Photoluminescence Spectroscopy

In Fig. 5.26a are reported the evolutions of the main photoluminescence peak measured from liquid helium temperature up to 300 K in the cases of a 1.5 nm thin GaN–$Ga_{0.9}Al_{0.1}N$ quantum well (top of the figure) and a 2.2 nm thick $Ga_{0.82}In_{0.18}$ N–GaN quantum well. Both samples were grown on sapphire after pre deposition of a GaN template. The blue shift of the photoluminescence between 2 and 60 K is about 12 meV for the GaN–$Ga_{0.9}Al_{0.1}N$, which we interpret in terms of exciton detrapping in the wide well regions of the sample. Then the photoluminescence follows the trend expected for GaN on sapphire when the excitonic mobility edge is reached

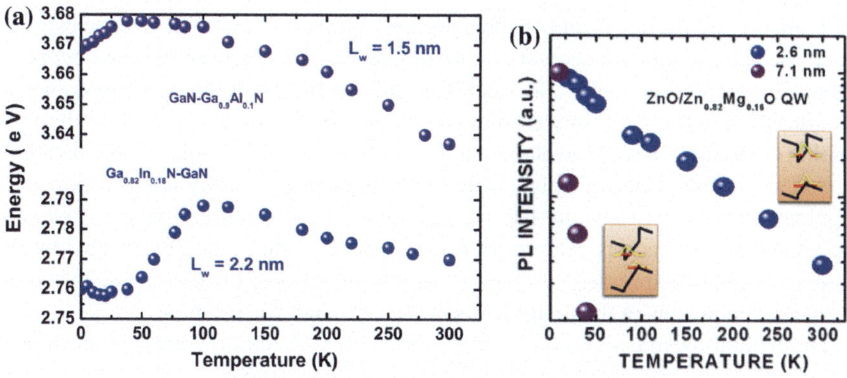

Fig. 5.26 **a** Temperature dependence of the photoluminesce energy for a 1.5 nm thin GaN–$Ga_{0.9}Al_{0.1}N$ quantum well (*top of the figure*) and a 2.2 nm thick $Ga_{0.82}In_{0.18}N$–GaN quantum well. **b** Temperature dependence of the photoluminescence intensity for ZnO−$Zn_{0.78}Mg_{0.22}O$ quantum wells of various thicknesses

by thermal heating of the exciton population. In the case of the $Ga_{0.82}In_{0.18}N$–GaN quantum well, the situation is more complex: first thermal induced detrapping of excitons from indium-rich regions is observed and the average energy of the excitonic population does-not really follow the evolution of the band gap of GaN. The excitonic mobility edge has not been reached and we measure photoluminescence based on both free and localized excitons. This behavior is particularly dependent on the structural properties of the sample and is called **s-shape dependence of the photoluminescencece energy**.

The evolution of the band gap versus temperature for such samples is generally fitted using an empirical laws of the kind:

$$E_{PL}(T{:}E_{PL}(0) + A\frac{T^2}{T+B} + C\frac{1}{T^2}$$

where A, B, and C are ad-hoc parameters. One can identify the description of the evolution of the photoluminescence energy in terms of the two-parameter (A and B) Varshni's equation plus a phenomenological term which accounts for the degree of localization in the sample via a third fitting parameter C. There is a large scattering for such values in the literature.

In Fig. 5.26b have been plotted the intensities of the luminescence intensity for a couple of ZnO−$Zn_{0.78}Mg_{0.22}O$ quantum wells of different well width. Clearly the photoluminescence intensity is less robust for the thick well than in case of the thin one. This can be first interpreted in terms of the influence of the electric field which strongly localized carriers at the hetero-interfaces rendering their lifetimes very sensitive to the presence of topological defects in the barrier layers. For thin wells the wave function of the carriers is not so strongly localized in specific regions of the crystal and the exciton is less sensitive to imperfections of the material used for realizing the barrier layers. To this argument linked to material science aspects has to be completed by a second one more general. The exciton binding energy is strong in case of thin wells and weak for wide wells. Then the robustness of the exciton being better for thin well they are not so efficiently dissociated by thermal heating or interaction with the phonon field even stronger is the impact of localization in case of GaInN–GaN quantum wells: the fluctuation of alloy composition is a supplementary effect reducing the average value of the electron-hole distance and the photoluminescence is even more robust with T in that case. These qualitative statements can be expressed more quantitatively in terms of the thermal distribution of exciton-polaritons in the reciprocal space as we shall see it below.

The evolution of the photoluminescence intensity with temperature is proportional to the ratio of the effective (τ) and radiative (τ_{rad})decay times:

$$I(T) = g \cdot \tau(T)/\tau_{rad}(T)$$

where g is a generation term which depends on the experimental context. The effective decay time (τ) is the sum of two different terms: the radiative part and the non radiative one τ_{nonrad}.

$$\tau(T)^{-1} = \tau_{rad}(T)^{-1} + \tau_{nonrad}(T)^{-1}$$

Then the photoluminescence intensity can be written:

$$I(T) = g \cdot \tau_{nonrad}(T)/[\tau_{rad}(T) + \tau_{nonrad}(T)]$$

or:

$$I(T) = g\frac{1}{1 + \frac{\tau_{rad}(T)}{\tau_{nonrad}(T)}}$$

This equation indicates that the photoluminescence intensity is rule by the ratio

$$\frac{\tau_{rad}(T)}{\tau_{nonrad}(T)}$$

One generally assumes that $\tau_{rad} \ll \tau_{nonrad}$ at low temperature in decent quality samples. In that case the photoluminescence intensity is maximum and it is postulated and alegated that the quantum efficiency equals unity.

$$I(T) = I(0)\frac{1}{1 + \frac{\tau_{rad}(T)}{\tau_{nonrad}(T)}}$$

The radiative decay time is an intrinsic quantity like will be shown later. The non radiative part is correlated to the crystalline quality of the sample and eventually to its design. In the simplest of the description one can writes it phenomenologically as:

$$\tau_{nonrad}(T) = \tau_{nr0} e^{E_a/k_B T}$$

where τ_{nr_0} is a characteristic constant and E_a is an activation energy. It is possible to define an activation temperature $T_a = E_a/k_B$.

The crystal growers have to target the best values for both τ_{nr_0} and E_a so that the non radiative decay time will be always larger than the radiative one so that the photoluminescence intensity is not influenced too much by increasing the temperature.

The radiative part of the decay time is much more complex to establish. **Let us begin by considering free excitons.** The spatially coherent coupling of the exciton (inside a crystal) and the electromagnetic field out of the crystal (a photon that we detect outside the crystal) has been described in Chap. 4 in terms of the exciton polariton picture. In bulk crystals, excitons are propagating waves which are dispersion relations of the center of mass of the electron-hole interacting pair. The motion of this center of mass expresses in terms of a wave number $\vec{K} = \vec{k_e} + \vec{k_h}$ and a translation mass $M(\vec{K}) = m_e(\vec{k_e}) + m_h(\vec{k_h})$ as $E(K) = \frac{\hbar^2}{2m_0}\frac{K^2}{M(\vec{K})}$.

The requirement of energy and momentum conservation in the radiative process imposes bounds on the momentum of those free excitons that can decay into a photon. For an infinite bulk crystal, only those states at the crossing of the exciton and photon dispersion relations can decay into a photon susceptible to cross the crystal-air interface and to be detected. This argument also holds for low dimensional systems with some specificities that we will not detail here (for quantum wells, the exciton is confined to the plane while the photon is not). One can demonstrate that, within a very good approximation, the radiative recombination rate is proportional to the amount of excitons with translation wave vector in a n-dimensional light cone:

$$\Gamma(T) = \Gamma_0 \frac{\int_{|\boldsymbol{K}|<\omega/c} d\boldsymbol{K} \exp\left(-\frac{\hbar^2 K^2}{2M(K)k_B T}\right)}{\int d\boldsymbol{K} \exp\left(-\frac{\hbar^2 K^2}{2M(K)k_B T}\right)}$$

where K is the wave-vector of an exciton center of mass, $M(K)$ is the translation mass of the exciton and k_B is as usual the Boltzmann constant. We have used the Boltzmann statistics instead of the Bose Einstein one in order to get analytical equations, but this does not impact the physics. The integration in the numerator is performed at k lying inside the n-dimensional light cone, while for the denominator one should integrate over the whole k-space. Γ_0 is a proportionality constant which depends on the material of the design of the nanostructure.

This equation is interesting because it offers the possibility, while changing the temperature to study the evolution of the radiative recombination rate with T and it also gives access while changing the dimension of the semiconductor in which excitons are free to travel ($d\boldsymbol{K} = 4\pi K^2 dK$ for a bulk material, $2\pi K dK$ for a quantum well and dK for a wire) to some trends specifics for bulk crystals, quantum wells and quantum wires. For an exciton confined in a quantum dot owing to the absence of free motion the decay is naturally equal to that of a single particle and the statistical calculation proposed above and detailed below does not hold.

Let us first consider the 3D case:

$$\Gamma^{\text{bulk}}(T) = \Gamma_0^{\text{bulk}} \times \frac{4}{\sqrt{\pi}} \left(\frac{\hbar^2}{2Mk_BT}\right)^{3/2} \times \int_0^{\omega/c} k^2 dk \exp\left(-\frac{\hbar^2 k^2}{2Mk_BT}\right)$$

$$\Gamma^{\text{bulk}}(T) = \Gamma_0^{\text{bulk}} \left[\text{erf}\sqrt{\frac{T_0}{T}} - \frac{2}{\sqrt{\pi}}\sqrt{\frac{T_0}{T}} \exp\left(-\frac{T_0}{T}\right)\right]$$

where erf(x) is the error function and $T_0 = \hbar^2\omega^2/(2Mk_Bc^2)$.

One can demonstrate that the radiative lifetime $\tau_R = 1/(2\Gamma)$.

Let $\tau_0^{\text{bulk}} = 1/(2\Gamma_0^{\text{bulk}})$, then, in the regions of high temperatures $T_0/T \ll 1$, **the radiative recombination lifetime in a bulk material writes:**

$$\tau_R^{\text{bulk}}(T) = \frac{3}{4}\sqrt{\pi}\tau_0^{\text{bulk}} \left(\frac{T}{T_0}\right)^{3/2} \propto T^{3/2}$$

The results for 2D and 1D cases can be obtained in a similar way. Let us now consider the 2D case:

$$\Gamma^{\text{QW}}(T) = \Gamma_0^{\text{QW}} \frac{\hbar^2}{Mk_BT} \int_0^{\omega/c} kdk \exp\left(-\frac{\hbar^2 k^2}{2Mk_BT}\right)$$

$$\Gamma^{\text{QW}}(T) = \Gamma_0^{\text{QW}} \left[1 - \exp\left(-\frac{T_0}{T}\right)\right]$$

Then the radiative lifetime at high temperatures for a quantum well is given by:

$$\tau_R^{\text{QW}}(T) = \tau_0^{\text{QW}} \frac{T}{T_0} \propto T$$

One can further calculate τ_0^{QW} using a 2D excitonic trial function with one variational parameter λ, (see more readings section of that chapter) that:

$$\tau_R^{\text{QW}} \approx \frac{2c^2 k_B \hbar^3}{(\hbar\omega_{LT})a_B^3} \left[\frac{1}{m_{eU} + m_{hU}} + \frac{1}{m_{eV} + m_{hV}}\right]^{-1} \frac{1}{(\hbar\omega)^3} \frac{\lambda^2}{I_{eh}^2} T$$

In the equation above, the quantum well plane orientation is indicated through translation masses in the U and V directions: $m_{eU} + m_{hU}$ and $m_{eV} + m_{hV}$ respectively, $\hbar\omega_{LT}$ is the longitudinal-transverse splitting energy (*this is a representation of the oscillator strength of each exciton in the bulk, it depends on the polarization conditions for detecting the light*), a_B is the Bohr radius in the bulk material. These two quantities are typical of the bulk semi conductor used as confining layer.

Finally the quantum confinement appears through the recombination energy $\hbar\omega$, the in-plane variational Bohr radius λ and the value of the overlap integral between confined electron and confined hole envelope functions I_{eh}.

These last three terms implicitly contain influences of band offsets, quantum well width, electric fields....

Of course more sophisticated theoretical calculation would be needed to get a better representation of τ_R^{QW} but the physics is given by the approach above and we can account for the different variations of the photoluminescence intensities reported for ZnO-ZnMgO quantum wells in Fig. 5.26b. For the 7.1 nm thick quantum well, the photoluminescence intensity more rapidly collapses than for the 2.6 nm thin well when T increases. From the 2.6 nm to the 7.1 nm situation the in-plane variational Bohr radius is increasing by a factor about 2–3 (see Fig. 5.5b), and the recombination energy varies from 3.4 to 2.9 eV. The dominating contribution to the collapse of intensities (or robustness) is linked to the many-decade variations of the overlap integral I_{eh} with well width (see Fig. 5.14b) which makes the radiative decay time to scale several decades between these two samples. The two samples being grown in identical conditions there is no great reason to face dramatic differences at the scale of the non-radiative recombination channels. **The longer the radiative decay time, the weaker the temperature robustness of the photoluminescence.**

Before to measure the non radiative and radiative components of the decay times for quantum wells, we browse the evolution of the radiative decay times with dimensionality.

For the 1D case (quantum wire):

$$\Gamma^{QWW}(T) = \Gamma_0^{QWW}\left(\frac{\hbar^2}{2\pi M k_B T}\right)^{1/2} \times \int_{-\omega/c}^{\omega/c} dk \exp\left(-\frac{\hbar^2 k^2}{2Mk_B T}\right)$$

$$\Gamma^{QWW}(T) = \Gamma_0^{QWW}\operatorname{erf}\sqrt{\frac{T_0}{T}}$$

The radiative lifetime for a quantum wire at high temperatures varies like:

$$\tau_R^{QWW}(T) = \frac{\sqrt{\pi}}{2}\tau_0^{QWW}\sqrt{\frac{T}{T_0}} \propto \sqrt{T}$$

To obtain an expression for $\tau_R^{QWW}(T)$ we consider a z-oriented cylindrical quantum wire of radius a with a strong lateral confinement. The Hamiltonian of the internal motion of exciton in z-direction reads

$$\mathcal{H}_z = -\frac{\hbar^2}{2\mu_{eh}}\frac{\partial^2}{\partial z^2} - \frac{e^2}{\varkappa\sqrt{z^2+d^2}}$$

where $z = z_e - z_h$, μ_{eh} is the exciton reduced mass in the wire direction (z), $\varkappa$ is the effective dielectric constant times $4\pi\epsilon_0$. The parameter d is introduced to eliminate the energy divergency arising in one-dimensional Coulomb problem. For simplicity we will take $d = a$. The wavefunction of Hamiltonian above can be found using variational technique with the following trial function $f(z)$:

$$f(z) = \frac{1}{\sqrt{a_w}} \exp\left(-\frac{|z|}{a_w}\right)$$

with a single variational parameter a_w. According to Kavokin and Ivchenko:

$$\Gamma_0^{\mathrm{QWW}} = \frac{\pi}{4} a_B^3 q^2 \omega_{\mathrm{LT}} \int\int d\rho d\rho' \Phi(\rho)\Phi(\rho')\mathrm{J}_0(q|\rho - \rho'|)$$

with J_0 being the Bessel function of zero index, q is the wave vector of light in the quantum wire, ρ and ρ' and the following wavefunction are electron and hole coordinates.

$$\Phi(\rho) = \frac{1}{\pi\sqrt{a_w a_e a_h}} \exp\left(-\frac{\rho^2}{\bar{a}^2}\right)$$

where a_e and a_h are in-plane effective radii of electron and hole and $\bar{a}$ is defined as:

$$\bar{a}^2 = \frac{2a_e^2 a_h^2}{a_e^2 + a_h^2}$$

The integration finally gives:

$$\Gamma_0^{\mathrm{QWW}} = \frac{\pi}{4} \omega_{\mathrm{LT}} \frac{q^2 a_B^3 \bar{a}^4}{a_w a_e^2 a_h^2} \exp\left(-\frac{1}{2} q^2 \bar{a}^2\right)$$

The physics underlying the temperature dependence of the radiative recombination rates for any dimensionality when increasing temperature is illustrate in Fig. 5.27. At low temperature, all excitons have wave vector contained within the light cone; the recombination rate is very high. When increasing the temperature, the probability the exciton has to have a large wave number and a higher energy is given by the Boltzmann statistics. This probability increases with T. The average population of excitons with wave vector inside the light cone decreases when T increases. The radiative recombination rate decreases with T and the radiative recombination time increases with T.

The problem of the radiative recombination for a localized center (impurity, quantum dot) has to be formulated differently. To get mathematical expression we simplify it as follows.

Let us consider a quantum dot with a strong confinement in all spacial directions (i.e. the separation between energy levels due to confinement is larger than the Coulomb energy).

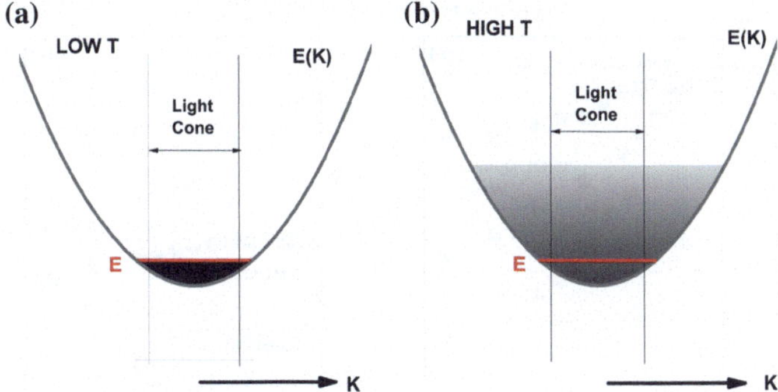

Fig. 5.27 **a** Plot of the energy of excitons against K and relative population of exciton in the light cone at low temperature (*shaded area*). Almost all of them have wave vectors inside the light-cone (its boundaries in K-space are represented using *vertical lines*). **b** Plot of the energy of excitons versus K and relative population of exciton in the light cone at high temperature (*shaded area*). Only part them have wave vectors inside the light-cone (its boundaries in K-space are represented using *vertical lines*)

The conventional way to derive radiative decays of excitons in different nanostructures is a non-local dielectric response approach. Γ_0^{QD} is obtained by taking the overlap between t a 3D Green function and the exciton wave-function $\Phi(\boldsymbol{r})$ at coinciding electron and hole coordinates:

$$\Gamma_0^{QD} = \frac{1}{6} q^2 a_B^3 \omega_{LT} \int\int d\boldsymbol{r} d\boldsymbol{r}' \frac{\sin q|\boldsymbol{r}-\boldsymbol{r}'|}{|\boldsymbol{r}-\boldsymbol{r}'|} \Phi(\boldsymbol{r})\Phi(\boldsymbol{r}')$$

where q is the light wave-vector in quantum dot material, a_B is the 3D exciton Bohr radius and ω_{LT} is the longitudinal-transverse splitting frequency. To further simplify our modeling, let us consider a spherical dot with parabolic confinement and effective radii $a_{e,h}$ for electron and hole. Then $\Phi(\boldsymbol{r})$ reads

$$\Phi(\boldsymbol{r}) = \frac{1}{\pi^{3/2}(a_e a_h)^{3/2}} \exp\left(-\frac{r_e^2}{2a_e^2}\right) \exp\left(-\frac{r_h^2}{2a_h^2}\right)$$

The integral can be calculated analytically and gives:

$$\Gamma_0^{QD} = \frac{1}{6} \omega_{LT} (q a_B)^3 \frac{\bar{a}^6}{a_e^3 a_h^3} \exp(-q^2 \bar{a}^2)$$

where again

$$\bar{a}^2 = \frac{2a_e^2 a_h^2}{a_e^2 + a_h^2}$$

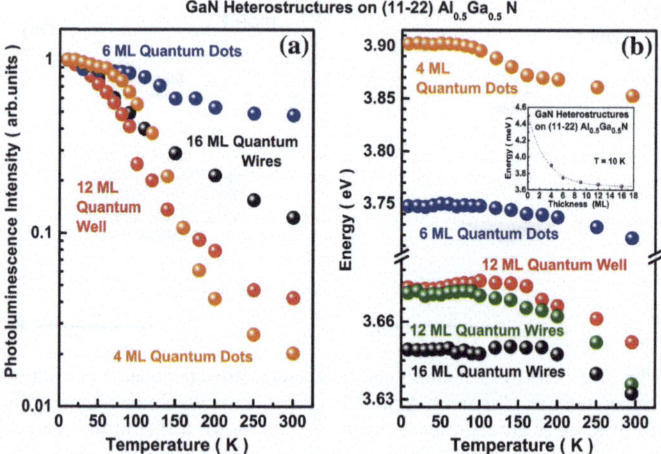

Fig. 5.28 The PL intensity (*left-hand part*) and energies (*right-hand part*) against T for a series of GaN– $Al_{0.5}GA_{0.5}N$ hetero-structures of varying dimensions. Small confining objects (4 ML *thin dots*) have been also chosen to indicate that thermal escape effects may superimpose to classical non-radiative recombination channels

For $a_e = a_h$ and $q\bar{a} \ll 1$ the dependence on the dot size vanishes yielding

$$\Gamma_0^{QD} = \frac{1}{6}\omega_{LT}(qa_B)^3$$

The important result is that the radiative decay time does not vary with T for a zero-dimensional object.

In summary, of a n-dimensional semiconductor, the temperature-dependent photoluminescence intensity can finally write:

$$I(T) = I(0)\frac{1}{1 + \frac{\tau_{rad_0}}{\tau_{nrad_0}} T^{\frac{n}{2}} e^{\frac{-T_a}{T}}}$$

In Fig. 5.28 are plotted the results of photoluminescence intensity measurements performed on a series of nano-objects of different dimensionalities: Quantum wells, quantum wires and Quantum Dots. The growth orientation is exotic is that case; it is a (11–22) reticular plane. The confining material is GaN, the barrier layers are made of $Al_{0.5}GA_{0.5}N$. the quantum confined Stark effect is very weak in such sample (see insert in the right-hand part of the figure). The left hand part of the figure indicates that photoluminescence intensity of 6 ML thick dots is more robust than the photoluminescence intensity of wires and wells. This is what is expected from the equations ruling the evolutions of decay times versus T and designs. However the photoluminescence intensity of the smallest dot sample is not very good although the evolution of the photoluminescence energy with T follows the general trend as reported in Fig. 5.28, right-hand part. This is an interesting asymptotic case where the delocalization of the envelope functions in the barrier layer is important for such sample so that thermal escape of the carriers out of the well is made very efficient. In that very specific confining system, the non radiative decay time is very short due to thermally activated escape of the carriers away from the dot region.

This is a first example of situations where one has to write:

$$\tau^{-1} = \tau_{rad}^{-1} + \tau_{nonrad}^{-1} + \tau_{escape}^{-1}$$

which complicates the equations above and requires for instance to define a second activation energy to describe the escape time. Data interpretation and understanding of what happens is as usual delicate.

5.4 Time-Resolved Photoluminescence

In this section we will address the determination of radiative, non radiative and escape contributions too the decay time. The principle of such spectroscopy is to shine the crystal with a laser pulse of high enough energy to create free electron hole pairs and then to record the emitted photoluminescence when time passes. The time- dependent photoluminescence may be recorded spectrally, spatially, both spatially and spectrally for detecting the spectral dependence of escape time. The photoluminescence intensity I(t) is correlated to a decay time τ (often named effective decay time) and expresses as:

$$I(t) \propto \frac{1}{\tau_{rad}} e^{-\frac{t}{\tau}}$$

The principle of such experiment is sketched in Fig. 5.29, in which the sketch of the laser pulse which is produced at a given repetition rate when time passes is plotted top in the figure. Then, middle in the figure is plotted a photoluminescence transient collected for laser repetition rate lower than the photoluminescence decay time so that decay time is correctly measured (photoluminescence intensity of created by laser pulse n vanishes at the impact of laser pulse n + 1). Bottom in the figure is plotted a decay time measured in condition of a laser repetition rate too high with respect to the decay time. then a photoluminescence background remains which prevents accurate measurement of the decay time. When a value is announced for a decay time it is mandatory to check the experimental conditions in which it was measured.

In Fig. 5.30 have been reported a series of low temperature decay time measurements collected on a series of GaN–AlGaN quantum wells corresponding to Fig. 5.10a. The low temperature photoluminescence of the 15 ML quantum well overlaps with the donor-bound exciton related one in the GaN buffer. The donor bound exciton decay time being shorter than the QW one, one can discriminate both contributions. We remark that the photoluminescence decay time increases with well width as expected for in case of presence of Quantum confined Stark Effect through

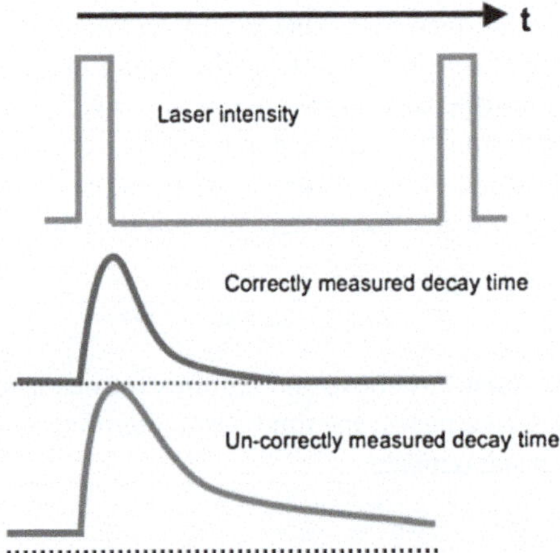

Fig. 5.29 *Top* Time variation of the laser intensity. *Middle* A photoluminescence transient collected for a good laser repetition rate: the photoluminescence intensity vanishes before the next laser pulse (the origin of photoluminescence intensity is represented by a *dotted line*). *Bottom* Uncorrect method to measure decay times. the photoluminescence does not vanish before the onset of the next laser pulse: the decay time will be underestimated

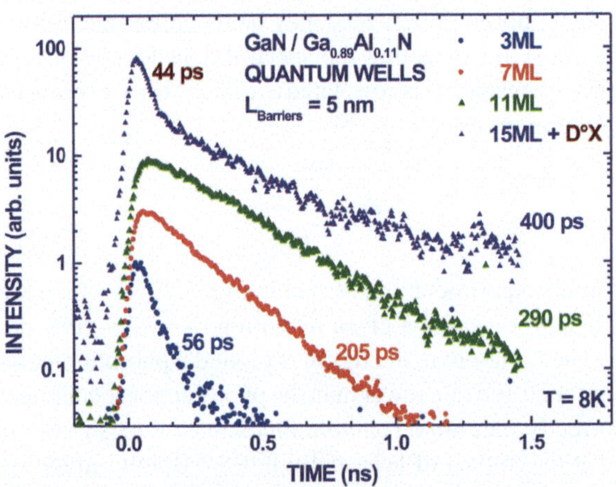

Fig. 5.30 Low temperature decay times for the multiple quantum well in Fig. 5.10a. For the 15 ML quantum well the donor-bound exciton related photoluminescence in the GaN buffer overlaps with the QW PL

quantities λ, I_{eh} and $\hbar\omega$ of the equation that gives it and that we remind here:

$$\tau_R^{QW} \approx \frac{2c^2 k_B \hbar^3}{(\hbar\omega_{LT})a_B^3}\left[\frac{1}{m_{eU}+m_{hU}}+\frac{1}{m_{eV}+m_{hV}}\right]^{-1}\frac{1}{(\hbar\omega)^3}\frac{\lambda^2}{I_{eh}^2}T$$

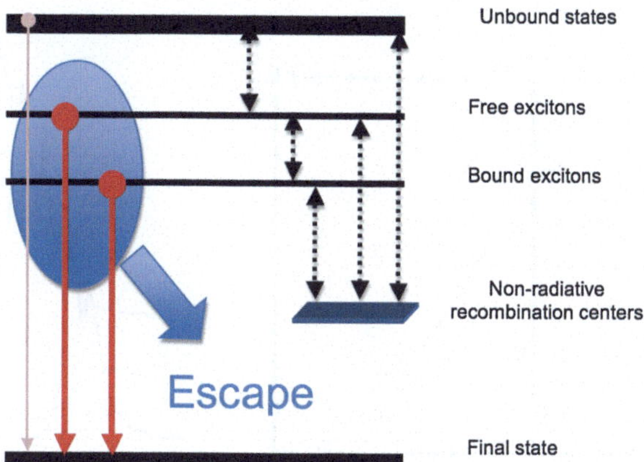

Fig. 5.31 Radiative recombination process (*full lines*) competing with non-radiative trapping and thermal-induced detraining (*dashed lines*) and eventually with thermal escape out of the active region of the sample (*thick arrow*)

When increasing the lattice temperature the decay times may change (and they generally do in most samples) due to complex thermalization processes between the radiative and non-radiative states as sketched in Fig. 5.31. Thermalization of the electron-hole pairs occur with a certain efficiency towards levels deeper in the band gap sometimes via non radiative recombination centers, which contributes to non radiative recombination channels in parallel to the recombination ones. Thermal induced-de-trapping of excitons (or electron hole pairs—may also occur which is a complementary mechanism to reduced the internal quantum efficiency. at the end the photoluminescence intensity and its decay time are balanced functions of all these processes.

We wish to emphasize the fact that we always have a mixture of bound and free excitons and that vanishingly small decay times are never measured at very low temperatures. Let N_{Loc} and Γ_{Loc} the areal density and radiative recombination rate of localized excitons, and let N_{Free} and Γ_{Free} the areal density and radiative recombination rates of free excitons. We know how to express Γ_{Loc} and Γ_{Free} from the quantum dot-related equations and from the quantum well related ones. They are different. The density of localized exciton expresses as:

$$N_{Loc} = N_D e^{\frac{E_{Loc}}{k_B T}}$$

where N_D is the areal density of localization centers and E_{Loc} is a localization energy.

The density of free excitons is:

$$N_{free} = (4 m_0 M k_B T)/(2\pi\hbar^2)$$

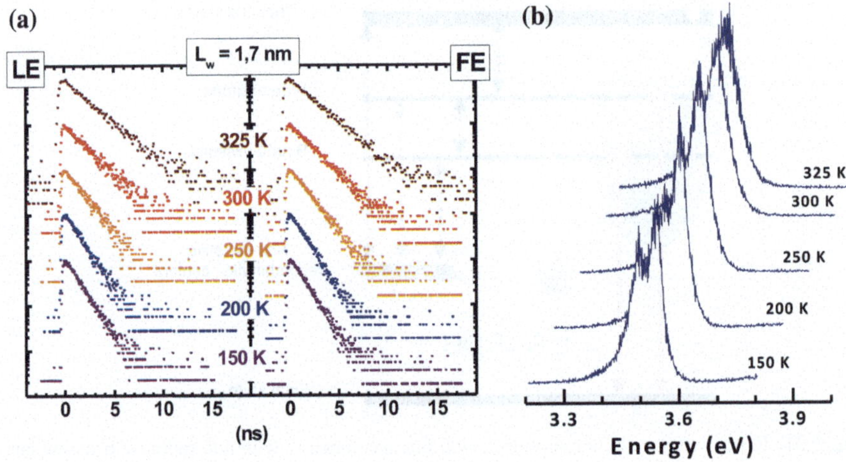

Fig. 5.32 **a** Photoluminescence decays collected for a ZnO–ZnMgO quantum well with localized and free exciton-related photoluminescence. **b** Evolution of the intensity of the PL for the ZnO–ZnMgO quantum well. After [73]

where

$$M = \sqrt{(m_{eU} + m_{hU})(m_{eV} + m_{hV})}$$

Then, when the population of free excitons N_{Free} exceeds the population of localized ones N_{Loc}; the PL decay times is:

$$\tau(T) = \tau_{Free}(T)\ (1 + \frac{N_D \Gamma_{Loc}}{N_{Free}\Gamma_{Free}})^{-1}$$

In Fig. 5.32a are reported as series of photoluminescence decays collected for a homoepitaxial, M-plane grown ZnO-ZnMgO quantum well. In Fig. 5.32b are ported the photoluminescence spectra which are constant in intensity up to 325 K which is an experimental proof of the lack of non-radiative recombination channels. The decay times are a direct measurement of the radiative recombination time. Both Localized and free excitons contribute to the photoluminescence are in thermal equilibrium: they have identical the decay times whatever the temperature is. The radiative decay time increases linearly with T at slope of 8 ps K^{-1}.

This text-book situation is not always met: in Fig. 5.33 is represented a series of PL transient at different temperatures (left), the temperature dependence of the PL intensity (middle) and the values of the decay times (effective, radiative and non radiative) of a 3 nm GaN–$Ga_{0.5}Al_{0.5}N$ quantum well (11–22)-grown so that QCSE is absent of the device. The decay times decrease with T as does the PL intensity. Processing these data leads to a rapidly decreasing non-radiative decay time and a radiative decay time linearly increasing with T at slope of 8 ps K^{-1}.

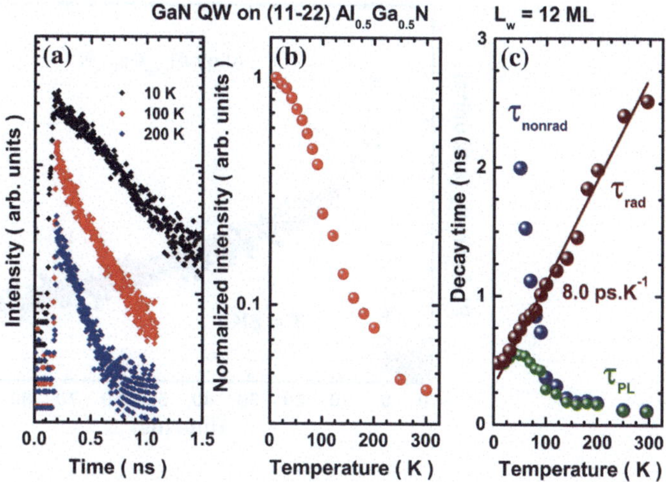

Fig. 5.33 PL transient at different temperatures (*left*), the temperature dependence of the PL intensity (*middle*) and the values of the decay times (effective, radiative and non radiative) of a 3 nm GaN–$Ga_{0.5}Al_{0.5}N$ quantum well (11–22)-grown so that QCSE is absent of the device. Data from courtesy Julien Brault and Daniel Rosales

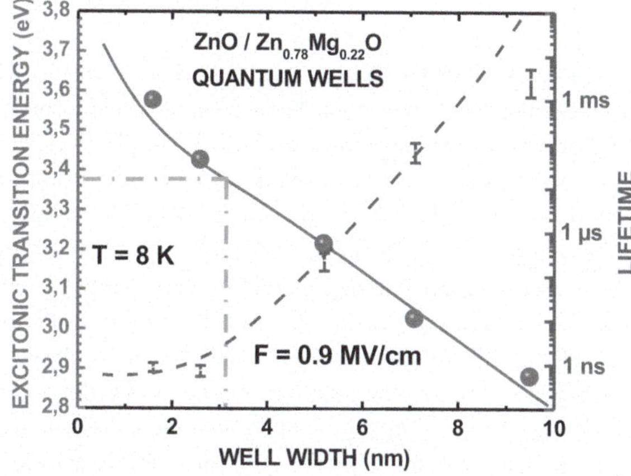

Fig. 5.34 PL energies (*left scale*) and PL decay times for polar ZnO–$Zn_{0.78}Mg_{0.22}O$ polar quantum wells. After [70]

It has been computed oscillator strengths susceptible to scale through more than 6 decades for quantum wells. This can be effectively found and is illustrated for ZnO-ZnMgO polar quantum wells in Fig. 5.34: the decay time is measured to scale through 8 decades for well thickness ranging from 1.6 up to 9.6 nm just due to Quantum Confined Stark Effect.

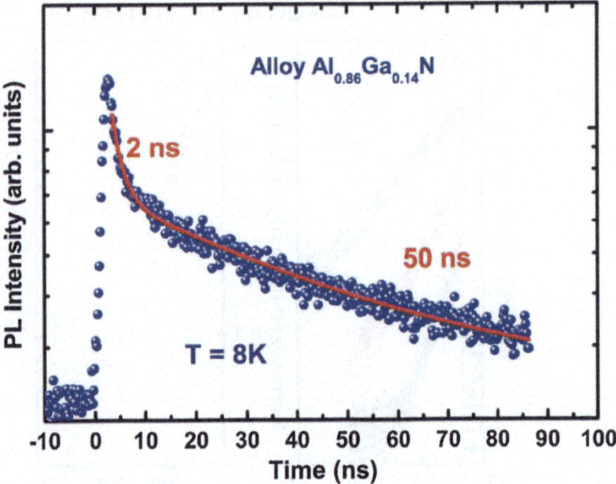

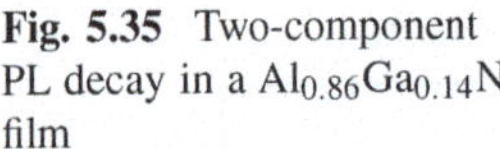
Fig. 5.35 Two-component PL decay in a $Al_{0.86}Ga_{0.14}N$ film

Photoluminescence transients presented so far in this section are mono-exponential decays of the intensity: a plot of the logarithm of the intensity versus time gives a straight line. It is often found more complex (multi exponential) behaviors which may be related to conditions of the experiment. Among them are the following ones:

- One excites photo carriers in the barrier regions of the sample when the laser energy is higher than the band gap of the barrier layers. When the barrier layer is made of a ternary or quaternary semiconductor alloy, the photo created carriers are not immediately trapped by the quantum well, and the generation of carriers into the well varies with T. The carrier diffusion time may be longer than the recombination time in the quantum well the secondary emission decay ruled by diffusion mechanisms of carriers from the barrier layers. Photon emitted by radiative recombination in the barrier layers may excite electron-hole pairs in the quantum well. This (weak) secondary emission is ruled by the radiative recombination time in the barrier layer. If it is longer than in the quantum well the time-dependent generation rate reveals a long decay time that superimposes to the short one as illustrated in Fig. 5.35 for a $Al_{0.86}Ga_{0.14}N$ film. This is easily measurable at delays longer than the quantum well effective photoluminescence decay time. If the recombination time in the barrier layer is shorter than the recombination time in the quantum well this is not observed or may be extracted by comparing the shape of the laser pulse with the photoluminescence rising time.
- Spin-triplet non-radiative excitons (excitons from Γ_6 symmetry in the bulk for instance) may be thermally de-trapped toward the bright exciton states. This contributes to the photoluminescence kinetics with a second longer decay time. This is a process which involves predictions of fundamental physics and which is thermally activated and which can be unambiguously detected and identified in single quantum dot spectroscopy.

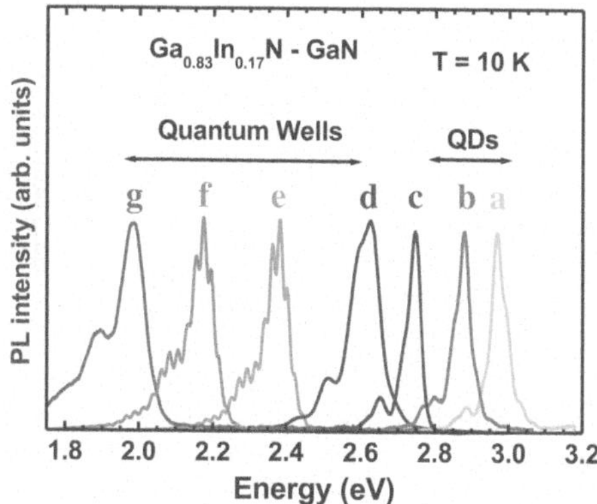

Fig. 5.36 Low temperature photoluminescence features reported on a series of $Ga_{0.83}In_{0.17}N$–GaNGaInN–GaN nano structures: polar quantum well giving the low energy photoluminescence lines and quantum dots to address the high energy regions

- Excitons may be trapped by defects in the well. There is a widely documented zoology for such defects among which are potential fluctuations of more complex defects. The typical situation for facing such behavior is the case of GaInN–GaN quantum wells.

In Fig. 5.36 are reported a series of low temperature photoluminescence features reported on a series of $Ga_{0.83}In_{0.17}N$–GaN nano structures: polar quantum well giving the low energy photoluminescence lines and quantum dots to address the high energy regions. There is no drastic evolution in terms of the shape of these photoluminescence features except the modification of the exciton-photonon coupling which increases when increasing the wavelength or said with an alternative phrasing when decreasing the photoluminescence energy. This is a homogeneous series of samples.

The photoluminescence decay is neither (and never) measured as a single exponential decay nor as a double exponential decay. It is a continuous non exponential decay, with at each time possibility to fit the decay in the neighborhood of this time one exponential which characteristics decay time is longer and longer when one measure it later and later after the excitation pulse. at the end the measurement is limited by the signal to noise ratio of the experimental setup. If we choose as an indicator the time τ_{10} necessary to reduced the photoluminescence intensity by a factor 10 after having reached its maximum, and if we plot the photoluminescence intensity as as function of the time expressed in units of τ_{10}, we obtain a universal plot, independently of the energy as indicated Fig. 5.37.

The interpretation of this complex behavior includes the account for existence of electric field together with a statistical distribution of localization centers that are going to act as electron traps or hole traps. As soon as they are created electron-hole pairs are split towards different well interfaces via diffusion through an inhomogeneous medium. Most of them recombine radiatively and their population decay can be phenomenologically described by a characteristics time, τ_{10}. The electric field

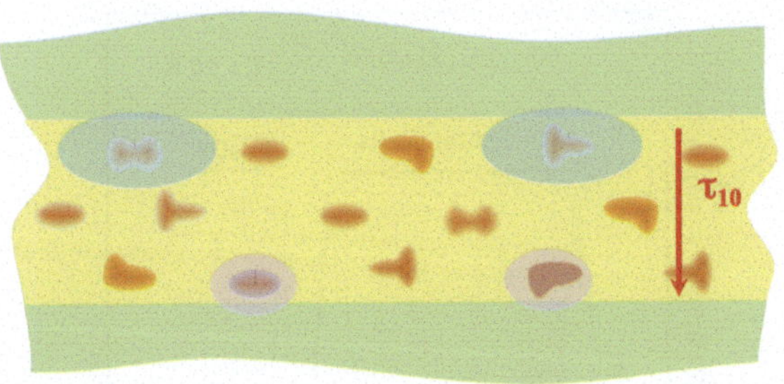

Fig. 5.37 The competitions between the electric field and the average well thickness that rule the value of τ_{10} and the random distribution of localization centers that both contribute to the continuous distribution of decay times

and the average well thickness rule the value of τ_{10} as indicated in Fig. 5.38 using an electric field of 2.5 MV/cm. A substantial amount of electron and hole will not recombine so fastly since they are localized in different parts of the crystal. They will statistically recombine at a given time which depends on their relative spacing in the well layer. The distance between such localization centers Is statistically and randomly distributed as sketched by the leopard skin in Fig. 5.57. The average distance between "efficient" localization centers is a growing function of well width, due to electric fields which split apart electron and holes and of the in-plane alloy disorder. Of course the intimate detail of the short range disorder is related to the whole alloy formation strategy by this qualitative description indicates the continuous distribution of decay times. These electron-hole configuration when producing a photon contribute to the long-lifetime photoluminescence. The localization of electron and hill to spatial potential fluctuations prevents the carrier diffusion towards non-radiative centers and reduces the spatial extension of electron and hole wave functions which produces and enhancement of the oscillator strength, a reduction of the radiative photoluminescence lifetime and an enhancement of the light emission intensity.

Time-resolved spectroscopy comes in very good complement to the coherently reinforce the interpretation of carrier localization represented in Fig. 5.25. To go even further in the comprehension of alloy disorder and of its impact time resolved photoluminescence experiment have been initiated and realized in the group of Professor Kawakami at the Kyoto University under conditions of very localized photo-excitation and fluorescence collection conditions. The principle of the method is sketched in Fig. 5.39. The sample is excited by a laser beam focused with a microscope objective. In the pinhole plane, one observes a magnified image of the photoluminescence light from the sample. With a pinhole, we can thus select points of observation on the sample and analyze their temporal behavior with a streak camera. In the example, the distance between points of excitation and detection is 900 nm.

A more sophisticated approach is sketched in Fig. 5.40 coming from the same group. Here the sample is illuminated at a very localized scale using a microscope objective (bottom objective in the figure) and the photoluminescence is collected

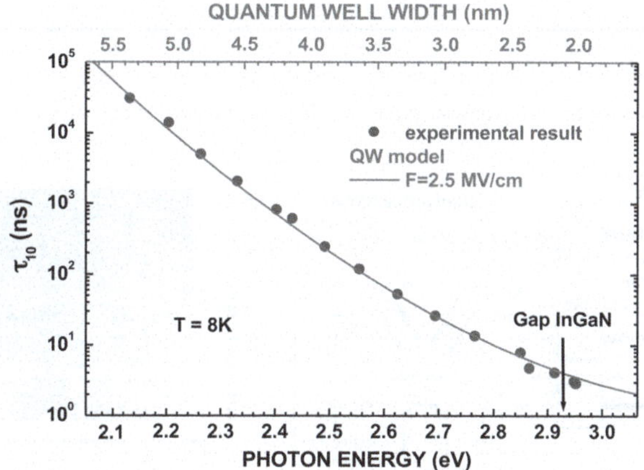

Fig. 5.38 The evolution of the indicative time τ_{10} versus well width or 8 K photoluminescence energy for $Ga_{0.83}In_{0.17}N$–GaN quantum wells

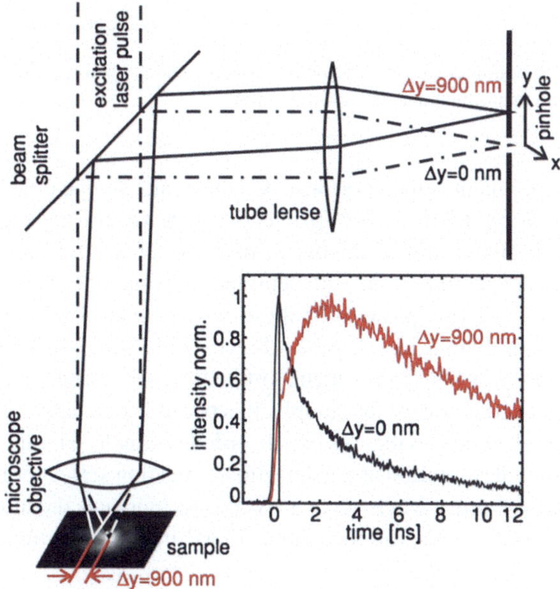

Fig. 5.39 Principle of measurement of carrier diffusions by means of photoluminescence spectroscopy. After [9]

using another microscope objective (middle in the figure). This gives a certain information in terms of spectral shape of the emitted light and of the decay times at the excitation position. The photoluminescence intensity in the I mode reflects radiative-nonradiative recombinations of diffused carriers, with a spatial resolution limited by the diffusion length of photogenerated carriers if it is larger than the aperture diam-

Applied Physics Express 3 (2010) 102102

Visualization of the Local Carrier Dynamics in an InGaN Quantum Well Using Dual-Probe Scanning Near-Field Optical Microscopy

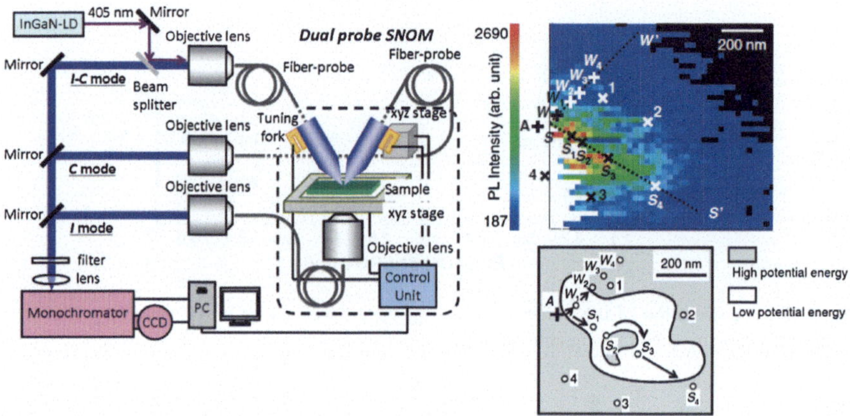

Fig. 5.40 Localized photoluminescence excitations and collections when performed in illumination and collection modes or in Illumination/collection mode

eter. A second experiment is then performed, a third microscope objective is used to excite and detect the light. This is the top objective in the picture and the Illumination-collection mode configuration of the experiment. Another information is obtained which gives another spectral shape of the emitted light and of the distribution of the decay times at the excitation position, in these experimental conditions. The spatial positions of the photoluminescence peaks directly reflect the potential energy. The decay times obtained in both experiments differ from the escape decay time away from the photo-excited region of the sample. It is possible to locally excite low band-gap regions and high band gap regions of a disordered sample. The spatial resolution depends on the spatial resolution at the detection/collection scales for that reason the Scanning Near-field Optical Microscopy (SNOM) technique is the most appropriate compared with far-field detection techniques to get the ultimate information about disorder.

After a long series of experiments in which excitation is made in both the high-energy and low-energy band gap regions of the quantum well, and collection is selectively recorded everywhere, authors come to the picture represented in Fig. 5.41 which validates the cathodoluminescent measurements reported in Fig. 4.35.

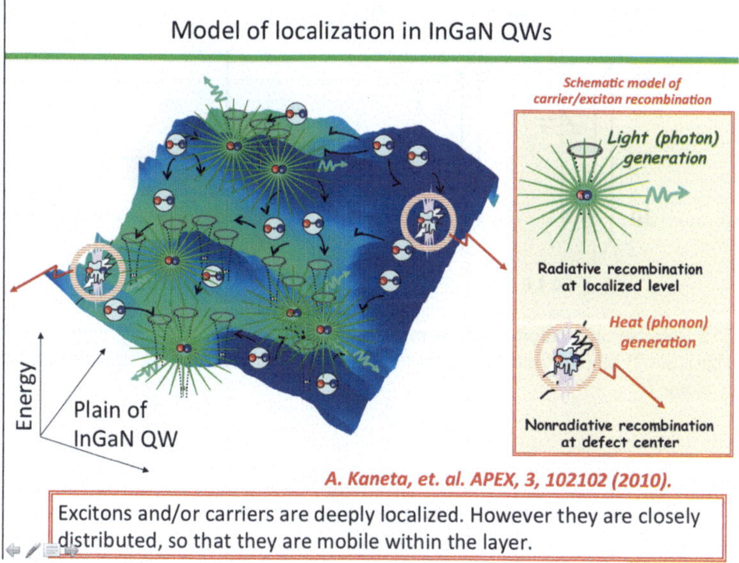

Fig. 5.41 Representation of alloy disorder as seen by Professor Kawakami, in 2013. Courtesy Yoichi Kawakami Kyoto University

5.5 Optical Properties of Non Polar Quantum Wells

The non-polar orientations are reticular planes with Miller indices of the $(hk-(h+k)0)$ kind. Then as demonstrated earlier in this chapter, group theory prescripts the lack of spontaneous and piezoelectric polarizations along the growth axis of such heterostructures. Quantum Confined Stark effect and its deleterious influence on device performances is thus by-passed. We will consider here the (10–10)-oriented M plane as text book case. Neither the non (11–20) A-plane nor other (hk·0) orientations will be discussed. The relative orientations of the M plane relatively to crystallographic axes $\overrightarrow{a_1}$, $\overrightarrow{a_2}$, $\overrightarrow{c}$, and respect to the unity vectors in the cartesian basis $\vec{x}$, $\vec{y}$, $\vec{z}$ is indicated in Fig. 5.42a. The realization of a M plane oriented quantum well breaks the wurtzitic translational symmetry along the $\vec{x}$ direction. This produces a macroscopic potential and the symmetry of the artificial crystal is reduced to C_{2v}: the only symmetry elements that are left are the two-fold symmetry axis along z and two symmetry planes generated by $(\vec{x}, \vec{z})$ and $(\vec{y}, \vec{z})$ as illustrated in Fig. 5.42b. The Bloch wave prevents to have mirror symmetry in the plane of the quantum well. C_{2v} is an abelian group.

The compatibility relations between the two-fold representation of C_{6v} $\Gamma_5(\vec{x}, \vec{y})$ and the irreducible representations of the abelian group C_{2v} are written here as:

$$\Gamma_5(\vec{x}, \vec{y}) = \Gamma_4'(\vec{x}) + \Gamma_2'(\vec{y})$$

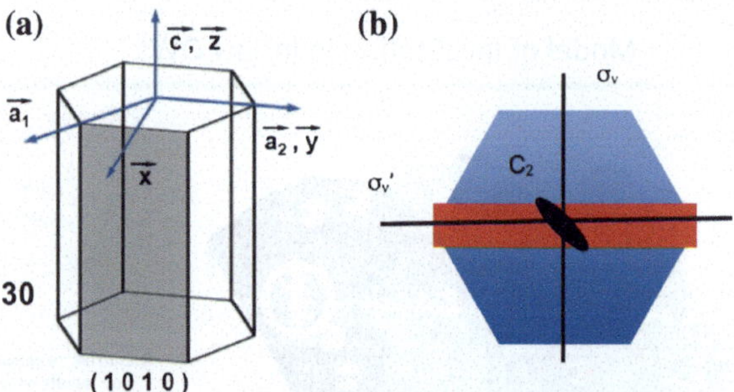

Fig. 5.42 **a** Relative orientation of the M plane with respect to the crystallographic axes and the cartesian basis set. **b** Macroscopic symmetry elements left after breaking the wurtzitic translational symmetry along $\vec{x}$

By analogy we have:

$$\Gamma_1(\vec{z}) = \Gamma_1'(\vec{z})$$

Group theory and the selection rules predict the $\Gamma_1'(\vec{z})$ excitons to be detected via radiative recombination and emission of a photon which escapes normal to the surface with its electric field $\vec{E}//\vec{z}$ and the $\Gamma_2'(\vec{y})$ excitons to be detected via radiative recombination and emission of a photon which escapes normal to the surface with its electric field $\vec{E}//\vec{y}$. In addition, the energies of these excitons are splitted by the short range spin exchange-related splitting between the Γ_1' and the Γ_2' excitons either for the quantum well or strained barrier layers. The optical response of the quantum well is polarized in its growth plane.

To interpret the polarization signature of the photoluminescence intensity we may restrict our model to the valence band structure and therefore work in the context of a band to band description of the recombination mechanisms which neglects the short range electron-hole spin exchange interaction but permits extraction of the physics of the light-matter interaction.

The Γ_9 and Γ_7 representations of the C_{6v} double group are all transforming as Γ_5' in C_{2v} symmetry and this indicates a 3×3 matrix representation of the valence band physics in contrast to the growth along the polar orientation which keeps the wurtzitic symmetry and lead to a 1×1 representation for the Γ_9 holes and a 2×2 representation for the Γ_7 ones.

Following the tradition, we treat the quantum confinement as a perturbation of the wurtzite crystal and we express the wave function of the valence band in terms of the standard $|J, m_J\rangle$ notation detailed in Chap. 4.

Here:

$$|3/2, \pm 3/2\rangle = |1, \pm 1\rangle \begin{pmatrix} \uparrow \\ \downarrow \end{pmatrix} = \frac{1}{\sqrt{2}}[x(\vec{r}) \pm iy(\vec{r})] \begin{pmatrix} \uparrow \\ \downarrow \end{pmatrix}$$

where $x(\vec{r})$ and $y(\vec{r})$ are two of the p-type Bloch states in a spinless description and $\uparrow$ and $\downarrow$ represent the 1/2 and $-1/2$ eigenstates of the spin operator.

Concerning the Γ_7 valence bands we choose as wave functions:

$$|1, \mp 1\rangle \begin{pmatrix} \uparrow \\ \downarrow \end{pmatrix} = -\frac{1}{\sqrt{2}}[x(\vec{r}) \mp iy(\vec{r})] \begin{pmatrix} \uparrow \\ \downarrow \end{pmatrix}$$

and:

$$|1, 0\rangle \begin{pmatrix} \uparrow \\ \downarrow \end{pmatrix} = z(\vec{r}) \begin{pmatrix} \uparrow \\ \downarrow \end{pmatrix}$$

where $z(\vec{r})$ is the third p-type Bloch states in a spinless description.

The 6×6 valence band Hamiltonian is obtained as two similar 3×3 hamiltonians which we write as follows:

$$\begin{bmatrix} |1, \pm 1\rangle \begin{pmatrix} \uparrow \\ \downarrow \end{pmatrix} & |1, \mp 1\rangle \begin{pmatrix} \uparrow \\ \downarrow \end{pmatrix} & |1, 0\rangle \begin{pmatrix} \uparrow \\ \downarrow \end{pmatrix} \\ \Delta_1 + \Delta_2 + \frac{\hbar^2}{2m_0}(A_2 + A_4)k_x^2 & -\frac{\hbar^2}{2m_0}A_5 k_x^2 & 0 \\ -\frac{\hbar^2}{2m_0}A_5 k_x^2 & \Delta_1 - \Delta_2 + \frac{\hbar^2}{2m_0}(A_2 + A_4)k_x^2 & \sqrt{2}\Delta_3 \\ 0 & \sqrt{2}\Delta_3 & \frac{\hbar^2}{2m_0}A_2 k_x^2 \end{bmatrix}$$

In this representation, Δ_1 depicts the matrix element which accounts for splitting of the p-type valence band states in a wurtzitic environment, the convention is to express the operator as $\Delta_1 L_z^2$, where L_z is the z-component of the spinless hole angular momentum. The quantities Δ_2 and Δ_3 are the matrix elements required to describe the anisotropic spin orbit interaction. The quantities A_2, A_4 and A_5 are the Luttinger parameters that are introduced in the $\vec{k} \cdot \vec{p}$ description of the dispersion relation in the valence band of wurtzitic semiconductors at Γ, m_0 represents the electron mass at rest, and k_x is the confined hole wave vector which takes discrete values thanks to the breaking of the translational symmetry in the $\vec{x}$ direction by alternating stacking of GaN and AlGaN layers [(10–10) growth plane].

We re-arrange this equation in the $x(\vec{r})$, $y(\vec{r})$ and $z(\vec{r})$ basis as:

$$\begin{bmatrix} |x(\vec{r})\rangle \begin{pmatrix} \uparrow \\ \downarrow \end{pmatrix} & |y(\vec{r})\rangle \begin{pmatrix} \uparrow \\ \downarrow \end{pmatrix} & |z(\vec{r})\rangle \begin{pmatrix} \uparrow \\ \downarrow \end{pmatrix} \\ \Delta_1 + \frac{\hbar^2}{2m_0}(A_2 + A_4 + A_5)k_x^2 & i\Delta_2 & -\Delta_3 \\ -i\Delta_2 & \Delta_1 + \frac{\hbar^2}{2m_0}(A_2 + A_4 - A_5)k_x^2 & -i\Delta_3 \\ -\Delta_3 & i\Delta_3 & \frac{\hbar^2}{2m_0}A_2 k_x^2 \end{bmatrix}$$

This re-arrangement shows that the three-band system behaves like three independent parabolic dispersion relations in the limit of high k_x.

From the form of the Hamiltonian and consistent with the expansions of the initial valence band basis in terms of $x(\vec{r})$, $y(\vec{r})$ and $z(\vec{r})$ p-type Bloch states, the

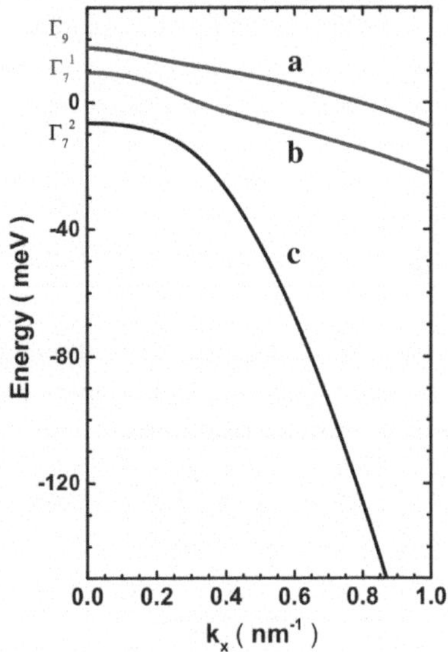

Fig. 5.43 Spectral dependence of the three valence band levels at $k_x \neq 0$ and $k_y = k_z = 0$. Note the initial anti-crossing of A and B levels at low values of k_x and the more pronounced one between B and C at higher values of k_x. For high k_x values levels A and B are quasi parallel and level C rapidly shifts apart the both of them

eigenvectors of the problem can be written in the most generic way as:

$$\Psi(\vec{r}) = \frac{\alpha}{\sqrt{\alpha^2+\beta^2+\gamma^2}}x(\vec{r}) + \frac{\beta}{\sqrt{\alpha^2+\beta^2+\gamma^2}}y(\vec{r}) + \frac{\gamma}{\sqrt{\alpha^2+\beta^2+\gamma^2}}z(\vec{r})$$

with $\alpha \neq \beta \neq \gamma$, due the orthorhombic symmetry.

Values of Δ_1, Δ_2 and Δ_3 are respectively 30.1, 6.2 and 5.5 meV for ZnO. Values of A_2, A_4 and A_5 are taken at −0.386, −2.089 and −2.059, respectively.

Plotted in Fig. 5.43 is the dispersion relation in ZnO for k_x values ranging between 0 and 0.05 nm^{-1}. We remark that starting from the Γ_9, Γ_7 and Γ_7 ordering for $k_x = 0$, increasing k_x produced a complex behavior with anti-crossings for high values of k_x to the relative ordering of the valence bands: x($\vec{r}$), z($\vec{r}$) and y($\vec{r}$) (spin component is neglected since hole levels are all twice spin-degenerate). We remark that y and z states are nearly parallel to each other for high values of k_x. A quantum well design allows discrete values of k_x which are the confined hole levels.

In terms of selection rules for the band to band transition the symmetry of the electric field of the photon may be Γ_1' ($\vec{E}//\vec{z}$), Γ_2' ($\vec{E}//\vec{y}$), or Γ_4' ($\vec{E}//\vec{x}$) in the C_2v orthorhombic environment. Group theory indicates that non-vanishing matrix elements for the band-to-band transitions transform as z^2 or y^2 or x^2.

In our experimental conditions, for photons collected with Pointing vectors parallel to (light emitted through the m-plane surface), transitions can be recorded for Γ_1' ($\vec{E}//\vec{z}$) and Γ_2' ($\vec{E}//\vec{y}$) polarizations of the photon. Thus, transition between the confined electron state of the band and the x-type confined hole state should not be allowed in our experimental configuration. Our calculations indicate that the first dipole allowed transition is z-type for high values of k_x (thin wells) and y-type for

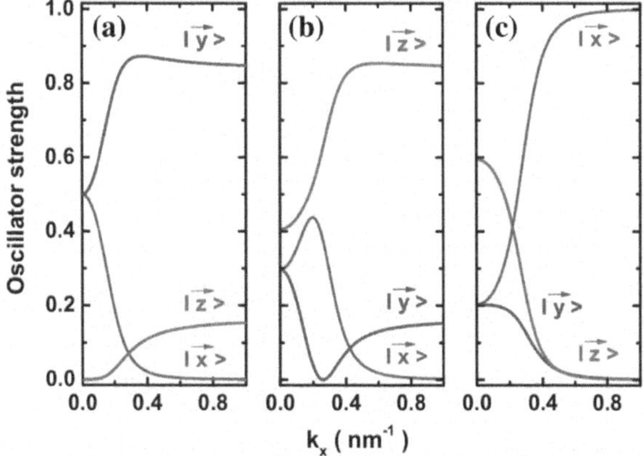

Fig. 5.44 Triptych representation of the expansion of **a**, **b** and **c** (*right*) levels against k_x for ZnO. At high values of k_x, the ground state valence band **a** is dominantly built from $|y\rangle$ Bloch state. The second level **b** is dominantly built from $|y\rangle$ Bloch state. For high values of, it appears the **c** is built of $|x\rangle$ Bloch state only. These expansion impact the selection rules for band to band transitions.

low values of k_x (wide wells). The oscillator strengths are obtained as the squared components of the coefficients of the eigenvectors projected over $|z\rangle$ and $|y\rangle$, respectively. The modifications of the oscillator strengths for z (π_z), y (π_y), and x (π_y) polarizations are plotted in Fig. 5.44 for the three band to band transitions between the first confined electron states and the three confined valence bands. The amplitudes of the different transitions are obviously very different from what is expected in case of samples grown along the polar c-plane direction but we remark that the lowest confined levels originating from the Γ_9 and the highest Γ_7 valence bands are both allowed in $\vec{E}//\vec{z}$ and $\vec{E}//\vec{y}$ polarizations.

In Fig. 5.45 are reported reflectivity features of a M-plane grown ZnO-ZnMgO quantum wells recorded in normal incidence conditions for π_z (top), and π_y polarization conditions. Note the bulk related transitions and transitions between the first confined electron and the two first confined hole states. The ordering of the quantum well related transitions and their amplitudes, agree with the predictions of Figs. 5.43 and 5.44 respectively.

To distinguish the situation of ZnO-based M plane quantum wells from GaN-based M plane quantum wells we plot in Figs. 5.46 and 5.47 the analogs of Figs. 5.43 and 5.44.

Of course, a strain-field in the quantum well layer (ZnO or GaN) may modify the results presented here and their interpretation, but they remain valid provided that the strain-field is added in the hamiltonian which produces complementary couplings. Finally taking into account the finite depths of the quantum well brings important refinements.

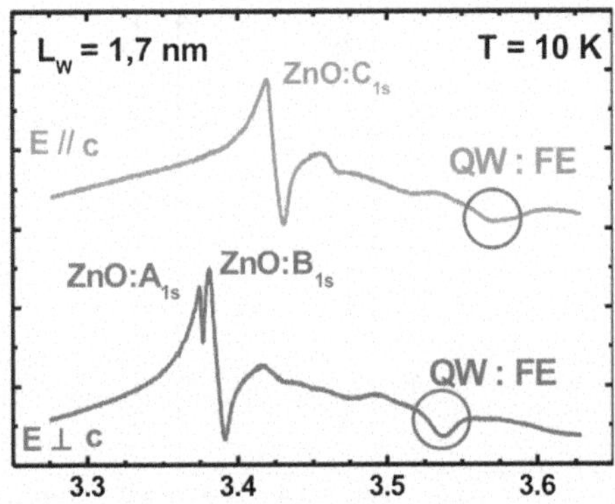

Fig. 5.45 Normal incidence reflectivity features of a M-plane grown ZnO–ZnMgO quantum wells in π_z (*top*), and π_y polarization conditions. Note the bulk related transitions and transitions between the first confined electron and the two first confined hole states

5.6 Optical Properties of Semipolar Quantum Wells

This is a situation typical of (hk.ℓ) reticular planes with simultaneously fulfilling: $h \neq 0, k = 0, \ell \neq 0$ or $hk\ell \neq 0$.

We know from chapter one that the relationship between the international basis set $\{\vec{x}, \vec{y}, \vec{z}\}$ and the $(\vec{U}, \vec{V})$ generate the $(hk \cdot \ell)$ reticular plane and $\vec{W}\}$ is orthogonal to it is:

The growth of the quantum well on the (hk$\cdot\ell$) reticular plane or along the $\vec{W}$ direction leads to a breaking of translational symmetry along this $\vec{W}$ direction. Therefore, values of the wave vector $\vec{k_W}$ are quantized.

$$\begin{bmatrix} \vec{U} \\ \vec{V} \\ \vec{W} \end{bmatrix} = \begin{bmatrix} -\frac{\sqrt{3}}{2}Ak & \frac{1}{2}A(2h+k) & 0 \\ -\frac{\sqrt{3}}{2}Aa\ell B_{ac}(2h+k) & -\frac{3}{2}Aa\ell B_{ac}k & 2c\frac{B_{ac}}{A} \\ AcB_{ac}(2h+k) & \sqrt{3}AcB_{ac}k & \sqrt{3}\ell aB_{ac} \end{bmatrix} \begin{bmatrix} \vec{x} \\ \vec{y} \\ \vec{z} \end{bmatrix}$$

For the conduction band it is obvious: we have to solve the Shrodinger equation in the context of the single band approximation, occasionally including an electric field.

For the valence band the situation is more complicated and the hamiltonian for the kinetic energy operator is the $6 \times 6\, \vec{k} \cdot \vec{p}$ hamiltonian:

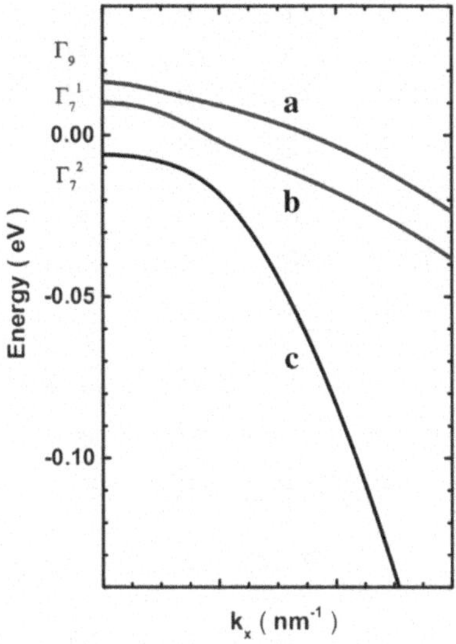

Fig. 5.46 Spectral dependence of the three valence band levels at $k_x \neq 0$ and $k_y = k_z = 0$. Note the initial anti-crossing of A and B levels at low values of k_x and the more pronounced one between B and C at higher values of k_x. For high k_x values levels A and B are quasi parallel and level C rapidly shifts apart the both of them

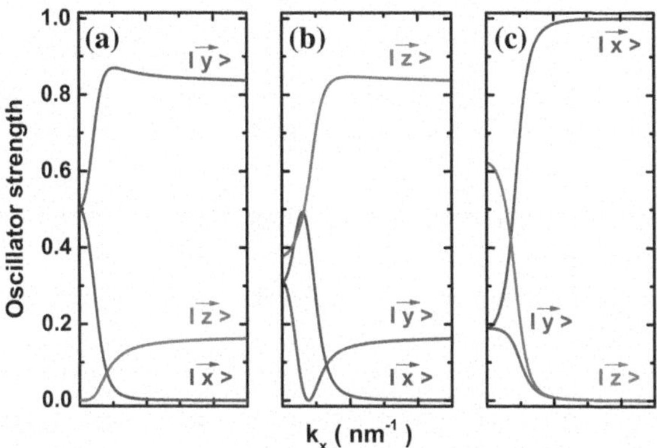

Fig. 5.47 Triptych representation of the expansion of A (*left*), B (*middle*) and C (*right*) levels against k_x for a GaN-based infinitely deep valence band well. At high values of k_x, the ground state valence band (A) is dominantly built from $|y\rangle$ Bloch state and from about 18 % of $|z\rangle$ state. The second level (B) is dominantly built from $|z\rangle$ Bloch state and from about 18 % of $|y\rangle$ state. Note the quasi similar expansion of A and B through $|y\rangle$ and $|z\rangle$, respectively. Last for high values of k_x, it appears the C is built of $|x\rangle$ Bloch state only. Values of Δ_1, Δ_2 and Δ_3 are respectively 10.1, 7.1 and 5.5 meV for GaN; values of A_2, A_4 and A_5 are -0.53, -2.51 and -2.51, respectively

$$\begin{bmatrix} |1,1\rangle\uparrow & |1,-1\rangle\downarrow & |1,1\rangle\downarrow & |1,0\rangle\uparrow & |1,-1\rangle\uparrow & |1,0\rangle\downarrow \\ H_{11} & 0 & 0 & H_{14} & H_{15}^{*} & 0 \\ 0 & H_{11} & H_{15}^{*} & 0 & 0 & H_{14}^{*} \\ 0 & H_{15} & H_{22} & \sqrt{2}\Delta_3 & 0 & H_{14} \\ H_{14}^{*} & 0 & \sqrt{2}\Delta_3 & H_{33} & H_{14} & 0 \\ H_{15} & 0 & 0 & 0 & H_{22} & \sqrt{2}\Delta_3 \\ 0 & H_{14} & H_{14}^{*} & 0 & \sqrt{2}\Delta_3 & H_{33} \end{bmatrix}$$

$$H_{11} = \Delta_1 + \Delta_2 + Ck_z^2 + (A+B)k_x^2$$
$$H_{22} = \Delta_1 - \Delta_2 + Ck_z^2 + (A+B)k_x^2$$
$$H_{33} = Dk_z^2 + E(k_x^2 + k_y^2)$$
$$H_{14} = \frac{1}{\sqrt{2}}Gk_z(k_x - ik_y)$$
$$H_{15} = \frac{1}{2}(A-B)(k_x^2 - k_y^2 + 2ik_xk_y)$$

with

$$k_x = AcB_{ac}(2h+k).\tilde{k}$$
$$k_y = \sqrt{3}AcB_{ac}k.\tilde{k}$$

and

$$k_z = \sqrt{3}\ell aB_{ac}.\tilde{k}$$

where $\tilde{k} = -i\frac{\partial}{\partial \vec{W}}$ represents (to a $\hbar^{-1}$ proportionality constant) the impulsion operator along the $\vec{W}$ direction.

Using the Luttinger parameters A_1, A_2, A_3, A_4, A_5 and A_6 we rewrite the matrix elements H_{ij} as follows:

$$H_{11} = \Delta_1 + \Delta_2 + \frac{\hbar^2}{2m_0}[3\ell^2a^2(A_1+A_3) + 4c^2(A_2+A_4)]B_{ac}^2.\tilde{k}^2$$
$$H_{22} = \Delta_1 - \Delta_2 + \frac{\hbar^2}{2m_0}[3\ell^2a^2(A_1+A_3) + 4c^2(A_2+A_4)]B_{ac}^2.\tilde{k}^2$$
$$H_{33} = \frac{\hbar^2}{2m_0}[3\ell^2a^2A_1 + 4c^2A_2]B_{ac}^2.\tilde{k}^2$$
$$H_{14} = -\frac{\hbar^2}{2m_0}A_6[2h + (1-i\sqrt{3})k]\ell acAB_{ac}^2.\tilde{k}^2$$
$$H_{15} = -\frac{\hbar^2}{2m_0}A_5[2h + (1-i\sqrt{3})k]^2a^2AB_{ac}^2.\tilde{k}^2$$

Which is then rewritten as:

$$\begin{aligned}
H_{11} &= \Delta_1 + \Delta_2 + \tilde{H_{11}}.\tilde{k}^2 \\
H_{22} &= \Delta_1 - \Delta_2 + \tilde{H_{22}}.\tilde{k}^2 \\
H_{33} &= \tilde{H_{33}}.\tilde{k}^2 \\
H_{14} &= \tilde{H_{14}}.\tilde{k}^2 \\
H_{15} &= \tilde{H_{15}}.\tilde{k}^2
\end{aligned}$$

with the following definitions:

$$\begin{aligned}
\tilde{H_{11}} &= \frac{\hbar^2}{2m_0}[3\ell^2a^2(A_1 + A_3) + 4c^2(A_2 + A_4)]B_{ac}^2 \\
\tilde{H_{22}} &= \frac{\hbar^2}{2m_0}[3\ell^2a^2(A_1 + A_3) + 4c^2(A_2 + A_4)]B_{ac}^2 \\
\tilde{H_{33}} &= \frac{\hbar^2}{2m_0}[3\ell^2a^2A_1 + 4c^2A_2]B_{ac}^2 \\
\tilde{H_{14}} &= -\frac{\hbar^2}{2m_0}A_6[2h + (1 - i\sqrt{3})k]\ell acAB_{ac}^2 \\
\tilde{H_{15}} &= -\frac{\hbar^2}{2m_0}A_5[2h + (1 - i\sqrt{3})k]^2a^2AB_{ac}^2
\end{aligned}$$

Thus we can write for the hamiltonian of the valence band the $6 \times 6\, \vec{k} \cdot \vec{p}$ operator valid for any $\vec{W}$ direction:

$$\begin{bmatrix}
|1,1\rangle\uparrow & |1,-1\rangle\downarrow & |1,1\rangle\downarrow & |1,0\rangle\uparrow & |1,-1\rangle\uparrow & |1,0\rangle\downarrow \\
\Delta_1+\Delta_2+\tilde{H_{11}}\tilde{k}^2 & 0 & 0 & \tilde{H_{14}}\tilde{k}^2 & \tilde{H_{15}}^*\tilde{k}^2 & 0 \\
0 & \Delta_1+\Delta_2+\tilde{H_{11}}\tilde{k}^2 & \tilde{H_{15}}^*\tilde{k}^2 & 0 & 0 & \tilde{H_{14}}^*\tilde{k}^2 \\
0 & \tilde{H_{15}}\tilde{k}^2 & \Delta_1-\Delta_2+\tilde{H_{11}}\tilde{k}^2 & \sqrt{2}\Delta_3 & 0 & \tilde{H_{14}}\tilde{k}^2 \\
\tilde{H_{14}}^*\tilde{k}^2 & 0 & \sqrt{2}\Delta_3 & \tilde{H_{33}}\tilde{k}^2 & \tilde{H_{14}}\tilde{k}^2 & 0 \\
\tilde{H_{15}}\tilde{k}^2 & 0 & 0 & 0 & \Delta_1-\Delta_2+\tilde{H_{11}}\tilde{k}^2 & \sqrt{2}\Delta_3 \\
0 & \tilde{H_{14}}\tilde{k}^2 & \tilde{H_{14}}^*\tilde{k}^2 & 0 & \sqrt{2}\Delta_3 & \tilde{H_{33}}\tilde{k}^2
\end{bmatrix}$$

To solve the problem requires to perform a multi-band envelope function approximation, but again a plot of the dispersion relations against the unknown $\tilde{k}$ may give valuable information regarding the valence band splittings and the wave functions of the different valence band levels. This is more complicated than the scope of the present book but the important message is the non vanishing values of matrix element H_{14} or H_{15} which lead to the vanishing of selection rules for optical transitions.

After making an ad-hoc unitary transform one can reduce the problem to the two 3×3 hamiltonians below:

$$\begin{bmatrix} \Delta_1 + \Delta_2 + \tilde{H_{11}}\tilde{k}^2 & \tilde{H_{15}}^*\tilde{k}^2 & \mp i\tilde{H_{14}}\tilde{k}^2 \\ \tilde{H_{15}}\tilde{k}^2 & \Delta_1 - \Delta_2 + \tilde{H_{11}}\tilde{k}^2 & \sqrt{2}\Delta_3 \mp i\tilde{H_{14}}\tilde{k}^2 \\ \pm i\tilde{H_{14}}^*\tilde{k}^2 & \sqrt{2}\Delta_3 \pm i\tilde{H_{14}}^*\tilde{k}^2 & \tilde{H_{33}}\tilde{k}^2 \end{bmatrix}$$

Let us consider a square well case (no electric field). We can write the wave functions in the well layer:

$$F(w) = \begin{bmatrix} C_1 e^{i\tilde{k_{11}}w} \\ C_2 e^{i\tilde{k_{11}}w} \\ C_3 e^{i\tilde{k_{33}}w} \end{bmatrix} + \begin{bmatrix} D_1 e^{-i\tilde{k_{11}}w} \\ D_2 e^{-i\tilde{k_{11}}w} \\ D_3 e^{-i\tilde{k_{33}}w} \end{bmatrix}$$

and

$$F^R(w) = \begin{bmatrix} R_1 e^{-\tilde{\kappa_{11}}w} \\ R_2 e^{-\tilde{\kappa_{11}}w} \\ R_3 e^{-\tilde{\kappa_{33}}w} \end{bmatrix}$$

and

$$F^L(w) = \begin{bmatrix} L_1 e^{\tilde{\kappa_{11}}w} \\ L_2 e^{\tilde{\kappa_{11}}w} \\ L_3 e^{\tilde{\kappa_{33}}w} \end{bmatrix}$$

in the right hand barrier layer and left-hand barrier layer respectively.

With our notations, the hole levels of the one band approximation are obtained as the solutions of the unidimensional Schrödinger equation:

$$[\tilde{H_{ii}}(w)\tilde{k}^2 + V_i(w) - E]\Psi_i(w) = 0$$

Where $V_i(w)$ is the valence band potential (neither including crystal field splitting nor spin-orbit interaction) in each of the materials combined to form the quantum well.

In the well layer, $\tilde{k_W} = \sqrt{2m_W(E - V_W)}/\hbar$; in the barrier layer, $\tilde{\kappa_B} = \sqrt{2m_B(V_B - E)}/\hbar$; and $V_i = V_B - V_W$.

We have to ensure the continuity of the wave function at both interfaces: $F(w_L) = F^L(w_L)$ at the left-hand interface and $F(w_R) = F^R(w_R)$ at the right -hand interface.

A second boundary conditions requires the continuity of the current of probability through the interfaces. This operator writes:

$$J = \begin{bmatrix} \tilde{H_{11}}\tilde{k} & \tilde{H_{15}}^*\tilde{k} & \mp i\tilde{H_{14}}\tilde{k} \\ \tilde{H_{15}}\tilde{k} & \tilde{H_{11}}\tilde{k} & \mp i\tilde{H_{14}}\tilde{k} \\ \pm i\tilde{H_{14}}^*\tilde{k} & \pm i\tilde{H_{14}}^*\tilde{k} & \tilde{H_{33}}\tilde{k} \end{bmatrix}$$

One has to define J^B in the barrier layers and J^W in the well layer using the parameters of the materials acting as barrier layers and then at the left-hand interface the continuity of the density of probability writes:

$$J^B F^L(w_L) = J^W F(w_L)$$

Similarly at the left-hand interface the continuity of the density of probability writes:

$$J^W F(w_R) = J^B F^R(w_R)$$

Then, given a well width the three eigen energies can be obtained and the twelve coefficients as well. The detailed calculation is out of the scope of this chapter either if square wells or triangular ones. The strain field may also be added via a strain-related hamiltonian.

5.7 Optical Properties of Quantum Dots

We begin by discussing of the optical properties of assemblies of quantum dots. Such dots are grown with a distribution of sizes and densities that can be changed by the growth conditions. The dots are in general fabricated by a Stranski-Krastanov (SK) growth. This process is classically described by a minimization of the system free energy (defined as the sum of the elastic and surface energies),which can have a strong influence on the nanostructure size and shape. In Fig. 5.49 are reported some typical Atomic Force Microscopy data taken in case of Indium nitride quantum dots deposited on a (0001) GaN template by MOVPE at different temperatures using Ammonia and trimethylindium as precursors. As indicated by the treatment of the AFM picture the average size of the dots (h_m is the average height of the dots) and their densities vary with the growth temperature (Fig. 5.48).

In Fig. 5.49 are reported some typical Atomic Force Microscopy data taken in case of Indium nitride quantum dots deposited on a (0001) GaN template by MOVPE at different pressure conditions of ammonia and trimethylindium. The growth conditions clear impact the sizes and densities of the dots.

Finally the gas precursor may impact the interaction of the chemical species with the surface and impact the growth as indicated in Fig. 5.50.

The symmetry of the growth surface may impact the shape of the dots as illustrated in Fig. 5.51a, b. In Fig. 5.51a is represented the AFM picture of a large indium nitride island grown on an AlN template. Clearly the shape of a truncated pyramid is found with a sixfold symmetry around the growth axis. The aspect ratio (the ratio between the average dot diameter d_m and the height h_m) is about 5.8: $d_m = 115$ nm and $h_m = 20$ nm. for such dots while it is generally obtained around 10 for InN dots grown on a GaN template. In Fig. 5.51b is represented the AFM picture of an indium nitride dot grown on a (111)-oriented silicon surface. Clearly the shape of a truncated pyramid is found with a three-fold symmetry around the growth axis. The dimension of this dot

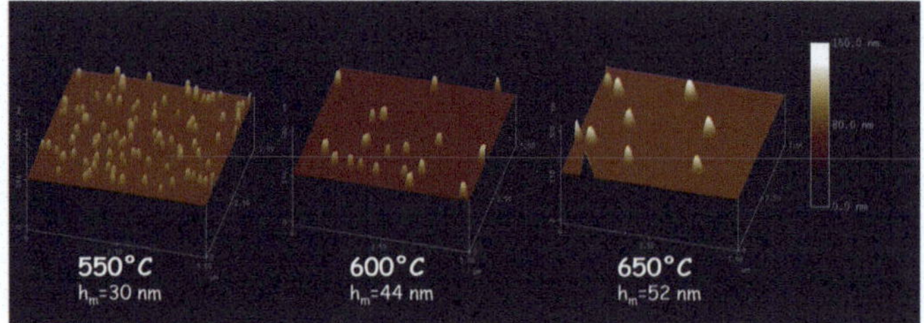

Fig. 5.48 Atomic force microscopy measurements of shapes and densities of indium nitride quantum dots deposited on a (0001) GaN template by MOVPE at different temperatures. Courtesy Olivier Briot. After [10]

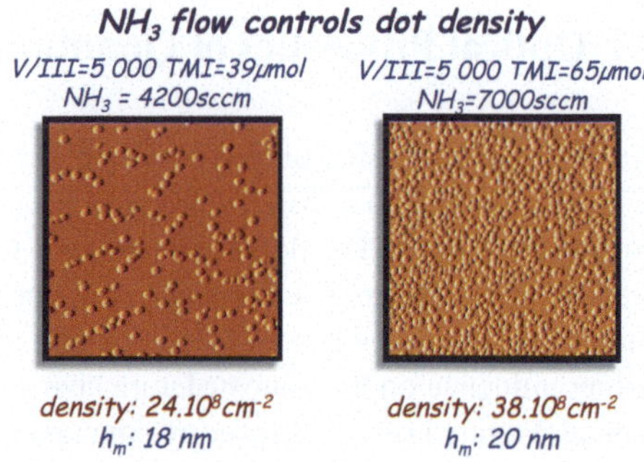

Fig. 5.49 Atomic force microscopy measurements of shapes and densities of indium nitride *quantum dots* deposited on a (0001) GaN template by MOVPE under different growth conditions. Courtesy Olivier Briot. After [10]

are: $h_m = 2$ nm; $d_m = 18$ nm and the aspect ratio $d_m/h_m = 9$. The aspect ratio is not a constant quantity, it is a complex function of the substrate/material association, of the growth method and it may change with the growth time. In particular for indium nitride dots grown on (111)-oriented silicon, we found it to range between 3 for very large dots to 9 for small dots. Sometimes dots are not formed as regular in-plane isotropic three-fold or six-fold symmetric pyramids. Reasons for this are multiple among which is the symmetry of the nucleation center which triggers the onset of the growth, the anisotropy of the composition of the environment, or the anisotropy of the template. The optical response of such oblate or prolate dots is in general anisotropic in the growth plane. Internal strain is not necessary isotropic too.

Another approach to growth dots with controlled anisotropic properties and shapes is to use a growth surface which is not the (0001) plane. M-plane, A plane have been used as well as non polar orientations. The growth of the dot is more or less strongly anisotropic and the concept of aspect ratio fails: one has to calibrate the dots using their three dimensions in the $(\vec{U}, \vec{V}, \vec{W})$ basis, eventually via expression of these

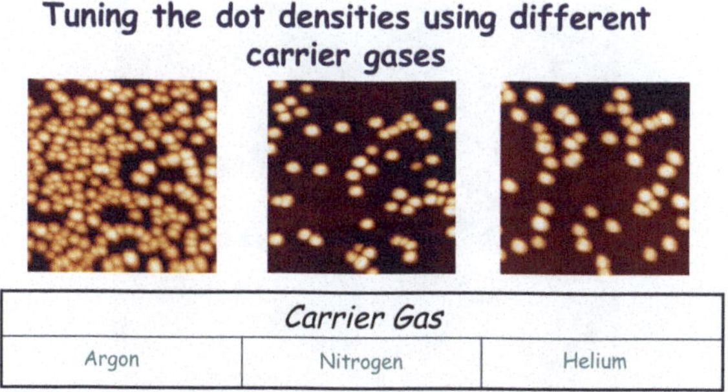

Fig. 5.50 Atomic force microscopy measurements of shapes and densities of indium nitride *quantum dots* deposited on a (0001) GaN template by MOVPE with different carrier gases. Courtesy Matthieu Moret, Olivier Briot and Sandra Ruffenach. After [11]

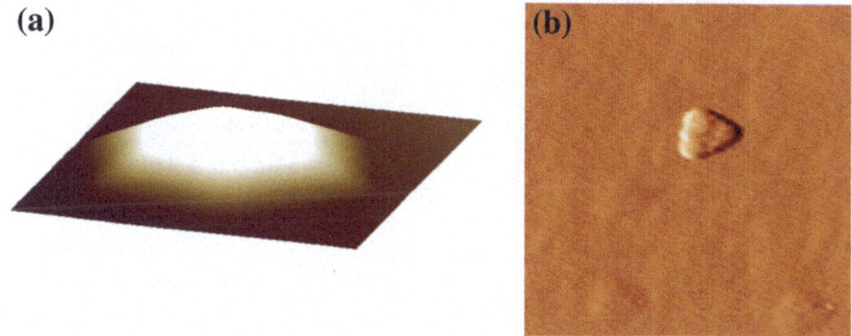

Fig. 5.51 *Left* **a** Indium nitride dot on (0001)-oriented AlN. *Right* **b** Indium nitride dot on (111)-oriented Silicon. Courtesy Matthieu Moret, Olivier Briot and Sandra Ruffenach

vector relatively to the international basis. In Fig. 5.52 are reported as series of AFM pictures collected for GaN quantum dots grown by Molecular beam epitaxy on a (11–22)-oriented $Al_{0.5}Ga_{0.5}N$ template. Clearly the anisotropy of the growth surface is found at the scale of the dots which grow anisotropically forming at the end nano chains and wires. The lateral size of the dots along the [1–100] direction increases permanently. Strong asymmetry of the grown dots has also been illustrated by the high-resolution transmission electron microscopy (TEM) measurements performed along two crystallographic directions in the dot base. They clearly indicate the formation of a continuous (undulating) film along the [−1−123] direction and more efficient coalescence of the dots along the [1–100] direction. The optical properties of such ensemble of and nano-objects are clearly anisotropic in the growth plane.

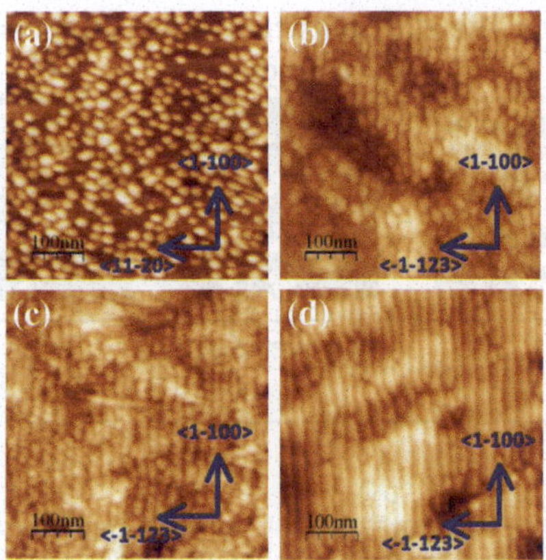

TABLE I: Lateral sizes and heights of the GaN nanostructures under examination

	4 MLs	6 MLs	9 MLs	12 MLs	16 MLs
$l_{[-1-123]}$ (nm)	18±2	19±3	20±4	21±4	21±4
$l_{[1-100]}$ (nm)	17±1	25±5	39±7	41±4	-
Height (nm)	1.1	2.3±0.5	2.84±0.88	3.3±0.6	4.31±0.83

Fig. 5.52 Atomic force microscopy measurements of shapes GaN *quantum dots* deposited on a (11–22) $Al_{0.5}Ga_{0.5}N$ template. by MBE showing the evolution of the dots morphology with the orientation of the $Al_{0.5}Ga_{0.5}N$ template. **a** GaN dots grown on (0001)-oriented template (**a**), and GaN nano structures: dots (**b**) and wires (**c, d**), grown on (11–22)-oriented template demonstrating alignment along the [1–100] direction. below are given the average sizes of the dots. Courtesy Abdelkarim Kahouli, Julien Brault and Benjamin Damilano. After [12]

5.7.1 Optical Properties of Polar Quantum Dots

In Fig. 5.53 are reported some photoluminescence spectra taken on two GaN quantum dot samples of different average heights of 5 and 3.7 nm. *Top* in the plots are plotted the cw spectra which indicate a redshift of the average PL bands when increasing the average size of the dots. Below are reported series of time-resolved photoluminescence spectra collected at different delays after a very short excitation pulse. The photoluminescence clearly red-shifts when time passes.

This behavior is complex and results of different contributions. Number one, there is Quantum confined Stark Effect in such samples. We excite a collection of dots of different sizes. When time passes, excitons confined in the small dots recombine first (their radiative decay time is short) and the exciton-populated dots redistribute in favor of large dots (with longer) decay times. This gives a red-shift of the PL energy which saturates to the energy of the most numerous dots. At long delays, only big dots being still populated with excitons and thus susceptible to contribute to the photoluminescence, the photoluminescence feature is washed out from the high energy contribution of small dots. The full-width at half maximum of this photoluminescence

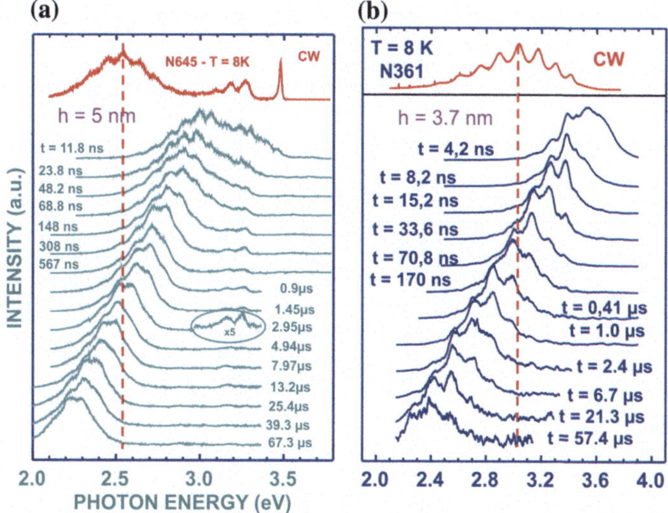

Fig. 5.53 CW PL spectrum of two quantum dot samples taken at $T = 8$ K (*top spectra*). TRPL spectra at different time delays (*bottom spectra*) at $T = 8$ K. The energy density per pulse is 2×10^4 J/cm^2. Even for the lowest excitation power density in a CW PL experiment, the spectrum is clearly *blue-shifted* compared with the TRPL spectra at large time delays. After [75]

has to decrease when time passes. The decrease of the exciton population is expected to be observed via a decrease of the spectrally integrated photoluminescence intensity when time passes. This is effectively found also. Finally, in terms of decay times, one expects to record a complex non exponential recombination dynamics at short delays when dots of different size recombine and a much more pronounced evolution towards a mono exponential behavior which is finally reached at long delays when the red-shift of the PL energy saturates. Then, one measures a significant PL energy and decay time which can be quantitatively correlated to an average design of the dots and may allow a quantitative modelization. These phenomena are summarized in Fig. 5.54.

This kind of experiments has been reproduced using a large collection of GaN quantum dots which height range from 1 to 4 nm. The decay times that are measured scale in these case from about 300 ps for a 4.2 eV emission up to 1 ms for a red (1.7 eV) light emitting dot. In that later case the decay time is long and the photon emission rate is small. the light emission is not very intense at the naked eye. The plot of the decay time versus energy, or dot size permits to estimate the average value of the electric field which is about 10 MV/cm for these GaN–AlN polar dots as sketched in Fig. 5.55. Values comparable can be obtained using GaInN as active material.

For the sake of the completeness, one has to indicate that the situation is even more complicated since the dots mays trap more than one electron-hole pair. Then these carrier create an electric field internal to the existing one which produces a blue-shift of the transition energy. The recombination rate decreases with time (the decay time

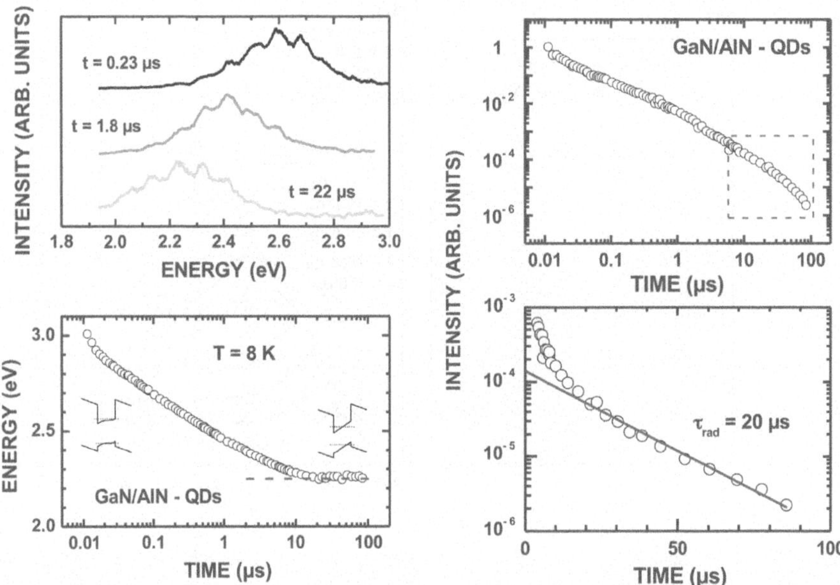

Fig. 5.54 *Left-hand top corner*: PL spectra of GaN/AlN QDs (height: 3.3 nm), taken 0.23, 1.8, and 22 μs after the excitation pulse. The fringes are due to interferences in the buffer layer and remain constant for all times. *Left-hand top corner*: PL peak energy versus the delay after excitation, obtained from a Gaussian fit to the spectrum. The error on the determination of the energy is estimated to ±50 meV. The *dashed line* is a guide for the eye. The sketches present the potential felt by electrons and holes, and their ground states, with (*left*) and without (*right*) screening of the electric field. *Right-hand top corner*: Spectrally integrated intensity of the PL (3.3 nm QDs again) versus the delay after excitation, on log-log scale. *Right-hand bottom corner*: the data from the *dashed square* are presented on a linear scale. The plain line is a mono-exponential fit of the decay between 20 and 80 μs. After [78]

increases) with progressive reduction of the excess charge density in the dot which also contributes to the complex recombination dynamics as sketched in Fig. 5.56. The measured radiative lifetime (circles) increases from the high photo injection injection density regime to low photo-injection conditions. Note the simultaneous saturation of the red-shift and the onset of the mono exponential PL dynamics when an average value of one electron hole pair per dot is reached. The exact measurement of the energy and decay time requires conditions of very low photo-excitation density.

GaN-based quantum dots have been also studied when fabricated along non polar direction or semi polar directions. In Fig. 5.57 are represented some photoluminescence spectra taken in case of GaN quantum dots deposited on a (11–22) $Al_{0.5}Ga_{0.5}N$ template. The photoluminescence energy does not reveal Quantum Confined Stark Effect: it does not shift below the energy of GaN when increasing the well-width compared with polar quantum dots.

Interesting to note too is the Full Width at Half Maximum of the photoluminescence which is much more smaller than for Polar quantum dots as illustrated in

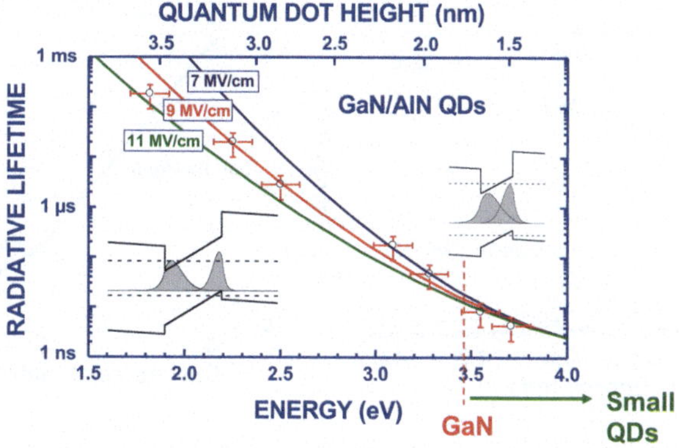

Fig. 5.55 Measured radiative lifetime (*circles*) versus measured energy of PL peak for the different QD samples. *Solid lines* show the calculated value of lifetime as a function of energy of the fundamental transition for an effective electric field of respectively 7, 9, and 11 MV/cm. The calculated QD height corresponding to the transition energies for a 9 MV/cm electric field is reported on the top axis. The electron and hole energy levels and wave functions are sketched in the regimes dominated by Stark effect (*left*) and quantum confinement (*right*). After [69]

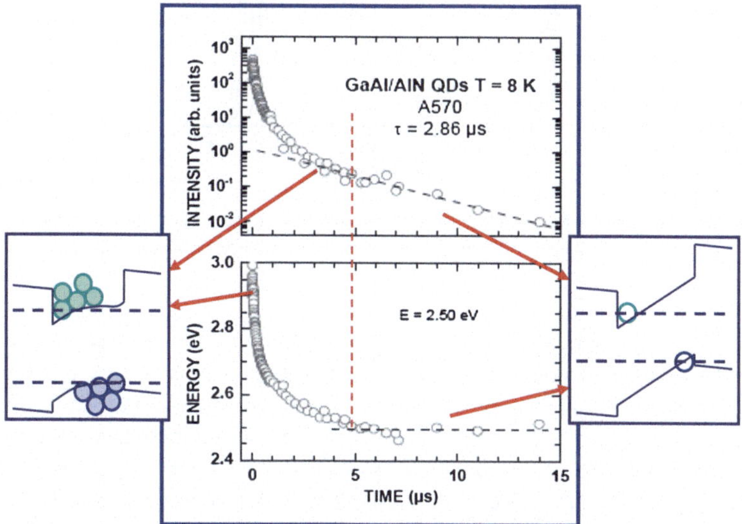

Fig. 5.56 Measured radiative lifetime (*circles*) versus time for a QD samples from the high photo injection injection density regime to low photo-injection conditions which indicate the simultaneous saturation of the red-shift and the onset of the mono exponential PL dynamics. After [76]

Fig. 5.58. The physics below is again the lack of Quantum Confined Stark Effect through the collection of dots. The dramatic energy shift with dot height is not measured when QCSE is absent. Then, the FWHM is just correlated to slight effects of confinement or strain fluctuations through the part of the sample shined by the laser.

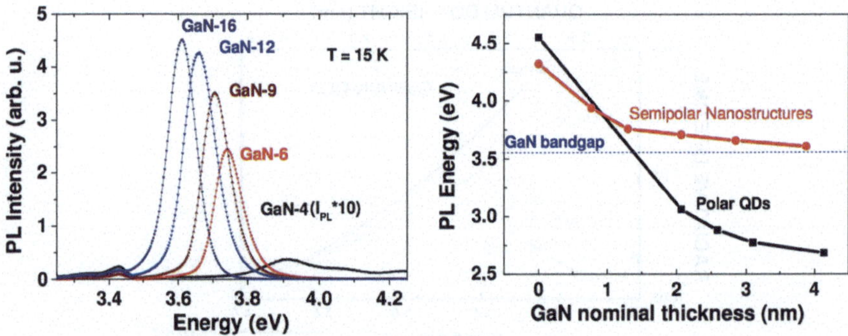

Fig. 5.57 *Left* Photoluminescence spectra collected for a series of GaN quantum dots deposited on a (11–22) $Al_{0.5}Ga_{0.5}N$ template. *Right* Comparison of the energy shift for polar GaN *quantum dots* and GaN quantum dots deposited on a (11–22) $Al_{0.5}Ga_{0.5}N$ template After [76]

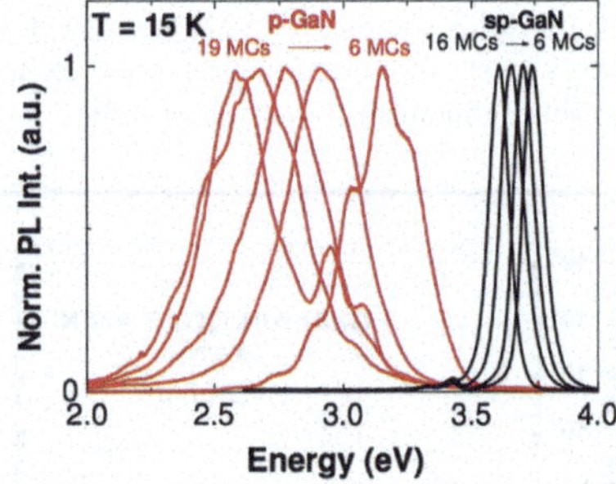

Fig. 5.58 Photoluminescence spectra collected for a series of polar GaN *quantum dots* and GaN *quantum dots* deposited on a (11–22) $Al_{0.5}Ga_{0.5}N$ template. Courtesy Mathieu Leroux

To measure the optical signature of a single quantum dot one has to drastically reduce the number of dots shined by the laser beam. It is necessary to excite the photoluminescence using a microscope objective and/or to restrict the excited region of the sample. Figure 5.59a represents a typical photoluminescence spectra for a single polar GaN–AlN quantum dot recorded at 8 K by Thierry Guillet and Richard Bardoux in our laboratory. The photoluminescence spectrum is composed of several sharp photoluminescence line which can be attributed to ground state recombination and excited states. We remark that these lines energy shift when time passes. This jitter is related to the fluctuation of charges in the crystal in the neighborhood of the shined dot illustrated in Fig. 5.59b. When the is environment changes, it produces an in-plane electric field and an associate in-plane Stark effect which changes the carriers confinements and the recombination energy. These fluctuations may be established as being similar to the random telegraph noise. They are very much often reported in case of single dot spectroscopy. In nitrides the amplitude of the jitter is bigger than it is in case of CdTe-based quantum dots, for which it is even bigger than for InAs. This is really an electrostatic effect.

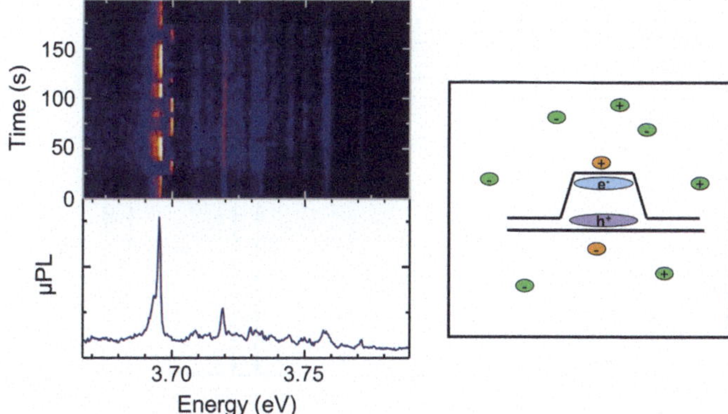

Fig. 5.59 *Bottom left* Photoluminescence spectrum of a single GaN–AlN *quantum dot*. *Top left* Two dimensional representation of the photoluminescence energy (x axis) against time (y axis) which shows the jitter. *right* Sketch of the electrostatic environment in the neighborhood of the dot. the electron and hole densities of probabilities are sketched to illustrate their relative position with respect to the apex of the dot. After [79]

5.7.1.1 Biexcitons

The exciton, an electron-hole pair bound by the attractive Coulomb force, which has been widely discussed so far is the fundamental optical transition of a semiconductor system or nanostructure which, similar to atoms in gas phase, may interact with other excitons to produce exciton molecules or biexcitons. Bi-excitons have been also observed in bulk materials including GaP:N in the nineteen sixties. The main criterion for observing bi-excitons from the experimental side was to saturate the nitrogen isoelectronic impurities with excitons so that the excellent electron goole pairs could form bi-excitons. Figure 5.60 reminds us the text-book experiment of James Merz, Roger Faulkner and Paul Dean in 1969 [81].

When one exciton recombines radiatively, the ground state is the crystal without exciton and the energy of the emitted photon is $h_{\nu_{exc}} = E_g - E_X$ where E_X the exiton binding energy (balanced direct electron-hole electrostatic interaction + exchange terms).

When one exciton of the bi-exciton recombine, the ground state is the exciton. The initial energy of the complex is $E_g - E_X + E_g - E_X - E_{X2} = 2E_g - E_X - E_{X2}$.

Where E_{X2} is the bi-exciton binding energy(balanced contributions of electron-electron repulsion + hole hole repulsion + electron-hole electrostatic interactions + exchange terms).

When a biexciton recombines radiatively, the ground state for the transition is a single exciton (which can itself also recombine radiately). The energy of the emitted photon in case of a bi-exciton to exciton recombination is: $h_{\nu_{biexc}} = E_g - E_X - E_{X2}$. The bi-exciton binding energy can be estimated from the difference between energy of the photon corresponding to the bi-exciton to exciton recombination and

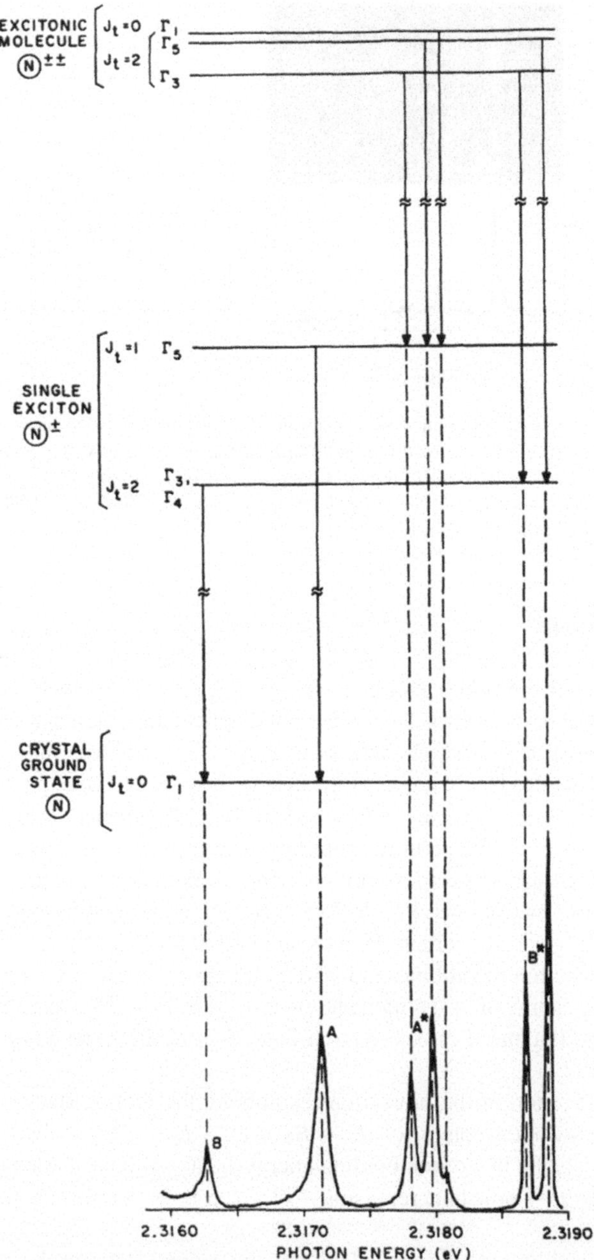

Fig. 5.60 Photoluminescence spectrum and energy-level diagram of the isoelectronic nitrogen trap in GaP:N. Lines *A* and *B* arise from the decay of the states of a single exciton bound to nitrogen split by short-range electron-hole exchange interaction. Lines *A** and *B** result from the radiative decay of the states of the excitonic molecule bound to nitrogen, the final states of the transitions being the *A* line and *B* line states, respectively, of the single exciton. After [81]

the energy energy of the photon corresponding to the exciton to crystal-ground state recombination.

In Fig. 5.60, the bi-exciton to exciton recombination appears at higher energy than the exciton to ground state recombination. Then the bi-exciton binding energy is negative in GaP:N.

A relevant requirement for the observation of the biexciton formation in a semiconductor quantum well is the high optical quality of the nanostructures. The in-plane structural disorder produces exciton localization which strongly produces inhomogeneous broadening of the exciton optical resonance which may masks the observation of the biexciton recombination (if formed). Nitride-based QWs offer the advantage of large Coulomb interaction, with an exciton binding energy estimated to be larger than 30 meV in low Al content narrow GaN/AlGaN QWs. However the interface roughness is not so flat as discussed earlier compared with the in-plane lateral extension of the exciton and bi-exciton observation remains exceptional (see for instance [13]). Single quantum dot spectroscopy may facilitates the observation of bi-excitons and in turn it really does. In Fig. 5.61 are reported micro-photoluminescence measurements recorded in Lausane in the Group of Nicolas Grandjean on Polar GaN–AlGaN quantum dots of different heights. The low temperature Micro-PL spectra of a single 8 ML high dot indicates a negative biexciton binding energy E_b. Measured at various power densities (P0 $\approx$ 1W/cm^2), the intensities of exciton and biexciton emission lines (labeled X and XX respectively indicate a slope of 1 and 2, respectively whIch probes the excitonic and biexcitonic nature of the two transitions. Similar experimentt performed using a 7-monolayer high for indicates a positive binding energy (the bi-exciton line is observed at higher energy than the excitonic transition).

The reason for this behaviour is correlated to several balancing phenomena. From the point of view of classical physics (neglecting exchange interactions), the bi-exciton can be sketched as in Fig. 5.62 with two electrons and two holes.

The two electrons are confined near the apex of the quantum dot while the hole have the possibility to dilute their density of probability in the wetting layer beneath the quantum dot. The gap mismatch between GaN and AlN is strong and the electrons are forced to be confined near the apex of the dots, although they strongly repulse each other. The smaller the dot the stronger this repulsion. For this reason, the exciton binding energy is positive in small dots and negative in big ones. This experimentally reported trend sketched in Fig. 5.63. At the time of writing; the bi-exciton binding energyE_{BXX} is correlated to the excitonic transition energy $E_X$by the empirical equation expressed in eV in polar GaN quantum dots:

$$E_{BXX}(eV) \approx 7.10^{-3}(E_X - 3.9)(eV)$$

Theoretical treatment is currently in hand in several groups.

Quantum dots constitute an excellent platform for manybody physics and quantum optics. There are results of paramount importance that were reported in that specific field among which are: [15–22]

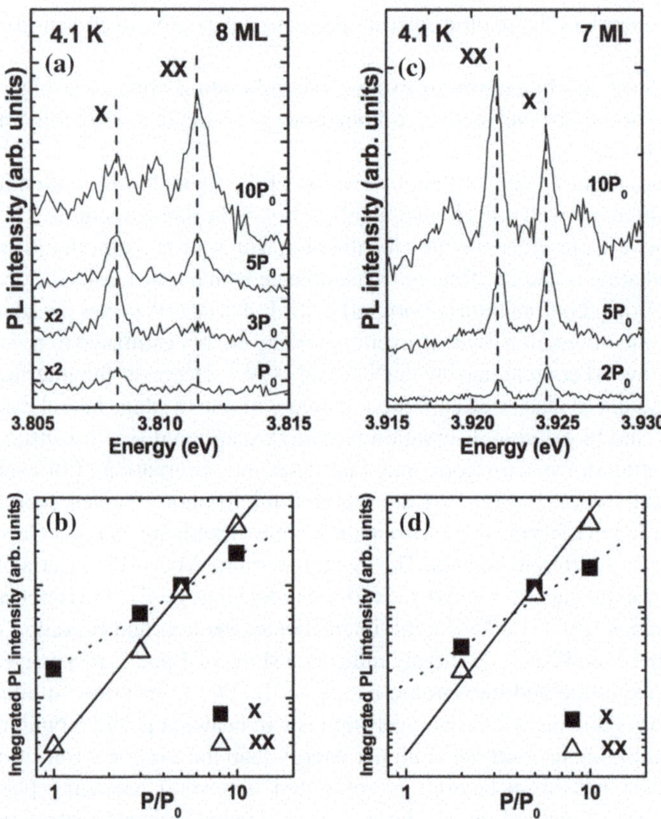

Fig. 5.61 Photoluminescence spectra indicating the relative energies of excitons and bi-excitons in polar GaN *quantum dots* (*upper figures*) and the intensity versus pump power (*down figures*). After [14]

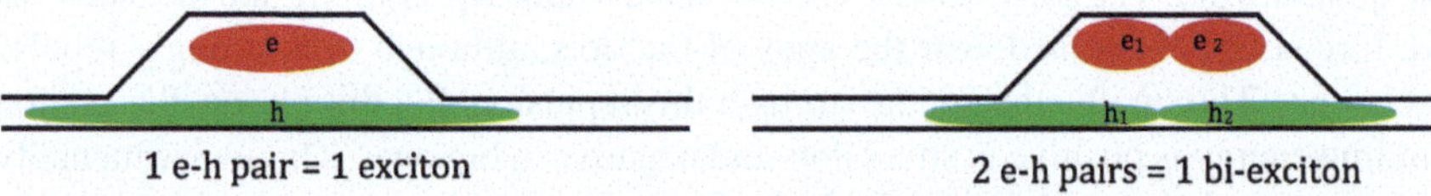

Fig. 5.62 Relative positions and electron and hole in GaN polar *quantum dots*. **a** One electron-hole pair forming one exciton. **b** Two electron-hole pairs forming a bi-exciton

There is a huge amount of interesting papers. I decided to select through the wealthy literature some typical papers and papers of my group which give via their lists of references a global overview of the state of the art in that field.

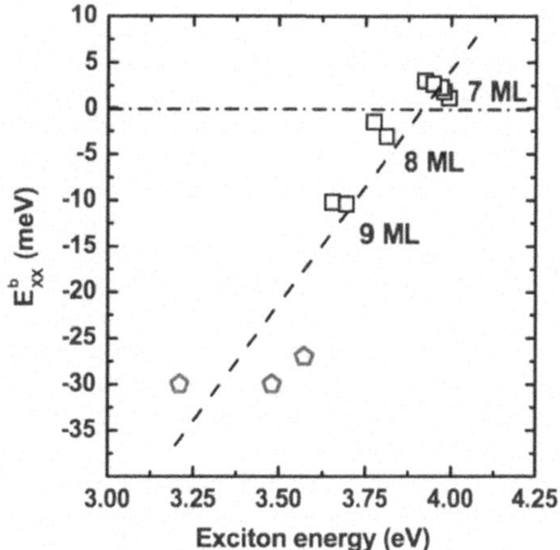

Fig. 5.63 Dependence of the biexciton binding energy on the exciton emission energy for GaN–AlGaN polar QDs of different vertical sizes. After [14]

References

1. M. Leroux, N. Grandjean, M. Laügt, J. Massies, B. Gil, P. Lefebvre, P. Bigenwald, Quantum confined stark effect due to built-in internal polarization fields in (Al, Ga)N/GaN quantum wells. Phys. Rev. B **58**, R13371 (1998)
2. M. Leroux, N. Grandjean, J. Massies, B. Gil, P. Lefebvre, P. Bigenwald, Barrier-width dependence of group-III nitrides quantum-well transition energies. Phys. Rev. B **60**, 1496 (1999)
3. N. Grandjean, J. Massies, M. Leroux, Self-limitation of AlGaN/GaN quantum well energy by built-in polarization field. Appl. Phys. Lett. **74**, 2361 (1999)
4. M. Gallart, A. Morel, T. Taliercio, P. Lefebvre, B. Gil, J. Allegre, H. Mathieu, N. Grandjean, M. Leroux, J. Massies, Scale effects on exciton localization and nonradiative processes in GaN/AlGaN quantum wells. Phys. Stat. Sol. (b) **180**, 127 (2000)
5. R. Butte, J.F. Carlin, E. Feltin, M. Gonschorek, S. Nicolay, G. Christmann, D. Simeonov, A. Castiglia, J. Dorsaz, H.J. Buehlmann, S. Christopoulos, G.B.H. von Hogersthal, A.J.D. Grundy, M. Mosca, C. Pinquier, M.A. Py, F. Demangeot, J. Frandon, P.G. Lagoudakis, J.J. Baumberg, N. Grandjean, Current status of AlInN layers lattice-matched to GaN for photonics and electronics. J. Phys. D—Appl. Phys. **40**, 6328 (2007)
6. B. Damilano, N. Grandjean, S. Vzian, J. Massies, J. Cryst. Growth **227–228**, 466 (2001)
7. T.J. Ochalski, B. Gil, P. Bigenwald, M. Bugajski, A. Wojcik, P. Lefebvre, T. Taliercio, N. Grandjean, J. Massies, Dual contribution to the stokes shift in InGaN-GaN quantum wells. Phys. Stat. Sol. (b) **228**, 111 (2001)
8. A. Hangleiter, J.S. Im, J. Off, F. Scholz, Optical properties of nitride quantum wells: how to separate fluctuations and polarization field effects. Phys. Stat. Sol. (b) **216**, 427 (1999)
9. J. Danhof, U.T. Schwarz, A. Kaneta, Y. Kawakami, Time-of-flight measurements of charge carrier diffusion in $In_xGa_{1-x}N$-GaN quantum wells. Phys. Rev. B **84**, 035324 (2011)
10. B. Maleyre, S. Ruffenach, O. Briot, B. Gil, Growth of InN quantum dots by MOVPEby. Phys. Stat. Sol. (c) **2**, 826 (2005)
11. S. Ruffenach, O. Briot, M. Moret, B. Gil, Control of InN quantum dot density using rare gases in metal organic vapor phase epitaxy. Appl. Phys. Lett. **90**, 153102 (2007)
12. D. Rosales, T. Bretagnon, B. Gil, A. Kahouli, J. Brault, B. Damilano, J. Massies, M.V. Durnev, A.V. Kavokin, Excitons in nitride heterostructures: from zero- to one-dimensional behavior. Phys. Rev. B **88**, 125437 (2013)
13. F.S. Cheregi, A. Vinattieri, E. Feltin, D. Simeonov, J.-F. Carlin, R. Butt, N. Grandjean, M. Gurioli, Biexciton kinetics in GaN quantum wells: time-resolved and time-integrated photoluminescence measurements. Phys. Rev. B **77**, 125342 (2008)

14. D. Simeonov, A. Dussaigne, R. Butt, N. Grandjean, Complex behavior of biexcitons in GaN quantum dots due to a giant built-in polarization field. Phys. Rev. B **77**, 075306 (2008)
15. S. Kako, K. Hoshino, S. Iwamoto, S. Ishida, Y. Arakawa, Exciton and biexciton luminescence from single hexagonal GaN/AlN self-assembled quantum dots. Appl. Phys. Lett. **85**, 64 (2004)
16. C. Santori, S. Gotzinger, Y. Yamamoto, S. Kako, K. Hoshino, Y. Arakawa, Photon correlation studies of single GaN quantum dots. Appl. Phys. Lett. **87**, 051916 (2005)
17. S. Kako, C. Santori, K. Hoshino, S. Gotzinger, Y. Yamamoto, Y. Arakawa, A gallium-nitride single-photon source operating at 200K. Nat. Mater. **5**, 887 (2006)
18. F. Rol, S. Founta, H. Mariette, B. Daudi, Le Si Dang, J. Bleuse, D. Peyrade, J.-M. Grard, and B. Gayral, Probing exciton localization in nonpolar GaN/AlN quantum dots by single-dot optical spectroscopy. Phys. Rev. B **75**, 125306 (2007)
19. J. Renard, R. Songmuang, G. Tourbot, C. Bougerol, B. Daudin, B. Gayral, Evidence for quantum-confined stark effect in GaN/AlN quantum dots in nano wires. Phys. Rev. B **80**, 121305(R) (2009)
20. C. Kindel, S. Kako, T. Kawano, H. Oishi, Y. Arakawa, G. Honig, M. Winkelnkemper, A. Schliwa, A. Hoffmann, D. Bimberg, Exciton fine-structure splitting in GaN/AlN quantum dots. Phys. Rev. B **81**, 241309 (2010)
21. S. Sergent, S. Kako, M. Burger, D. As, Y. Arakawa, Narrow spectral linewidth of single zinc-blende GaN/AlN self-assembled quantum dots. Appl. Phys. Lett. **103**, 151109 (2013)
22. M.J. Holmes, K. Choi, S. Kako, M. Arita, Y. Arakawa Room-temperature triggered single photon emission from a IIINitride site-controlled nanowire quantum dot. Nano Lett. (2014)

Further Reading

23. *Wave mechanics applied to semiconductor heterostructures* by G. Bastard (Les Editions de Physique, Paris, 1990).
24. *Nitride Semiconductor Devices "Principle and simulation"* edited by J. Piprek (Wiley VCH, 2007).
25. *Introduction to nitride semiconductor blue lasers and light-emitting diodes*, edited by S. Nakamura, S. F.Chichibu (CRC Press, London, 2000).
26. *Nitride phosphors and Solid State lighting* edited by R.-J. Xie, Y. Q. Li, N. Hirosaki, H. Yamamoto, (CRC Press, Taylor and Francis, London, 2011).
27. *Self-assembled quantum dots*, edited by Z. M. Wang (Springer, 2008).
28. *Rare Earth-doped III-Nitrides for Optoelectronic and Spintronic applications*, edited by K. O'Donnell, V. Dierof, (Springer, 2010).
29. *Light-emitting diodes*, by E. F. Schubert (Cambridge University Press, 2006).
30. *GaN-based materials and Devices* edited by M. S. Shur, R. F. Davis (World Scientific, 2004).
31. *Wide energy band gap electronic Devices*, edited by F. Ren, J. C. Zolper (World Scientific, 2003).
32. *Variational treatment of the exciton binding energy problem in low dimensional systems with one marginal potential*, B. Gil, and P. Bigenwald, Solid State Communications, 94, 883 (1995).
33. *Excitonic properties and resonance widths in biased*, (Ga, In)As-GaAs double quantum wells, P. Bigenwald, and B. Gil, Phys. Rev. B **51**, 9780 (1995).
34. *Optical Spectroscopy in ZnCdSe-ZnSe Graded Index Separate Confinement, Heterostructures*, L. Aigouy, V. Mathet, F. Liaci, B. Gil, N. Briot, T. Cloitre, O. Briot, M. Averous, and R. L. Aulombard, Phys. Rev. B **53**, 4708 (1996).
35. *Recombination dynamics of free and localized excitons in GaN-AlGaN quantum wells*, P. Lefebvre, J. Allegre, B. Gil, A. Kavokine, H. Mathieu, H. Morkoc, W. Kim, A. Salvador, A. Botchkarev, Phys. Rev. B 57, 9447 (R) (1998).
36. *Effect due to built-in internal polarization fields in, AlGaN/GaN quantum wells*, M. Leroux, N. Grandjean, M. Laügt, J. Massies, B. Gil, P. Lefebvre, and P. Bigenwald, Phys. Rev. B **58**, 13371(R) (1999).

37. *Observation of long-lived oblique excitons in GaN/AlGaN multiple quantum wells*, B. Gil, P. Lefebvre, J. Allegre, H. Mathieu, N. Grandjean, M. Leroux, J. Massies, P. Bigenwald, and P. Christol, Phys. Rev. B 59, 10246 (1999).
38. *Time resolved photoluminescence as a probe of internal electric fields in GaN/AlGaN quantum wells*, P. Lefebvre, J. Allegre, B. Gil, H. Mathieu, N. Grandjean, M. Leroux, J. Massies, and P. Bigenwald, Phys. Rev. B 59, 15563 (1999).
39. *Photoreflectance Investigation of the bowing parameter in AlGaN alloys lattice-matched to GaN*, T. Ochalski, B. Gil, P. Lefebvre, N. Grandjean, M. Leroux, J. Massies, S. Nakamura, and H. Morko, Appl. Phys. Lett. 74, 3353 (1999).
40. *Slow spin relaxation observed in InGaN/GaN multiquantum wells*, M. Julier, A. Vinatierri, M. Colocci, P. Lefebvre, B. Gil, D. Scalbert, C. A. Tran, R. F. Karlicek Jr., and J. P. Lascaray, Physica Status Solidi 216, 341 (1999).
41. *Confined Excitons in GaN-AlGaN Quantum Wells*, P. Bigenwald, P. Lefebvre, T. Bretagnon and B. Gil, Physica Status Solidi 216, 37, (1999).
42. *Reduction of oscillator strength due to piezoelectric fields in* $GaN - Al_xGa_{1-x}N$ *quantum wells* by Jin Seo Im, H. Kollmer, J. Off, A. Sohmer, F. Scholz, and A. Hangleiter, Phys. Rev. B 57, R9435, (1998).
43. *Theoretical study of orientation dependence of piezoelectric effects in wurtzite strained GaInN/GaN heterostructures and quantum wells*, by T.Takeuchi, H. Amano, and I. Akasaki, Japanese Journal of Applied Physics 39 413, (2000).
44. *Quantum-confined stark effect due to piezoelectric fields in GaInN strained quantum wells* by T. Takeuchi, S.Sota, M. Katsuragawa, M. Komori, H. Takeuchi, H. Amano, and I. Akasaki, in Japanese Journal of Applied Physics, 36, L382, (1997).
45. *Photoreflectance spectroscopy as a powerful tool for the investigation of GaN/AlGaN quantum well structures*, T. Ochalski, B. Gil, P. Lefebvre, N. Grandjean, M. Leroux, J. Massies, Solid State Communications 109, 567, (1999).
46. *Barrier width dependence of group III nitrides quantum well transition energies*, M. Leroux, N. Grandjean, J. Massies, B. Gil, P. Lefebvre, and P. Bigenwald, Phys. Rev. B 60, 1496, (1999).
47. *Highly photo-excited nitride quantum wells: threshold for exciton bleaching*, P. Bigenwald, A. Kavokin, B. Gil, and P. Lefebvre, Physica Status Solidi 216, 481, (1999).
48. *Electron-hole plasma effect on excitons in GaN-GaAlN quantum wells*, P. Bigenwald, A. Kavokin, B. Gil, and P. Lefebvre, Phys. Rev. B 61, 15621 (2000).
49. *Exclusion principle and screening of excitons in, GaN/AlxGa1-xN quantum wells*, P. Bigenwald, A. Kavokin, B. Gil, and P. Lefebvre. Phys. Rev. B **63**, 035315 (2001)
50. *Excitons and trions confined in quantum systems: from low to high injection regimes*, P. Bigenwald, A. Kavokin and B. Gil, Physica Satus Solidi (a) 195, 587 (2003).
51. *Observation and modeling of the time-dependent de-screening of internal electric field in a wurtzite GaN-*$Al_0.15Ga_0.85N$ *quantum well after high photo-excitation*, P. Lefebvre, S. Kalliakos, T. Bretagnon, P. Valvin, T. Taliercio, B. Gil, N. Grandjean, and J. Massies, Phys. Rev. B 69, 035307 (2004).
52. *Photoreflectance spectroscopy investigation of GaN-AlGaN quantum well structures*, T. J. Ochalski, B. Gil, P. Lefebvre, N. Grandjean, J. Massies, and M. Leroux, Physica Status Solidi 216, 221, (1999).
53. *Dynamics of excitons in GaN-AlGaN MQWs with varying depths, thicknesses and barrier layers*, P. Lefebvre, M. Gallart, T. Taliercio, B. Gil, J. Allegre, H. Mathieu, N. Grandjean, M. Leroux, J. Massies, and P. Bigenwald, Physica Status Solidi 216, 261 (1999).
54. *High internal electric field in a graded-width InGaN/GaN quantum well: Accurate determination by time-resolved photoluminescence spectroscopy*, P. Lefebvre, A. Morel, M. Gallart, T. Taliercio, J. Allegre, B. Gil, H. Mathieu, B. Damilano, N. Grandjean, and J. Massies, Appl. Phys. Lett. 78, 1252 (2001).
55. *Effects of GaAlN barriers and of dimensionality on optical recombination processes in InGaN quantum wells and quantum boxes*, P. Lefebvre, T. Taliercio, A. Morel, J. Allgre, M. Gallart, B. Gil, H. Mathieu, B. Damilano, N. Grandjean, and J. Massies, Appl. Phys. Lett. 78, 1538 (2001).

56. *Extremely sharp dependence of the exciton oscillator strength on quantum well width in GaN-AlGaN system: the polarization effect*, M. Zamfirescu, B. Gil, N. Grandjean, G. Malpuech, A. Kavokin, P. Bigenwald, and J. Massies, Phys. Rev. B 64 (R), 121304 (2001).
57. *Large size dependence of exciton longitudinal phonon coupling in nitride-based quantum wells and quantum boxes*, S. Kalliakos, X. B. Zhang, T. Taliercio, P. Lefebvre, B. Gil, N. Grandjean, B. Damilano, and J. Massies, Appl. Phys. Lett. 80, 428, (2002).
58. *Donor Acceptor-like behavior of Electron-hole recombinations in In, GaN-GaN low dimensional systems*, A. Morel, P. Lefebvre, S. Kalliakos, T. Taliercio, T. Bretagnon, and B. Gil. Phys. Rev. B **68**, 045331 (2003).
59. *Donor binding energies in group-III-nitride-based quantum wells, Influence of internal electric fields*, A. Morel, P. Lefebvre, T. Taliercio, M. Gallart, B. Gil, and H. Mathieu. Materials Science and Engineering B **82**, 221 (2001).
60. *Effects of the optical pumping intensity and temperature on excitons in GaN-based quantum wells*, P. Bigenwald, A. Kavokin, P. Christol, P. Lefebvre, and B. Gil, Materials Science and Engineering B 82, 185 (2001).
61. *Temperature enhancement of the exciton binding energy in nitride quantum structures*, P. Bigenwald, B. Gil, A. Kavokin, and P. Christol, Physica Status Solidi (a) 183, 125 (2001).
62. *Blue-light emission from* $GaN - Al_{0.5}Ga_{0.5}N$ *quantum* dots by T. Huault, J. Brault, F.Natali, B.Damilano, D. Lefebvre, L.Nguyen, M.Leroux, J Massies, in applied Physics letters, 92, 051911 (2008).
63. *The dual contribution to the stokes-shift in InGaN-GaN quantum wells*, T. J. Ochalski, B. Gil, P. Bigenwald, M. Bugajski, A. Wojcik, P. Lefebvre, T.Taliercio, N. Grandjean, and J. Massies, Physica Status Solidi (b) 228, 111 (2001).
64. *Carrier dynamics in group III nitride low dimensional systems : localization versus quantum confined Stark effect*, P. Lefebvre, T. Taliercio, S. Kalliakos, A. Morel, X. B. Zhang, M. Gallart, T. Bretagnon, and B. Gil, Physica Status Solidi (b) 228, 65 (2001).
65. *Experimental and theoretical tools for the study of exciton properties versus disorder in nitride-based heterostructures*, B. Gil, M. Zamfirecu, P. Bigenwald, G. Malpuech, and A. Kavokin, Physica Status Solidi (b) 228, 471 (2001).
66. *The effect of localization and of electric fields on LO-phonon-exciton coupling in InGaN-GaN quantum wells and quantum boxes*, S. Kalliakos, P. Lefebvre, X. B. Zhang, T. Taliercio, B. Gil, N. Grandjean, B. Damilano, and J. Massies, Physica Status Solidi (b) 190, 149 (2002).
67. *Barrier Composition Dependence of the Internal Electric Field in ZnO/Zn1-xMgxO Quantum Wells*, T. Bretagnon, P. Lefebvre, T. Guillet, T. Taliercio, B. Gil, and C. Morhain, Appl. Phys. Lett. 90, 201912 (2007).
68. *Polarized emission from GaN-AlN quantum dots: single dot spectroscopy and symmetry-based theory*, R. Bardoux, T. Guillet, B. Gil, P. Lefebvre, T. Bretagnon, T. Taliercio, S. Rousset, and F. Semond, Phys. Rev. B 77, 235315 (2008).
69. *Visualization of the Local Carrier Dynamics in an, InGaN Quantum Well Using Dual-Probe Scanning Near-Field Optical Microscopy*, A. Kaneta, T. Hashimoto, K. Nishimura, M. Funato, and Y. Kawakami, Applied Physics Express **3**, 102102 (2010).
70. *Internal electric field in wurtzite ZnO/Zn0.78Mg0.22O quantum wells.*, C. Morhain, T. Bretagnon, P. Lefebvre, X. Tang, P. Valvin, T. Guillet, B. Gil, T. Taliercio, M. Teisseire-Doninelli, B. Vinter, and C. Deparis, Phys. Rev. B 72, 241305 (R) (2005).
71. *Short-range spin exchange interaction in ZnO-ZnMgO quantum wells*, B. Gil, P. Lefebvre, T. Bretagnon, T. Guillet, J. A. Sants, T. Taliercio, and C. Morhain. Phys. Rev. B **74**, 153302 (2006).
72. *Low temperature reflectivity study of nonpolar, ZnO/(Zn, Mg)O quantum wells grown on M-plane ZnO substrates*, L. Baur, T. Bretagnon, C. Brimont, T. Guillet, B. Gil, D. Tainoff, M. Teisseire, J. M. Chauveau, Appl. Phys. Lett. **98**, 101913 (2011).
73. *Exciton radiative properties in nonpolar homoepitaxial, ZnO/(Zn, Mg)O quantum wells*, L. Baur, T. Bretagnon, B. Gil, A. Kavokin, T. Guillet, C. Brimont, D. Tainoff, M. Teisseire, and J.-M. Chauveau, Phys. Rev. B **84**, 165312 (2011).

74. *Phonon-assisted exciton formation in, ZnO/(Zn, Mg)O single quantum wells grown on C-plane oriented substrates*, T. Bretagnon, L. Baur, T. Guillet, C. Brimont, M. Gallart, B. Gil, P. Gilliot, and C. Morhain, Journal of Luminescence. **136**, 355 (2013).
75. *Photoluminescence energy and linewidth for stackings of GaN-AlN quantum dot planes*, S. Kalliakos, T. Bretagnon, P. Lefebvre, T. Taliercio, B. Gil, N. Grandjean, B. Damilano, A. Dussaigne, and J. Massies, J. Appl. Phys. 96, 180 (2004).
76. *Time dependence of the photoluminescence of GaN/AlN quantum dots under high photo-excitation*, T. Bretagnon, S. Kalliakos, P. Lefebvre, P. Valvin, B. Gil, N. Grandjean, A. Dussaigne, B. Damilano, and J. Massies, Phys. Rev. B 68, 205301 (2003).
77. *Nontrivial carrier recombination dynamics and optical properties of over-excited GaN/AlN quantum dots*, S. Kalliakos, T. Bretagnon, P. Lefebvre, S. Juillaguet, T. Taliercio, T. Guillet, B. Gil, N. Grandjean, B. Damilano, A.Dussaigne, and J. Massies. 5th International Symposium on Blue Lasers and Light-Emitting Diodes, Physica Status Solidi (b) 241, 2779 (2004).
78. *Radiative lifetime of a single electron-hole pair in GaN/AlN quantum dots*, T. Bretagnon, P. Valvin, R. Bardoux, P. Lefebvre, T. Guillet, T. Taliercio, B. Gil, N. Grandjean, F. Semond, B. Damilano A. Dussaigne, and J. Massies, Phys. Rev. B 73, 113304 (2006).
79. *Photoluminescence of single GaN/AlN hexagonal quantum dots on, Si(111): Spectral diffusion effects*, R. Bardoux, T. Guillet, P. Lefebvre, T. Bretagnon, T. Taliercio, S. Rousset, B. Gil, and F. Semond, Phys. Rev. B **74**, 195319 (2006).
80. *Time resolved photoluminescence study of ZnO-ZnMgO quantum wells*, T. Bretagnon, P. Lefebvre, P. Valvin, B. Gil, C. Morhain, and X. D. Tang, Journal of Crystal growth **287**, 12 (2006).
81. *Excitonic Molecule Bound to the Isoelectronic Nitrogen Trap in GaP*, by J. L. Merz, R. A. Faulkner, and P. J. Dean, Phys. Rev. 188, 1228 (1969).

Index

B. Gil, *Physics of Wurtzite Nitrides and Oxides*, Springer Series in Materials Science 197, DOI: 10.1007/978-3-319-06805-3,

The manufacturer's authorised representative in the EU is Springer Nature Customer Service Centre GmbH, Europaplatz 3, 69115 Heidelberg, Germany. If you have any concerns regarding our products, please contact ProductSafety@springernature.com

Printed and bound by CPI Group (UK) Ltd, Croydon, CR0 4YY
15/07/2026
02167635-0002